Machine Learning and Statistics

Sixth-Generation Computer Technology Series

Branko Souček, Editor
Star Service Sp.A.

Neural and Massively Parallel Computers: The Sixth Generation
Branko Souček and Marina Souček

Neural and Concurrent Real-Time Systems: The Sixth Generation
Branko Souček

Neural and Intelligent Systems Integration: Fifth and Sixth Generation
Branko Souček

Neural and Intelligent Systems Integration: Fifth and Sixth Generation Integrated Reasoning Information Systems
Branko Souček and the IRIS Group

Dynamic, Genetic, and Chaotic Programming: The Sixth Generation
Branko Souček and the IRIS Group

Fuzzy, Holographic, and Parallel Intelligence: The Sixth Generation Breakthrough
Branko Souček and the IRIS Group

Fast Learning and Invariant Object Recognition: The Sixth Generation Breakthrough
Branko Souček and the IRIS Group

Modular Learning in Neural Networks: A Modularized Approach to Neural Network Classification
Tomas Hrycej

Integrating Rules and Connectionism for Robust Commonsense Reasoning
Ron Sun

Frontier Decision Support Concepts: Help Desk, Learning, Fuzzy Diagnoses, Quality Evaluation, Prediction, Evolution
Vito Leonardo Plantamura, Branko Souček, and Giuseppe Visaggio (eds.)

Evolution and Optimum Seeking
Hans-Paul Schwefel

Foundations of Distributed Artificial Intelligence
G. M. P. O'Hare and N. R. Jennings (eds.)

Machine Learning and Statistics: The Interface
G. Nakhaeizadeh and C. C. Taylor (eds.)

Machine Learning and Statistics

The Interface

Edited by

G. NAKHAEIZADEH
C. C. TAYLOR

A Wiley-Interscience Publication

JOHN WILEY & SONS, INC.

New York - Chichester - Brisbane - Toronto - Singapore - Weinheim

This text is printed on acid-free paper.

Library of Congress Cataloging in Publication Data:

Machine learning and statistics: the interface / G. Nakhaeizadeh & C.C. Taylor, editors.
p. cm. — (Sixth-generation computer technology series)
Papers from a workshop which followed the European Conference in Machine Learning (ECML-94).
"A Wiley-Interscience publication."
ISBN 0-471-14890-3 (cloth : alk. paper)
1. Mathematical statistics—Data processing—Congresses. 2. Machine learning—Congresses. I. Nakhaeizadeh, Gholamreza. II. Taylor, C. C. (Charles C.) III. Series.
QA276.4.M23 1996
006.3′1′015195—dc20 96-4668
CIP

Printed in the United States of America

10 9 8 7 6 5 4 3 2 1

CONTENTS

CONTRIBUTORS

C. BAROGLIO
Dipartimento di Informatica
Università di Torino
C.so Svizzera 185 10149, Torino
Italy

I. BRUHA
Department of Computer Science and Systems
McMaster University
Hamilton, Ontario L8S4K1
Canada

M. CHIOGNA
Dipartimento di Scienze Statistiche
Università degli Studi di Padova
via San Francesco 33
35121 Padova
Italy

A. CORNUÉJOLS
Laboratoire de Recherche en Informatique
Universite de PARIS-SUD, Orsay
91405 Orsay Cedex
France

F. ESPOSITO
Dipartimento di Informatica
Università degli Studi
via Orabona, 4
I-70124 Bari
Italy

A. GIORDANA
Dipartimento di Informatica
Università di Torino
C.so Svizzera 10149, Torino
Italy

D. J. HAND
Department of Statistics
The Open University
Milton Keynes MK 76AA
UK

R. J. HENERY
Department of Statistics and Modelling Science
University of Strathclyde
Glasgow G1 1XH
UK

W. E. HENLEY
N. M. Rothschilds & Sons, Ltd.
New Court
St. Swithins Lane
London EC4P 4DU
UK

W. Z. LIU
Intelligent Computer Systems Centre
Faculty of Computer Studies and Mathematics
University of the West England
Frenchay Campus
Bristol BS16 1QY
UK

G. LO BELLO
Dipartimento di Informatica
Università di Torino
C.so Svizzera 10149, Torino
Italy

D. MALERBA
Dipartimento di Informatica
Università degli Studi
via Orabona, 4
I-70124 Bari
Italy

M. MOULET
Laboratoire de Recherche en Informatique
Universite of Paris-Sud, Orsay
91405 Orsay Cedex
France

W. MÜLLER
Fraunhofer-Institute for Information and Data Processing
10117 Berlin
Germany

G. NAKHAEIZADEH
Daimler-Benz Research and Technology
8 9013 Ulm
Germany

R. PIOLA
Dipartimento di Informatica
Università di Torino
C.so Svizzera 185, 10149, Torino
Italy

M. RICHELDI
Centro Studi E Laboratori Telecomunicazioni S.p.A
Artificial Intelligence Unit
via G. Reiss Romoli 274, 10148 Torino
Italy

M. ROSSOTTO
Centro Studi E Laboratori Telecomunicazioni S.p.A.
Artificial Intelligence Unit
via G. Reiss Romoli 274 10148 Torino
Italy

B. SCHULMEISTER
Fraunhofer-Institute for Information and Data Processing
10117 Berlin
Germany

G. SEMERARO
Dipartimento di Informatica
Università degli Studi
via Orabona, 4
I-70124 Bari
Italy

C. C. TAYLOR
Department of Statistics
University of Leeds
Leeds LS2 9JT
UK

M. WESTPHAL
Daimler-Benz and SGZ Bank
76133 Karlsruhe
Germany

A. P. WHITE
School of Mathematics and Statistics
University of Birmingham
Birmingham B15 2TT
UK

F. WYSOTSKI
Fraunhofer-Institute for Information and Data Processing
10117 Berlin
Germany

ABOUT THE EDITORS

Gholamreza Nakhaeizadeh is a senior scientist at the Daimler-Benz Research Center in Ulm, Germany. He is also a Professor at Karlsruhe University. During 1990–1993 he directed the Machine Learning Project *StatLog*, supported by the European Union. His research fields include symbolic and statistical learning and their industrial and commercial applications.

Charles Taylor is Senior Lecturer in the Department of Statistics at the University of Leeds, England. His particular interests include nonparametric density estimation methods related to classification, and statistical methods in image analysis.

PREFACE

Many problems which are encountered in everyday life can be considered as classification or prediction tasks—"Is it safe to cross the road now?", "Is it worth taking my umbrella when I go out?", "Is it the right time to buy shares in this company?" These are all examples of decisions which many people make regularly. The problem can often be reformulated as one of classification. For example, if I cross the road now (given all the available information) will I cause an accident? For most people, the decision-making process is internal and the basis for a particular judgment, including the utilities and prior beliefs, is not normally qualified. Typically the "correct" decision is based on a correct classification and will be based on many factors and the relationship between them. An objective solution to such problems requires appropriate rules which are often not available in advance. Sometimes these "rules" are just plain common-sense or they can be provided by experts, but in the first instance they could often be learned by induction, gathering experience from a number of examples. In industrial applications, this is a time-consuming approach, and so a lot of work has been devoted to the automation of the inductive learning process. Broadly speaking, there have been workers from two distinct disciplines (Statistics, and a branch of Artificial Intelligence known as Machine Learning) working on these problems.

In the statistical community, historical roots can be traced back to R. A. Fisher's work in the 1930s on the discrimination of groups by using linear (line, planes etc.) functions. The idea of Automatic Interaction Detection (AID) was proposed by Sonquist and Morgan; this was further developed by Friedman who developed a recursive partitioning decision rule for nonparametric classification and has led to the well-known *Classification and Regression Trees* (CART) of Breiman and co-workers. In parallel to these attempts, a group of Machine Learning researchers had been developing algorithms to automate the induction process and consequently classification procedures. The well-known contribution of this group is the idea of Concept Learning Systems (CLS) due to Hunt et al. This was used in the implementation of the ID3 algorithms described by Quinlan in 1979.

Although there were almost independent developments in ML and Statistics, since the end of the 1980s the researchers of both communities have realized that their disciplines can help each other's development in a mutual manner. For example:

- Statistical and probabilistic approaches have been applied by ML-researchers in the development of different learning algorithms.
- Statistical tests have been used in optimization, pruning, and in particular, the evaluation of different ML-algorithms.
- Concepts like Case-Based Reasoning (CBR) and Instance-Based Learning (IBL) which are rather new in the ML-community have a significant interface with Statistics. The core of both approaches is the concept of the k-Nearest Neighbour (k-NN) algorithm which can be traced back to the early 1950s and during the last years has been further investigated by the statistical community.

On the other hand, decision tree and rule based algorithms as well as the Case-Based Reasoning approach have also been used to adapt statistical methods and, above all, Machine Learning has provided a strong challenge that rules or methods should be valuable and understood by humans. However, since statistics is concerned with the analysis and interpretation of data, and ML deals primarily with learning rules or structure, the mutual benefit of these two disciplines is only beginning to be realized. This volume is aimed at continuing this process by exploring areas of common interest and bringing about further cross-fertilization.

Much of the interface between ML and statistics is concerned with prediction and classification tasks and the papers in this book reflect this. Broadly, the methods encompass statistical approaches, Machine Learning trees and rules and neural network methods as well as a variety of hybrids and combinations. Sometimes simulations are used to demonstrate the effectiveness of an approach but more often it is on real datasets that comparisons are made. There are many applications and examples, including credit scoring, medical diagnoses, forecasting financial time series, control, and document understanding and analysis. In addition there are two more recent developments at the interface between Machine Learning and Statistics. One is concerned with learning by analogy using the *Minimum Description Length Principle.* The other is a fast-growing area of research known as *Knowledge Discovery in Databases* or *Data Mining* which attempts to make more use of the rapidly growing numbers of large databases which are being stored around the world.

Most of the papers contained in this book were written as a result of a workshop which followed the European Conference in Machine Learning (ECML-94), which took place in Catania, Sicily. The papers have been extended to make the material accessible to the reader who is not an expert in this area. It is also intended that the Introduction will provide an introduction for the novice, as well as guidance for further reading.

Ulm, Germany G. NAKHAEIZADEH
Leeds, United Kingdom C. C. TAYLOR

Machine Learning and Statistics

Introduction

G. NAKHAEIZADEH
C. C. TAYLOR

The intersection of, and interaction between Machine Learning (ML) and statistics is a rapidly growing area of interest. There are several areas of common research, the main ones being classification and prediction tasks. However, since statistics is concerned with the analysis and interpretation of data, and ML deals primarily with learning rules or structure, the mutual benefit of these two disciplines is only beginning to be realized. This volume attempts to explore some areas of common interest, particularly in classification (or discrimination), since this is chiefly where cross-fertilization is taking place.

The papers contained in this book were written as a result of a workshop that followed the European Conference in Machine Learning (ECML-94). The papers have been extended to make the material accessible to the reader who is not an expert in this area. It is also intended that this Introduction will help the novice, as well as providing guidance for further reading.

The next section deals with classification and prediction, which is where the subject areas primarily overlap. The section following that discusses more recent developments which have arisen as a consequence of the ever-increasing number and size of databases around the world. If the many pitfalls are avoided, both statistics and machine learning will play a crucial role in knowledge discovery in such systems. The final section discusses the prospects of learning rules in an "Occam's Razor" fashion, in which the simplest representation is used to learn by analogy.

Machine Learning and Statistics, The Interface, Edited by G. Nakhaeizadeh and C. C. Taylor.
ISBN 0-471-14890-3

CLASSIFICATION AND PREDICTION

Background

The term "classification" could cover any context in which some decision or forecast is made on the basis of currently available information. A more limited interpretation—also known as pattern recognition, discrimination, or supervised learning—involves the construction of a procedure applied to a continuing sequence of examples. In this case, each new example must be assigned to one of a set of predefined *classes* on the basis of observed *features* or *attributes*. A second usage of the term classification involves learning the classes from the data; this is also known as clustering or unsupervised learning. Here, we use the term in the first sense only.

Attempts to group discrimination procedures can be made on the basis of history (*classical* vs. *modern*, etc.), origin (ML vs. statistics) or type (rule-based vs. density estimation, etc.), but each approach leads to a number of exceptions or anomalies. Here we give brief introductions to give a flavor of some of the methods; full details can be found elsewhere (Hand, 1981; Hertz et al., 1991; Weiss and Kulikowski, 1991; McLachlan, 1992; Michie et al., 1994).

Statistical Approaches

In the statistical community, historical roots can be traced back to Fisher's (1936) work on linear discriminant analysis, of which there are now several derivatives. A simple way to describe this is in terms of maximum likelihood. Suppose that observations belonging to class k have features x_i, $i = 1, \ldots, n_k$, where x_i is a p-dimensional vector of attributes, and that the x_i have distribution given by $f_k(\mathrm{x})$. Most statistical approaches use a set of *training* data to estimate $f_k(\mathrm{x})$, $k = 1, \ldots, q$, by $\hat{f}_k$, say. New observations, say x^*, in a set of *test* data, are then assigned to the class l where

$$l = \arg \max_{k=1,\ldots,q} \hat{f}_k(\mathrm{x}^*)$$

The only remaining question is: What is the best way to estimate f? There are *parametric* approaches in which f is assumed to come from a known family of distributions, and the parameters are estimated from the data, and there are *nonparametric* methods in which no form of distribution is assumed and f is estimated directly. Both methods have their place, depending on the distribution of the data.

If the distribution is multivariate normal with mean μ_k and covariance matrix Σ_k, that is,

$$f_k(\mathrm{x}) = \frac{1}{\sqrt{(2\pi)^p |\Sigma_k|}} \exp\left\{-\frac{1}{2}(\mathrm{x} - \mu_k)^T \Sigma_k^{-1} (\mathrm{x} - \mu_k)\right\} \tag{1}$$

then the parameters μ_k and Σ_k can be estimated from the data. This gives a *quadratic* discriminant rule, since the boundaries between the classes (i.e. where $f_{k_1}(x) = f_{k_2}(x)$) are quadratic hyperplanes. If we further assume that the covariance matrices in equation (1) are all equal with $\Sigma = \Sigma_1 = \cdots = \Sigma_q$ then this gives Fisher's (1936) *linear* discriminant rule, since the boundaries are now linear hyperplanes.

More recently, increasingly flexible classes of models have been introduced, many of which attempt to provide an estimate of the joint distribution of the features within each class, which can, in turn, provide a classification rule. Recent advances have broadened the scope of linear discriminant-type analysis to encompass more flexible models, such as allowing for mixtures of Gaussian distributions (see Hastie et al., 1993a, b) Another recent approach, known as adaptive regression splines (Friedman, 1991) can also be interpreted as a generalization of decision trees (see below).

Nonparametric approaches to the estimation of f include kernel density estimation and k-nearest neighbor. The general form is to use

$$\hat{f}_k(\mathbf{x}) = \frac{1}{n_k} \sum_{i=1}^{n_k} K(\mathbf{x}, \mathbf{x}_i, \lambda_n) \tag{2}$$

where $K(\mathbf{x}, \mathbf{x}_i, \lambda_n)$ are kernel functions. For example we could use

$$K(\mathbf{x}, \mathbf{x}_i, \lambda_n) = \frac{1}{(\lambda_n\sqrt{2\pi})^p} \exp\left\{-\frac{1}{2}\sum_{j=1}^{p}\left(\frac{x_j - x_{ji}}{\lambda_n}\right)^2\right\},$$

with λ_n a smoothing parameter; if λ_n is small then we will get a very erratic estimate, and if λ_n is large then the estimate of $f(\mathbf{x})$ is very smooth. In this case, no distributional assumptions on the form of f are made, so if the data are multimodal, or very skewed, this method will often outperform methods based on normal distribution assumptions. However, one of the drawbacks is that the choice of λ_n is crucial to the performance. This can be tackled by cross-validation (see Habbema et al., 1974, for details).

The k-nearest neighbor approach considers estimating $f_h(\mathbf{x})$ by $k_h(\mathbf{x})/k$, in which the k observations nearest to $\mathbf{x}$ are such that $k_h(\mathbf{x})$ belong to class h with $k = \Sigma\, k_h(\mathbf{x})$. This leads immediately to the k-nearest neighbor classification rule: classify $\mathbf{x}^*$ to class l where

$$l = \arg \max_{h=1,\ldots,q} k_h(\mathbf{x}^*)$$

This method is also very useful when there are departures from the normal distribution. The main drawbacks are similar to those for the kernel method: The choice of k (which is akin to the choice of λ_n in equation (2)), and the choice of distance metric—i.e., what do we mean by "nearest"?

Trees

Decision Trees. Machine learning aims to generate classifying expressions simple enough to be understood easily by the human; they must mimic human reasoning sufficiently to provide insight into the decision process. To this end there has been substantial work in the machine learning community on methods to learn rules and trees from a set of training data.

There are a variety of classification trees—for example, CART (Breiman et al., 1984), ID3 (Quinlan, 1986), and C4.5 (Quinlan, 1993a)—but there is a common structure which we now outline. Most algorithms recursively partition the data, usually using a binary split. The partition boundaries are usually parallel to one of the attribute axes. Figure 1 shows an example of recursive partitioning of two attributes together with a corresponding decision tree. The basic differences in the algorithms lie in two areas.

1. *The splitting method.* CART uses the Gini index to measure the impurity at a node, and then choose the split to maximize the reduction in impurity; ID3 and C4.5 use a measure of information, and the CHAID decision tree algorithm (Kass, 1980) uses a splitting criterion based on a χ^2 test statistic.

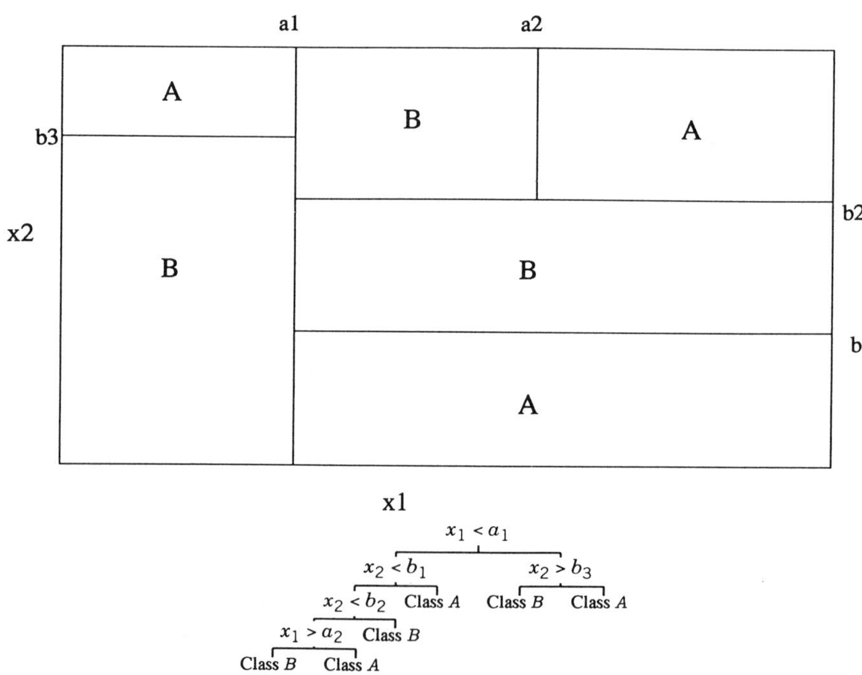

Figure 1. *Recursive partition (*top*) and corresponding decision tree (*bottom*).*

2. *The pruning method.* If the data are "noisy" then a very large tree would be required to classify correctly all of the examples in the training set. However, this *over-training* would lead to very poor performance when the test data are classified. In a sense this is equivalent to choosing a very small smoothing parameter in equation (2). To overcome this problem most algorithms "overgrow" the tree and then prune it back to conform with some cross-validation criteria. It should be noted that very large trees not only give large error rates, but also lead to poor understandability, so would not really qualify for the machine learning label. In practice, a small amount of accuracy would often be sacrificed for transparency.

In this volume, White and Liu's chapter (Chapter 1) traces the history of tree-based classification techniques from two points of origin: one in psychology and computer science and the other in mathematical statistics. They explain that tree-based classification is really a method of non-parametric discriminant analysis and the similarities with kernel-based methods are described briefly. The application of resampling techniques to obtain unbiased estimates of classification error rates is also discussed. They recommend the use of simulation experiments in the assessment of performance of these tree-based methods, because the utilization of many samples drawn from the same population allows the easy calculation of standard errors, which, in turn, facilitates the comparison of the power of different algorithms. Classification bias is also introduced, and they warn that classification error rate may not be the best measure of performance in circumstances where the class frequencies are unequal and classification bias is operating. (This problem of bias is also taken up by Henery in Chapter 7.) The problem relating to the fairness of variable selection in multi-branched trees when the variables have different numbers of possible discrete values is mentioned, and it is shown that there is a simple statistical solution. Finally, the correspondence between binary classification trees and logit models is explained and exemplified.

Chapter 2, by Müller and Wysotzki, investigates the use of a statistical splitting subroutine in a new recursive partitioning algorithm known as CAL5. The trees are automatically pruned during the tree construction with the help of a threshold for the estimated class probabilities in an interval. This pruning approach is completely different to those of the common approaches, which first build a complete decision tree (delivering an error rate of about 0% for the reclassification of the learning data set using this tree), and after that the tree has to be pruned by a sophisticated post pruning routine to gain some generalization feature for it. The node-splitting procedure of CAL5, which is presented in detail in their chapter, is based on hypothesis testing using the Binomial distribution for the class probabilities to be estimated. Its main differences to the splitting procedure of CART, as well as its advantages and disadvantages are discussed. A comparison of the CAL5 algorithm with selected statistical approaches (Discriminant Analysis and Sequential Probability Ratio Test, SPRT) is carried out by means of an artificial example of two classes with Gaussian probabilities in the feature space. For that the rejection regions of these three algorithms are computed, discussed, and compared. Finally, some real-world applications of CAL5 handled within the ESPRIT II Project *StatLog* and the German National Project WISCON (supported by the Federal Ministry for Research and Technology) are presented.

Probability Trees and Belief Networks. Often there is some expert knowledge which can be used as an aid to classification. There is a lot of research in artificial intelligence about the best way to use this knowledge for example, using symbolic reasoning. In this situation, we may want to combine this knowledge (which is often difficult to quantify) with information provided by the data, as used in most classification procedures. *Causal networks*, such as MUNIN (Jensen et al., 1987), CASTLE (Acid et al., 1991), and BIFROST (Højsgaard et al., 1992) are often used to encapsulate knowledge, and describe graphically the relationships between the variables. We illustrate the notion of a causal network with an example taken from Lauritzen and Spiegelhalter (1988). The structure of a knowledge base is represented as a *Directed Acyclic Graph* (*DAG*) in Figure 2. The arrows represent the direction of causality, meaning, for example, that a visit to Asia may cause tuberculosis. Note also that the graph contains logical links so that a positive X-ray and dyspnea do not discriminate between tuberculosis and lung cancer.

A causal network, consisting of nodes and edges, with a probability distribution defined for each node, can be written as a product of conditional distributions of the nodes given the *parents* of each node (see Neapolitan, 1990, for a proof). So, once a causal network has been established, it is straightforward to perform probabilistic inference. There are essentially two approaches

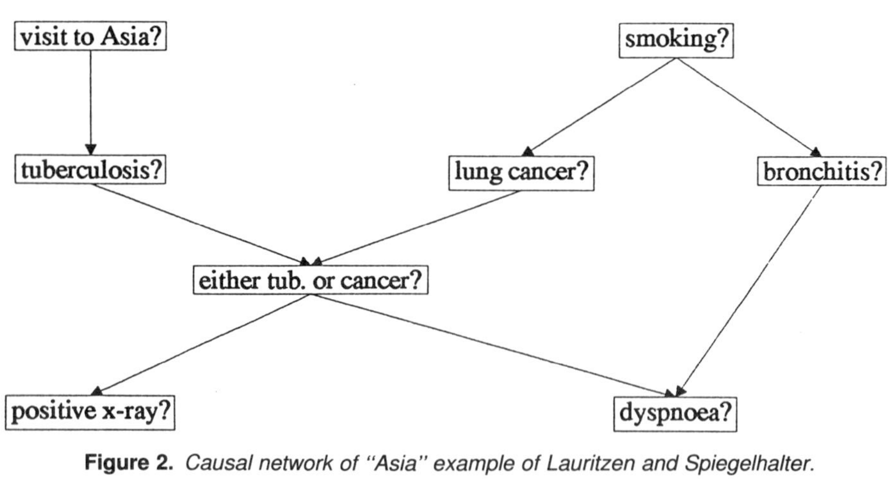

Figure 2. *Causal network of "Asia" example of Lauritzen and Spiegelhalter.*

to forming a causal network but these can be used together. One method uses only the data to learn the structure; the other can use expert knowledge, which can be described in probabilistic terms. It is often the case that the knowledge is expressed as a *prior* distribution, which can then be updated by the data to produce a *posterior* distribution and, subsequently, used for classification purposes, as well as a way of modeling the relationships between all the attributes. A good introduction, from a statistical perspective, to selection of undirected graphical representations, can be found in Chapter 8 of Whittaker (1990). Pearl (1988) is devoted to the construction of directed graphs, from an artificial intelligence perspective.

Any accepted method should possess 1) some theoretical, or heuristic justification, and 2) a reasonable level of performance—perhaps in terms of prediction accuracy, or interpretability. It seems that graphical techniques offer the most natural bridge between statistical and ML solutions to a classification task. In Chapter 3, by Chiogna, class-probability trees and belief networks are compared with respect to the problem of building predictive systems for the diagnosis of congenital heart diseases in newborn babies. Considering the mixed nature of the diagnostic features, the presence of missing values and of high-order interactions among the variables, the problem of building a classifier can be considered a fairly complex task to solve. The comparison is focused on numerical assessment of the accuracy of the classifiers and on the transparency of the models' structures. To assess accuracy, a cross-validation experiment is organized. Using the existing temporal stratification of the data, three training samples and their respective test samples are identified; data-derived classifiers are built for each of the compared techniques on the training samples and their numerical performance assessed on the test samples. Ability to gain insight into the problem is explored by comparing the classifiers' structures with equivalent expert-derived models. The experiment shows that class-probability trees outperform belief networks in terms of numerical accuracy of the derived predictive systems, in contrast with the potential for belief networks to produce a better insight into the problem. This emphasizes the difficulties in finding the right balance between good numerical performance of the classifiers and transparent understanding of the true underlying model generating the diagnostic classes. However, a closer look at the structures of the classifiers and at the misclassifications produced by them highlights that the two different types of classifiers perform comparably on homogeneous areas of the sample space; that is, on areas characterized by observations sharing the same diagnosis. This suggests that performance of the belief networks might be improved by refining the structure of the network on the heterogeneous areas of the sample space. If this could be achieved, belief networks could be a better model overall for supporting the diagnostic task.

In multiple-concept learning from examples it is possible to distinguish two situations: that in which concepts to be learned are mutually independent and that in which such an independence assumption does not hold. Chapter 4 by Malerba, Semeraro, and Esposito, presents two first-order learning systems, namely RES/H and RES/CSL, which integrate statistical and conceptual methods in order to learn independent and dependent multiple concepts, respectively. The concepts are represented by Directed Acyclic Graphs, in which the zeroth order is learned first. In RES/H the results of Fisher's linear discriminant analysis is mapped into a new descriptor which is considered, together with other symbolic features, by a conceptual learning system that performs a separate-and-conquer search strategy in the space of the inductive hypotheses. The integration aims at obtaining a more powerful and efficient learning system, since it is possible to cope with a heterogeneous set of features (qualitative and quantitative), to reduce the effect of noise, to describe structured objects, as well as to define and exploit the user's background knowledge. However, the independence assumption made in RES/H may raise some problems, such as non-learnability of concepts, decrease of predictive accuracy of learned concepts, and increase of learning time. On the contrary, dropping such an assumption raises other issues, such as the definition of the order in which concepts are to be learned. In Chapter 4, Malerba and colleagues investigate these issues and show that it is possible to use statistical techniques in order to infer concept dependencies before starting the learning process. Such dependencies are actually exploited to define the right order in which concepts have to be learned. Several statistical methods are presented and tested, some of which are general and appropriate for attribute-based domains, while another one has been implemented in RES/CSL and is appropriate for structural domains. Experimental results on an application in document understanding show that the statistical techniques improve the classification accuracy and simultaneously reduce the learning time for multiple dependent/independent concepts.

Rules

Note that decision trees can be written as a set of rules and there are some algorithms that aim to construct a set of concise, comprehensible rules. Often decision rules are generated for each class in turn, for each class assigning a *universal* rule, which assigns all examples to the current class. Specializations are then repeatedly formed and tested until all rules consistent with the data are found. Each rule must correctly classify at least a prespecified percentage of the examples belonging to the class in question with as few as possible *negative* examples (from other classes) covered.

An example of a rule learning algorithm is CN2 (Clark and Niblett, 1988), which is based on Michalski's (1969) AQ algorithm. Michalski introduced some terminology in rule induction: an attribute-test is called a *selector*; a conjunction of selectors is called a *complex*; and a disjunction of complexes is called a *cover*. If a rule is true of an example, we say that it *covers* the example. Thus CN2 works by generating a *complex* (a conjunct of attribute tests) which *covers* (is satisfied by) a subset of the training examples. This complex forms the *condition* part of a production rule: "**if** condition **then** class = c", where *class* is the most common class in the training set which satisfies the condition. The condition is a conjunction of selectors, each of which represents a test on the values of an attribute such as "weather = wet". To test significance, CN2 uses the entropy statistic, which is given by

$$2 \sum_{i=1}^{n} p_i \log(p_i/q_i),$$

where the distribution $p_1, p_2, \ldots, p_n$ is the observed frequency distribution of examples among classes satisfying a given complex and $q_1, q_2, \ldots, q_n$ is the expected frequency distribution of the same number of examples under the assumption that the complex selects examples randomly. This is taken as the $N = \Sigma p_i$ covered examples distributed among the classes with the same probability as that of examples in the entire training set. This statistic provides an information-theoretic measure of the (non-commutative) distance between the two distributions. The user provides a threshold of significance below which rules are rejected.

Decision trees *can* be readily converted into rules in a straightforward manner, but there are more sophisticated ways to carry out reorganization; see Quinlan's (1993a) trees-into-rules algorithm. This is especially useful when the decision tree is too big to be comprehensible, and classifiers generated in this way are usually much better in "mental-fit" even if there is a modest increase in error rate.

Decision-making systems exploiting unordered decision rules may be accompanied by some conflict situations, particularly when several rules associated with different classes match a selected object. Therefore, it is useful within certain classification procedures that each decision rule be accompanied by a numerical factor which can express its properties and characterize a measure of belief in correctness of the rule, its power, predictability, reliability, likelihood, and so forth. A collection of these properties is symbolized by a function commonly called the *rule quality*. There have been several attempts to define formulas for rule quality, the majority of which represent an intuitive or ad hoc approach. Chapter 5, by Bruha, presents two approaches to the definition of the rule quality: empirical and statistical. General characteristics of a formula which may be used as rule quality are pointed out, namely, its monotony toward the consistency and completeness of decision rules. These two characteristics can be combined to yield a single number: the rule quality, the meaning of which should be more or less understandable to human experts. A common policy is to use a weighted sum or multiplication of the above components.

Bruha goes on to discuss the potential use of three statistical techniques: contingency tables, information theory, and confidence sets. A characteristic called "independence property" is defined and it is indicated that all statistical formulas yield this property. Also, it is indicated that all the formulas are non-increasing functions of consistency and completeness. As soon as a formula for rule quality is selected and an unordered set of decision rules is induced, several different classification schemes can be used. Three of them are discussed by Bruha: ordering an unordered set of rules; weighted sum of qualities; and recalculating quality for combined rules. Recalculating combined rule qualities and additional schemes are experimentally tested on several well-known AI databases and compared from the empirical viewpoint. The covering learning algorithm CN4 is used as an inductive learning vehicle. The experiments reveal the classification characteristics of various quality formulas and classification schemes. The conclusions take into account both the experimental results and some theoretical observations. It is emphasized that the more complex classification schemes should be used in the case of redundant (multiple) rules with a large amount of overlap.

Neural Networks

The field of neural networks has arisen from diverse sources, and there are a broad class of techniques under this heading. Generally, neural networks consist of layers of interconnected nodes, each node producing a non-linear function of its input, which may come from other nodes or directly from the data. It has been argued that neural networks mirror the behavior of networks of neurons in the brain, but they are now viewed as useful tools in their own right.

Figure 3 shows an example of a multilayer perceptron with one hidden layer. The links between the units represent *weights*, so there is a set of weights, $w_{ij}^{(HI)}$, connecting the (input) observations to the hidden layer, and a set of weights, $w_{ij}^{(OH)}$, connecting the hidden layer to the output layer. The output, $y_i^{(O)}$, is then given by

$$y_i^{(O)} = f_{(O)}\left(\sum_k w_{ik}^{(OH)} f_{(H)}\left(\sum_j w_{kj}^{(HI)} x_j\right)\right)$$

where the univariate $f_{(.)}$ are usually both given by the transfer function

$$f(x) = \frac{1}{1 + e^{-x}}$$

The number of hidden layers and the number of nodes in each layer are often user-determined choices, and sometimes this is a difficult problem. The network is trained, that is, the weights are selected, to minimize the error sum of squares determined by the output values. The optimization of weights is done by gradient descent, either first-order or second-order. One of the difficulties lies in how to avoid *over-training* the network, in which the error is minimized on the training data to the extent that the network is unable to generalize, thus

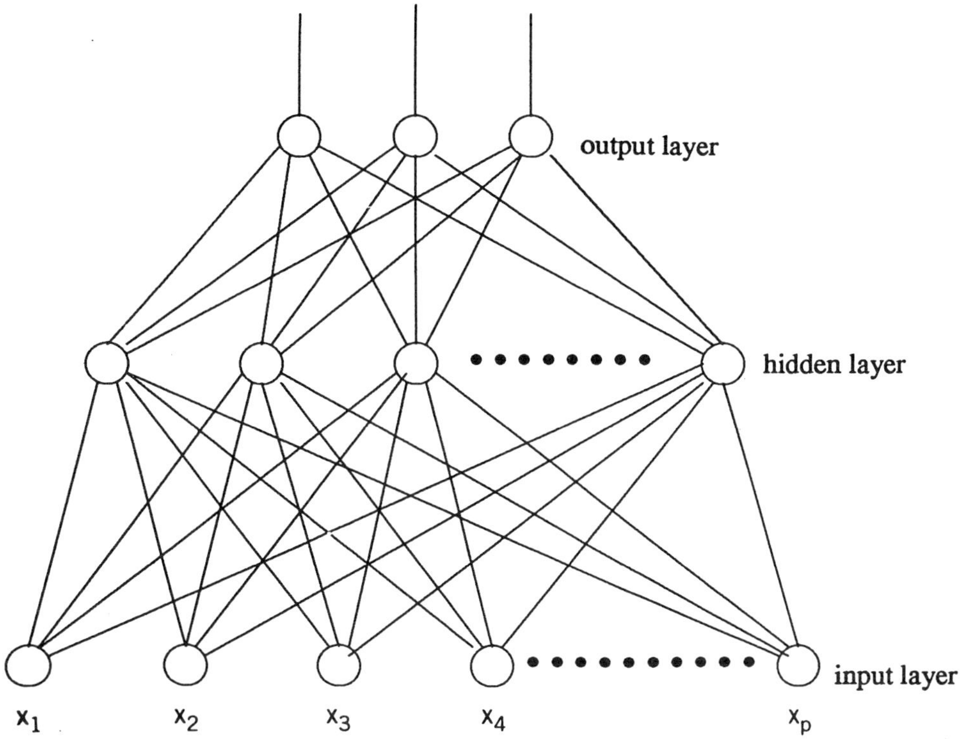

Figure 3. *Example of a neural network with one hidden layer.*

producing a large error rate for the unseen test data. Again, this problem is akin to that of choosing the smoothing parameter in the kernel density classifier or, to a lesser extent, in choosing the size of classification tree. However, it should be noted that neural networks have almost no user-transparency for any size of network.

There are close links between neural nets and flexible regression techniques, which are extensions of linear discrimination (see Ripley, 1944, and the references therein).

Another new algorithm, known as DIPOL92, is outlined in Chapter 6 by Schulmeister and Wysotzki. This is essentially a hybrid method, which uses methods from statistics as well as neural net approaches. In the first step of the algorithm, initial positions of the discriminating hyperplanes are determined by linear regression for each pair of classes. To optimize these positions, depending on the misclassified patterns, an error criterion function is defined. The second step involves minimizing this function by a gradient descent procedure for each hyperplane separately, corresponding to learning the weights of an artificial neuron. As an optional third step in the case of non-convex class regions, a clustering procedure decomposing the classes into appropriate subclasses can be applied. From the point of view of neural nets this corresponds to a reasonable fixing of the number and initial weights of neurons *a priori*. Thus a serious and actual problem of neural nets is solved in a pragmatic way using a statistical approach. The optimization of the positions of the decision hyperplanes represents an accumulated correction, that is, learning by epoch (of course, learning by epoch can be replaced by incremental learning). By using the classifier in the classification process the positions of the decision hyperplanes have to adapt to changing class distributions on the basis of environmental factors in time (not too rapidly, as compared with the initial conditions in the off-line learning process). For this they introduce on-line incremental supervised learning from new patterns, which are misclassified by the developed classifier.

Combining Methods

An increasingly asked question is whether methods can be *combined* in some manner to produce an improvement over any individual approach. This may lead to a loss of transparency in the final rule, but could lead to substantial reduction in error rates. Chapter 7, by Henery, discusses various ways to combine classification procedures. His common theme is the use of predicted class conditional probabilities as a basis for improving performance of algorithms, either individually or in combination. The first consideration is the processing of the output of several classifiers with the aim of finding similarities among the classifiers and clustering in the observations. Correspondence Analysis (see Hill, 1982, for a description) is normally used for analyzing contingency tables, but can be used more generally to analyze tables with positive entries and probabilities in particular. If several classification procedures are applied to a given data set, and the predicted class conditional probabilities are collected together, then the probabilities can be analyzed via correspondence analysis. Henery also considers the possibility of combining several classifiers through *stacked generalization* (Wolpert, 1992): here the aim is to create a better classifier by combining the best from several classifiers. This involves using estimated class-conditional probabilities of classifiers as attributes for alternative classifiers. So, in this context, a prediction is taken to be a predicted class-conditional probability, rather than the predicted class. This is motivated by appealing to an observation made by Fukunaga and Ando (1977), to the effect that class-conditional probabilities are the optimal Bayes features. These ideas are illustrated on a practical example from industry. Finally, he considers applying one classifier to the output of another in what might be called serial order (Henery, 1994). Here the suggestion is more specific: look for a particular form of bias in the output of the classifier, and use logistic regression techniques to remove this bias if it exists. The aim is to build a model for the true class probabilities as a function of the estimated class probabilities, using logistic regression as a post-processor to reduce the bias. This approach precludes classification methods that do not produce class probabilities, in particular, the nearest neighbor method (Fix and Hodges, 1951), or some of the older decision tree programs (Weiss and Kulikowski, 1991).

Statisticians are often more concerned with accuracy in predicting future observations than with understanding the structure between the attributes and their relationship to the class. On the face of it, it might appear that ML and statistics approaches merely have a different emphasis. However, an important criterion for a classification method to qualify under the ML heading is that the derived rule should be meaningful to humans and evaluable in the head. On this basis, many statistical algorithms, and especially neural net algorithms, would apparently not qualify. However, as described in Chapter 8 by Taylor, this may be more of a problem in presentation, rather than with the methods themselves. He considers cases-in-point of the linear discriminant function, the nearest neighbor algorithm, and a neural net method (LVQ), and shows how, with a little adaptation, they could qualify as ML methods. His suggested framework—of a recursive "distance" decision tree—is then adapted and extended in its own right, resulting in the potential for a superset algorithm in the sense that it can take the best from each learning method, and still result in an evaluable, and understandable rule. The classification rules are thus described in terms of "likeness", for example, that this observation is most like one of a number of prototypes whose class is known. Another formulation of this is known as *Case Based Reasoning*.

Chapters 7 and 8 deal with combining methods for the purposes of classification, whereas Chapter 9, by Baroglio, Giordana, Lo Bello and Piola, describes an application in control which seeks to combine neuro and fuzzy approaches. This multistrategy method combines statistical and connectionist learning techniques in order to synthesize a fuzzy controller from examples of correct behavior supplied by a human teacher. Fuzzy controllers are universal function approximators (Wang, 1992), which share properties both with neural networks and with symbolic rule-based systems, and then are suitable for combining symbolic and sub-symbolic learning algorithms. The learning procedure consists of several steps. First, an approximated control function is learned from examples using the algorithm CART (Breiman et al., 1984), which produces a regression tree. Then the regression tree is transformed into a network of symbolic rules, which becomes the layout of a fuzzy controller. The rule network can be simplified using post-pruning techniques. Finally, the fuzzy controller so obtained is refined off-line (or on-line), using a modification of the back-propagation algorithm, until the required accuracy is achieved. The method is evaluated and compared with alternative ones, using several case studies, including the classical test of Mackey Glass series, and a real control problem where a robot KUKA-IR 361 has to be controlled in order to accomplish the "peg-into-hole" task. In all cases, fuzzy controllers generated automatically proved to be better than other neural networks, such as multilayer perceptron and Time Delay Neural Network, in controlling the robot. Moreover, the resulting method was easy to use and computationally efficient. Finally, Baroglio and colleagues discuss the problem of generating the examples and show how it is possible to generate a controller better than the teacher itself by properly selecting the training events.

Applications

Many banks, building societies, and consumer credit companies are investigating the potential of credit scoring in assessing applications for credit. An empirical study of the point-awarding approach is made by Häußler (1979, 1981a, b). Fahrmeir and colleagues (1984) compare the results of a point awarding approach with the results obtained by the linear discriminant. In von Stein and Ziegler (1984) the authors use the k-nearest neighbor approach to analyze the problem of prognosis and surveillance of corporate credit risk. Linear discriminant is applied by Bretzger (1991) to early risk recognition in disposition credits. In a comprehensive study of corporate credit granting reported in Srinivisan and Kim (1987), the authors evaluate various approaches including parametric, nonparametric, and judgmental classification procedures. Within the nonparametric approaches they use a "recursive partitioning" method based on the decision tree concept. Their results show that this "recursive partitioning" approach performs better than the others.

Two recent developments in statistical credit scoring are discussed in Chapter 10, by Hand and Henley. They consider classical statistical approaches such as linear regression, logistic regression, and discriminant analysis, as well as more recent nonparametric methods such as k-nearest neighbor classifier. The first part of their chapter describes some of their work in applying such methods to the credit-scoring problem. Hand and Henley look at the choice of metric, the choice of k, and compare the performance of the method with regression approaches on a particular data set. Naturally, in a competitive world, credit-scoring agencies wish to improve their systems. Such improvements must be based on the application forms and on the true good/bad status, when this is known. While the status of accepted applicants is ultimately known, the true status of rejected applicants remains unknown, although the application form data is available for all applicants. In the second part of their chapter they review methods which have been suggested for using all of these available data and show that the application form data for the rejected applicants can only be used if additional assumptions are made. This section also has particular relevance to other applications (particularly in medicine) where the class is not observed.

Forecasting of future development of time series is a very important task with various applications in technical and commercial areas. Especially, it finds a lot of applications in finance. The use of statistical approaches to the forecasting of time series has a long tradition. Box and Jenkins (1976), Granger and Newbold (1977), and Theil (1971) are examples for such approaches.

Recently, a lot of work has been devoted to the application of machine learning and neural networks to forecasting. Graf and Nakhaeizadeh (1994), Bol et al. (1995) and Weigend and Gershenfeld (1993) are some examples of recent work. The application of case-based reasoning to forecasting is discussed in Nakhaeizadeh (1994).

In Chapter 11, by Westphal and Nakhaeizadeh, various forecasting procedures are applied to predict the development of the German Stock Price Index. Besides the single procedures based on the work of Quinlan (1992), Aha et al. (1991), and traditional k-nearest neighbor, they examined the performance of some combined forecasting procedures, among them an approach suggested recently by Quinlan (1993b). In this approach different model-based procedures are combined with an instance-based method. Their empirical results show that the combined approaches are, in many cases, more appropriate in the forecasting of German Stock Price Index. The authors also discuss, at the beginning of their contribution, the theoretical aspects of the applied single and combined approaches.

TOWARD KNOWLEDGE DISCOVERY IN DATABASES

The problem of *feature selection* has been widely studied for some time. In this case there is usually a model which has a *prespecified known form*, and the problem is to estimate parameters and select the variables that are useful in the context of the model. However, in practice, *model selection* is often much more important. Recent computing power has enabled a large number of models to be considered, and this is becoming common practice in many fields. As an example, suppose we want to predict the value of a response variable y_i by a (potentially) large number, p, of explanatory variables $\mathbf{x}_i$, $i = 1, \ldots, n$, with $\mathbf{x}^T = (x_1, x_2, \ldots, x_p)$. Even if we assume a linear regression model, there are pitfalls in using standard forward and/or backward selection methods, but if we extend the search space to include models of the form

$$y = b_0 + \sum_{j=1}^{p} b_j x_j^{c_j}$$

then the problem becomes almost intractable. To make the problem even harder we could extend the model space to allow for any of the variables to take the place of y, or consider also possible ratios and/or products of the explanatory variables. In such situations *model selection*, feature selection, and *model checking* are difficult. Statistical approaches to handle these uncertainties often make use of Bayesian methods (Chatfield, 1995, and Draper, 1995).

Existing learning algorithms often fail to discover valuable knowledge in real-world problems because of the abundance and redundancy of attributes (features) and of the high interaction between domain attributes. The representation of data often requires many features, only a few of which may be related to the target concept. Irrelevant features degrade the performance of concept learners both in speed and in predictive accuracy. The feature selection phase applies a set of algorithms to the input raw data that can detect and inspect feature intercorrelations. As a result, irrelevant features can be eliminated and the remaining features ranked according to their relevance to the concept description. In this framework, Richeldi and Rossotto propose in Chapter 12 ADHOC (Automatic Discoverer of Higher-Order Correlations), a new statistical algorithm that can discover the deep structure of data by detecting first- and higher-order correlations, partial correlations, and dependencies between data features. This knowledge makes machine learners employed in the concept-discovery phase focus on the significant concepts only, improving the efficiency and effectiveness of analysis. The new algorithm has been developed to handle missing values and attributes that lie on a nominal, ordinal, and interval scale, since most of the real-world problems do contain all these kind of data. Tests on algorithm performance has been conducted both on artificial data and on industrial domains (estimation of costs of software projects; classification of digital signal; and others), drawn from the telecommunication domain. This work is part of a research project aimed at integrating statistical and machine learning techniques to develop efficient and effective tools for the analysis of data drawn from telecommunication domains. More specifically, they developed a three-phase methodology for the process of knowledge and data discovery, in which the concept-discovery phase is driven by the knowledge on data dependencies collected in the feature-selection phase. (ADHOC is one of the approaches used to grasp dominant relationships among data.) The learned concepts are then represented using a notation easily understood by non-expert users in the concept-representation phase.

In data analysis, one major task is to induce a numerical law from the input experimental data. There are two extreme situations: in the first, a lot of prior information is available: the context of the data, the laws previously discovered, and so on: in summary the universe is quite well formalized. In this case, a simple solution is first to define a list of possible models (or forms) of the law, to apply regression on them, and to select the best law. In the opposite case, where the problem is totally new and unknown, the variables must be defined and the measuring devices chosen in order to set up hypotheses on the law and on its model; this requires much patience. Chapter 13, by Moulet, makes a comparison of numerical law discovery systems, which tackle the second problem and tend to assist the user by discovering the model of the law. Among these systems, a first approach, that she calls *heuristic*, deals with finding heuristics able to guide this search efficiently, and tries to simulate human experts solving similar problems. The initial ones were discovered by Langley et al. (1986) and implemented in the system BACON. These simple heuristics are able to reduce significantly the size of the search space. Her system ARC.2 (Moulet, 1992, 1993) is a direct descendant of BACON.3 (Langley, 1979) and of ABACUS (Falkenhainer and Michalski, 1986). It integrates heuristically polynomial regression in an iterative search. In this first family of systems, the main principle is to discover the law step by step, like a statistician usually does. A second and more recent approach does not discover the law by decomposing it, but explores exhaustively a solution space bounded by some parameters. This method, which she calls *exhaustive*, has no cognitive explanation, since it cannot be tractable by hand, but, on the contrary, relies on brute computational force and assumes that the law is more probably simple and rapidly accessible. Two main systems belong to this family: *E*quation *F*inder (EF) (Zembowicz and Zytkow, 1991, 1992) looks for the best adequate solution among many possible relations between only two variables x and y, and LAGRANGE (Dzeroski and Todorovski, 1993) applies linear regression on all products of the initial variables and proposes any satisfactory relation. These two approaches consider the discovery task in an inductive way, that is, only from the given data, with very little additional knowledge, and discover the model and the law at the same time. A third approach is guided by some background knowledge helping to constrain the search space. E* (Schaffer, 1990) or Discover (Wu and Wang, 1991) consider, for instance, a list of prototypes, which represent all possible models the system can test. In this approach, the goal is not to reach as many models as possible, but to propose an efficient search by exploring a smaller space. The comparison of all discovery systems would be both difficult and irrelevant, as the methods used, the scope of the systems, and their practical role could not be directly compared. She has therefore restricted her comparison to the two inductive approaches, through the comparison of three systems: on one hand, ARC.2 is representative of the *heuristic* approach; on the other hand, EF and LAGRANGE are two main representatives of the *exhaustive* approach. Although the two inductive approaches have a similar goal, they may be opposed for many reasons. Instead of proposing a list of goals for quantitative discovery as did Falkenhainer and Michalski (1986), she proposes to compare the three systems EF, LAGRANGE, and ARC.2, according to the following four main characteristics, which are important both on a theoretical and a practical point of view: 1)—the type of data they can deal with, 2)—the constraints limiting the application domain, 3)—the reachable laws, and 4)—the complexity of the search. Moulet shows in her chapter that the main constraint of ARC.2 concerning the organization of the data allows the automatic extraction of some information that is usually explicitly asked of the user and which helps to reduce the search significantly. On the other hand, exhaustive methods accept any kind of data, but, since they always apply the same strategy, may present a complexity much greater than the complexity of the law.

ANALOGY

In all of the above sections, an accurate prediction or classification will usually require quite a lot of observations, particularly in the presence of noise. However, it is possible to develop a rule on the basis of only a few, or indeed, only one, example. The essential ingredient is to view the rule as a transformation, and to describe the transformation using a measure of complexity. The selected rule is the transformation which is the simplest, that is, has minimum complexity. This approach has some affinity with Occam's Razor Principle, which has been used by scientists for centuries.

Chapter 14, by Cornuéjols, deals with learning transformations by minimizing description length. He approaches this by appealing to analogy, which is seen as a special form of induction in that 1) it works from the specific to the specific, apparently without relying on vast numbers of observations; and 2) it does not necessarily involve general rules that would apply both to the "source" of the analogy and to its "target." It thus appears quite remote from the realm of statistics. A survey of the current computational approaches to analogy shows that they all embrace the "analogy as a matching process" perspective, differing only on constraining factors and search mechanisms. An appropriate analogy is accordingly one that makes a good match between the descriptions of the analogs, allowing induction of new properties for the ill-known target. These matching-based theories give way to some dissatisfaction however. Mainly, 1) deep reasons, other than intuitive ones, for considering analogy as a matching process is lacking; 2) the proposed mechanisms are very dependent on the knowledge representation; 3) they do not explain well some properties of analogy, namely, non-symmetry and non-idempotency of analogy, or the propensity to learn making good analogies; and 4) these mechanisms exploit but do not explain why analogy preferably transfers abstract concepts rather than more literal ones. This chapter proposes a new point of view on analogy that offers possible answers to these queries. Analogy is seen as a realization of an economy principle that minimizes the complexity of the transformation from the source to the target, as measured by description length. These description lengths, in turn, are dependent upon statistical properties of the concepts and abstractions used to account for the analogy. This is where learning can take place, so as to facilitate further analogies in the same domain. Tests on a classical domain task confirm that the application of this principle correctly predicts the best analogies. The relevance of this research to scientists not directly interested in analogy is twofold. First, it provides a new distance measure between objects that can naturally take into account complex and symbolic properties. Second, it shows that algorithmic complexity measures can usefully be applied to problems involving large collections of objects, as well as to problems involving few objects. In both cases, underlying parameters (which represent useful summaries) can be learned to tune the system to its environment.

REFERENCES

Acid, S., do Campos, L. M., d., González, A., Molina, R., and Pérez de la Blanca, N. (1991). *CASTLE: Causal Structures from Inductive Learning*, Release 2.0. Rep. No. 91-4-3. University of Granada.

Aha, D. W., Kibler, D., and Albert, M. K. (1991). Instance-based learning algorithms. *Mach. Learn.* **6**, 37–66.

Bol. G., Nakhaeizadeh, G., and Vollmer, K. H., eds. (1995). *Finanzmarktanwendungen neuronaler Netze und Oekonometrischer Verfahren.* Physica-Verlag, Heidelberg.

Box, G. E. P., and Jenkins, G. M. (1976). *Time Series Analysis, Forecasting and Control,* 2nd ed. Holden-Day, San Francisco

Breiman, L., Friedman, J. H., Olshen, R. A., and Stone, C. J. (1984). *Classification and Regression Trees.* Wadsworth & Brooks, Monterey, CA.

Bretzger, T. M. (1991). Die Anwendung statistischer Verfahren zur Risikofruherkennung bei Dispositionskrediten. Ph.D. Thesis, Universität Hohenheim.

Chatfield, C. (1995). Model uncertainty, data mining and statistical inference (with discussion). *J. Roy. Stat. Soc., Ser. A* **158**, 419–466.

Clark, P., and Niblett, T. (1988). The CN2 induction algorithm, *Mach. Learn.* **3**, 261–283.

Draper, D. (1995). Assessment and propagation of model uncertainty (with discussion). *J. Roy. Statist. Soc., Ser. B* **57**, 45–97.

Dzeroski, S., and Todorovski, L. (1993). Discovering dynamics. In *Proceedings of the Tenth International Conference on Machine Learning*, Amherst, MA, pp. 97–103. Morgan Kaufmann, San Mateo, CA.

Fahrmeir, L., Haussler, W., and Tutz, G. (1984). Diskriminanzanalyse. In *Multivariate statistische Verfahren* (L. Fahrmeir and A. Hamerle, eds.). de Gruyter, Berlin.

Falkenhainer, B. C., and Michalski, R. S. (1986). Integrating quantitative and qualitative discovery in the ABACUS system. *Mach. Learn. J.* **3**.

Falkenhainer, B. C., and Michalski, R. S. (1990). In *Machine Learning: An Artificial Intelligence Approach* (Y. Kodratoff and R. S. Michalski, (eds.), Vol. 3. Morgan Kaufmann; San Mateo, CA.

Fisher, R. A. (1936). The use of multiple measurements in taxonomic problems. *Ann. Eugen. (London)* **7**, 179–188.

Fix, E., and Hodges, J. L. (1951). *Discriminatory Analysis, Nonparametric Estimation: Consistency Properties*, Rep. No. 4, Proj. No. 21-49-004. USAF School of Aviation Medicine, Randolph Field, TX.

Friedman, J. (1991). Multivariate adaptive regression splines (with discussion). *Ann. Stat.* **19**, 1–141.

Fukunaga, K., and Ando, S. (1977). The optimum nonlinear features for a scatter criterion in discriminant analysis. *IEEE Trans. Inf. Theory* **IT-23**, 453–459.

Graf, J., and Nakhaeizadeh, G. (1994). Recent development in solving the credit-scoring problem. In *Logistic and Learning for Quality Software, Management and Manufacturing* (V. L. Plantamura, B. Soucek, and G. Visaggio, eds.). Wiley, New York.

Granger, C., and Newbold, P. (1977). *Forecasting Economic Time Series.* Academic Press, New York.

Habbema, J. D. F., Hermans, J., and Van der Burght, A. T. (1974). Cases of doubt in allocation problems. *Biometrika* **61**, 313–324.

Hand, D. J. (1981). *Discrimination and Classification.* Wiley, Chichester, UK.

Hastie, T., Buja, A., and Tibshirani, R. (1993a). Penalized discriminant analysis.

Hastie, R., Tibshirani, R., and Buja, A. (1993b). Flexible discriminant analysis by optimal scoring.

Häussler, W. M. (1979). Empirische ergebnisse zur diskriminationsverfahren bei kredit-scoringsystemen. *Z. Oper. Res., Ser. B* **23**, 191–210.

Häussler, W. M. (1981a). Methoden der punktebewertung für kreditscoringsysteme. *Z. Oper. Res.* **25**, 79–94.

Häussler, W. M. (1981b). *Punktebewertung bei Kreditscoringsystemen.* Knapp, Frankfurt.

Henery, R. (1994). Reducing bias in statistical discrimination. *Appl. Stat.* (submitted for publication).

Hertz, J., Krogh, A., and Palmer, R. (1991). *Introduction to the Theory of Neural Computation.* Addison-Wesley, Reading, MA.

Hill, M. (1982). Correspondence analysis. In *Encyclopaedia of Statistical Sciences* (S. Kotz, N. L. Johnson, and C. B. Read, eds.), Vol. 2, pp. 204–210. Wiley, New York.

Højsgaard, S., Skjøth, F., and Thiesson, B. (1992). *User's Guide to BIFROST,* Tech. Rep. R-92-2001. Institute for Electronic Systems, Aalborg University, Aalborg, Denmark.

Jensen, F. V., Andersen, S. K., Kjaerulff, U., and Andreassen, S. (1987). MUNIN: On the case for probabilities in medical expert systems—a practical exercise. In *Proceedings of the First Conference of European Society for Artificial Intelligence in Medicine* (J. Fox, M. Fieschi, and R. Engelbrecht eds.), pp. 149–160. Springer-Verlag, Heidelberg.

Kass, G. V. (1980). An exploratory technique for investigating large quantities of categorical data. *Appl. Sta.* **29**, 119–127.

Langley, P. (1979). Rediscovering physics with BACON. 3. In *Proceedings of the Sixth International Joint Conference on Artificial Intelligence*, pp. 505–507. Morgan Kaufmann, San Mateo, CA.

Langley, P., Zytkow, J., Simon, H., and Bradshaw, G. L. (1986). The search for regularity: Four aspects of scientific discovery. In *Machine Learning: An Artificial Intelligence Approach* (R. S. Michalski, J. G. Carbonell, and T. M. Mitchell, eds.), Vol. 2. Morgan Kaufmann, Palo Alto, CA.

Lauritzen, S. L., and Spiegelhalter, D. J. (1988). Local computations with probabilities on graphical structures and their application to expert systems (with discussion). *J. R. Stat. Soc., Ser. B* **50**, 157–224.

McLachlan, G. J. (1992). *Discrimination and Statistical Pattern Recognition.* Wiley, New York.

Michalski, R. S. (1969). On the quasi-minimal solution of the general covering problem. In *Proceedings of the Fifth International Symposium on Information Processing*, pp. 125–128. Bled, Slovenia.

Michie, D., Spiegelhalter, D. J., and Taylor, C. C., eds. (1994). *Machine Learning, Neural and Statistical Classification.* Ellis Horwood, Chichester, UK.

Moulet M. (1992). ARC.2 Linear regression in ABACUS. In *Proceedings of the ML 92 Workshop on Machine Discovery, Aberdeen.*

Moulet M. (1993). ARC: Decouverte empirique de lois numériques ou ABACUS Revu et Corrige. Thèse d'Université, Paris 11.

Nakhaeizadeh, G. (1994). Learning prediction of time series. A theoretical and empirical comparison of CBR with some other approaches. In *Topics in Case-Based Reasoning* (S. Wess, K. D. Althoff, and M. M. Richter, eds.). Springer-Verlag, Berlin.

Neapolitan, E. (1990. *Probabilistic Reasoning in Expert Systems.* Wiley, New York.

Pearl, J. (1988). *Probabilistic Reasoning in Intelligent Systems: Networks of Plausible Inference.* Morgan Kaufmann, San Mateo, CA.

Quinlan, J. R. (1986). Induction of decision trees. *Mach. Learn.* **1**, 81–106.

Quinlan, R. (1992). Learning with continuous classes. In *Proceedings of the Fifth Australian Joint Conference on Artificial Intelligence.* pp. 343–348. World Scientific Press, Singapore.

Quinlan, J. R. (1993a). *C4.5: Programs for Machine Learning.* Morgan Kaufmann, San Mateo, CA.

Quinlan, J. R. (1993b). Combining instance-based and model-based learning. In *Proceedings of the Tenth International Conference on Machine Learning, Amherst, MA*, pp. 236–243. Morgan Kaufmann, San Mateo, CA.

Ripley, B. D. (1994). Neural networks and related methods for classification (with discussion). *J. R. Stat. Soc., Ser. B* **56**, 409–456.

Schaffer, C. (1990) Domain-independent scientific function-finding, Tech. Rep. No. LCSR-TR-149, Ph.D. Dissertation, Rutgers University, New Brunswick, NJ.

Srinivisan, V., and Kim, Y. H. (1987). Credit granting: A comparative analysis of classification procedures. *J. Finance* **42** 665–681.

Theil, H. (1971). *Applied Economic Forecasting*, North-Holland, Amsterdam.

von Stein, J. H. and Ziegler, W. (1984). The prognosis and surveillance of risks from commercial credit borrowers. *J. Bank. Finance* **8**, 249–268.

Wang, L.-X. (1992). Fuzzy systems are universal approximators. In *Proceeding of the 1992 IEE International Conference on Fuzzy Systems*, pp. 1163–1170.

Weigend, A., and Gershenfeld, N., eds. (1993). *Predicting the Future and Understanding the Past.* Addison-Wesley, Redwood City, CA.

Weiss, S. M., and Kulikowski, C. A. (1991). *Computer Systems that Learn: Classification and Prediction Methods from Statistics, Neural Networks, Machine Learning and Expert Systems.* Morgan Kaufmann, San Mateo, CA.

Whittaker, J. (1990). *Graphical Models in Applied Multivariate Statistics.* Wiley, Chichester, UK.

Wolpert, D. (1992). Stacked generalization. *Neural Networks* **5**, 241–259.

Wu, Y. H., and Wang, S. (1991). Discovering functional relationships from observational data. In *Knowledge Discovery in Databases* (G. Piatetsky-Shapiro and W. J. Frawley, eds.). AAAI Press/MIT Press, Cambridge, MA.

Zembowicz R., and Zytkow, J. M. (1991). Constructive induction of numerical terms in automated discovery of empirical equations. In *Methodologies for Intelligent Systems.* (Z. Ras, ed.). Springer-Verlag, Berlin.

Zewbowicz R., and Zytkow, J. M. (1992). Discovery of equations: Experimental evaluation of convergence. In *Proceedings of the Tenth National Conference on Artificial Intelligence.* MIT Press, Cambridge, MA.

CHAPTER 1

Statistical Properties of Tree-based Approaches to Classification

A. P. WHITE
W. Z. LIU

1. TREE-BASED CLASSIFICATION TECHNIQUES IN COMPUTER SCIENCE AND MATHEMATICAL STATISTICS

The principle of classification by induced trees is well known. Basically, an appropriate algorithm is set to work on a data set (called the 'training set') consisting of a number of cases, each of which comprises a class membership indicator and a number of attributes. The induced tree branches on the more important attributes (as judged by the information they convey about class membership) and, at the terminal nodes, predicted class membership is indicated. The tree may then be used to classify further cases from outside the training set that was used to build the tree. In the case of noisy data, the attributes do not provide an unambiguous indication of class membership and *clashes* will occur, in which cases from more than one class possess the same attribute vector. In these circumstances, prediction will be necessarily probabilistic, rather than definite and, hence, subject to error. However, the classification performance of the tree on an independent data set (drawn from the same population as the training set) serves to provide an unbiased measure of classification accuracy.

A powerful inductive algorithm should be able to generate classification trees that are valid beyond the training set, that is, to build trees that correctly classify not only as many objects as possible from the training set, but other (unseen) objects as well. This is the main objective of the induction task. In

Machine Learning and Statistics, The Interface, Edited by G. Nakhaeizadeh and C. C. Taylor.
ISBN 0-471-14890-3

order to achieve this goal, the classification tree must capture some *intrinsic* associations between class and attribute values. At the same time, it should, as far as possible, be insensitive to those aspects of the data set which are based on noise.

The study of tree-based classification techniques seems to have two points of academic origin. The first of these is the study of inductive learning. The influence of computer science in this field did not develop until the second half of this century.[1] Hunt was one of the early pioneers who modeled a psychological theory of human concept learning using computer programs. He developed a series of algorithms called 'concept learning systems' [CLS-1 to CLS-9; described in Hunt et al. (1966)]. An important descendent of these programs was Quinlan's ID3 algorithm (Quinlan, 1979). Basically, this was a procedure for discriminating between two classes in domains which were entirely free of uncertainty. (In fact, ID3 was developed initially for performing chess end-game analysis, discriminating between winning and non-winning positions.)

The introduction of uncertainty has the consequence that, generally speaking, the terminal nodes contain *probabilities* of class membership, rather than definite class assignments. The adaptation of this type of algorithm to deal with noisy domains took place much more recently. Quinlan et al. (1986) extended the ID3 system, producing the C4 algorithm, to deal with this type of problem. It is interesting to note that, initially, computer scientists were unaware of the penalties of constructing over-large trees. However, Quinlan et al. rediscovered the problem, and dealt with it by incorporating a pruning phase into the algorithm. Thus, an over-large tree was grown to begin with and then cut back to protect against overfitting.

The second point of origin lies in the discipline of mathematical statistics. It is interesting to note that, among mathematical statisticians, there has been a minority interest in these techniques for the last thirty years. However, this interest did not really come to the fore until the last ten years or so, prompted by the work of Breiman et al. (1984) and the associated CART software for performing classification and regression using binary trees. In the last year or two, the incorporation of tree-based classification and regression techniques into multi-purpose statistical software packages has begun to appear. The statistical package 'S', recently developed for use by statisticians themselves and described by Clark and Pregibon (1992), includes a set of procedures for conducting classification and regression tasks by the use of binary trees. Similarly, the well-known package SPSS has recently become available with CHAID (CHi-squared Automatic Interaction Detector). This is based on earlier work by Kass (1980), which uses multiple, rather than binary, branching.

[1]A more detailed review of these techniques from a computer science perspective is given in Liu and White (1991).

As machine learning algorithms are adapted to deal with uncertainty in the subject domain, the study of the operation and performance of these algorithms is really a matter with which the discipline of mathematical statistics is better equipped to deal. However, it must be admitted that, as in the field of expert systems, statisticians have been rather slow to offer their expertise in this area. Nevertheless, it is obvious that statistical matters such as overfitting avoidance, the estimation of classification error rates and probabilities of class membership, significance testing and measures of association are all involved and that, consequently, a proper study of this area cannot proceed in ignornace of statistical considerations.

2. TREE-BASED CLASSIFICATION AS NON-PARAMETRIC DISCRIMINANT ANALYSIS

From a statistical point of view, the operation of tree-based classification techniques on domains characterized by uncertainty is really a type of non-parametric discriminant analysis. It is instructive to compare and contrast the way in which tree-based techniques and kernel-based techniques partition the data space.

As explained by Breiman et al. (1984), a binary tree partitions the data space into multi-dimensional rectangles. This can be seen most easily in the case of an example with just two attributes and two classes, as illustrated in Figure 1.1 and Figure 1.2. The data space, *S*, is partitioned by the tree into five regions,

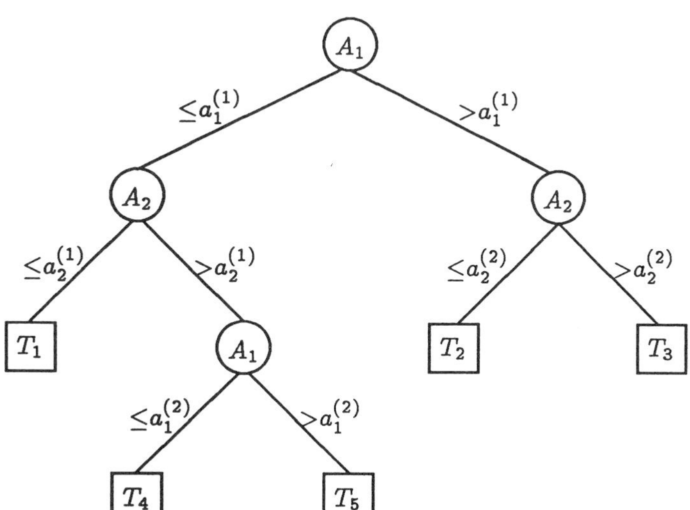

Figure 1.1. *A simple classification tree. See text for further explanation.*

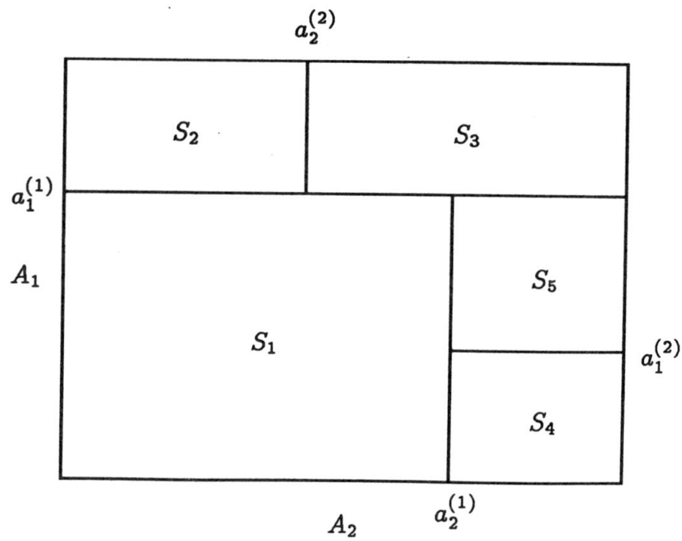

Figure 1.2. *A data space partitioned according to the classification tree shown in Figure 1.1.*

S_1 to S_5, which are mutually exclusive and jointly exhaustive. At each terminal node in the tree, there will be a number of cases from each class. These numbers form the basis of the probability estimates of class membership that are assigned to new test cases, when they are classified by the tree. More precisely, if a particular terminal node, T_i, in the tree has k_i cases, comprising $k_i^{(1)}$ from class 1 and $k_i^{(2)}$ from class 2, then these probability estimates will be simply $k_i^{(j)}/k_i$ for class j. The corresponding region of the data space is S_i. The process of classifying a new case from region S_i of the data space means that the probability estimates derived from T_i will be assigned to it.

By contrast, kernel-based methods involve the definition of *kernels* in the data space. These kernels are regions of the space that surround the case under classification and which include a number of other observations, from which the posterior probabilities of class membership of the particular case are estimated. There are two main types of kernel method. One type involves the use of spherical or ellipsoidal kernels of specified constant radius. The other type (usually called the k-nearest neighbor algorithm) utilizes spherical kernels of variable radius, which, for each case, enclose a region of space which is just sufficient to include the k nearest neighbors of the point under consideration.

It is interesting to compare and contrast the tree-based technique with the k-nearest neighbor (k-NN) algorithm. The two methods operate on the data space in different ways. The basic k-NN algorithm utilizes *all* the attributes in the data space and operates as follows. When a new case is presented for classification, a kernel is formed in the data space, centered on the observation concerned. The kernel has the shape of a hypersphere and is just large enough to contain the k nearest neighbors of the case under consideration, using some appropriate distance metric. Classification is then conducted according to the class probabilities estimated from the k nearest neighbors found in the kernel. Because, in general, the tree-based approach utilizes a *subset* of the available independent variables, it is less prone to overfitting than the nearest-neighbor technique, particularly if care is taken not to generate an over-large tree. With data spaces of high dimensionality (i.e., a large number of independent variables), the nearest-neighbor method tends to produce sub-optimal results because of overfitting, irrespective of the value of k. In these circumstances, the performance difference between the two techniques is likely to be particularly marked if the independent variables differ in the information that they convey about class membership. In such a situation, the tree-based method is particularly efficient at partitioning the data space on just the important variables and thus avoids overfitting. A comparison of the performance of these two techniques is given in Liu and White (1995).

3. ESTIMATION OF CLASSIFICATION ACCURACY

In theory, to test the performance of a classification tree, one should draw randomly another very large independent set of cases from the same population as the training set Ł was drawn from; then use the classification tree to classify each of these cases and compare the actual class of each such case with its predicted class. The proportion of correctly classified cases will be an unbiased estimate of the classification accuracy.

To make the exposition more precise, let us assume that there are N cases in the training set Ł and N' cases in the set Ł′ used to test the classification accuracy of the induced tree. Cases are defined as couples $(\mathbf{x}, c)$ where $\mathbf{x}$ is an attribute vector and c is a class label. Let d denote the classification tree. If a case $(\mathbf{x}_i, c_i)$ is classified by the induced tree d as class c_j, then we represent this application of classification as $(d(\mathbf{x}_i) = c_j)$. Obvously, if $c_j = c_i$ then the classification of object $\mathbf{x}_i$ is correct. Let us define the indicator function $I(.)$ to be unity if the logical statement inside the parentheses is true, otherwise zero. Then the above defined classification accuracy $R(d)$ is:

$$R(d) = \frac{1}{N'} \sum_{i=1}^{N'} I(d(\mathbf{x}_i) = c_i)$$

where $(\mathbf{x}_i, c_i) \in$ Ł′.

In actual problems, however, it is often difficult to obtain another large sample of classified cases in addition to the set Ł. Thus, under many circumstances, this set has to be used both to build the classification tree and to estimate the classification accuracy of the tree. Such estimates are referred to as *internal estimates* by Breiman et al. (1984). There are basically three types of internal estimate:

1. The first, least accurate, and most commonly used is the *resubstitution* estimate. This estimate uses Ł as the training set to construct the classification tree first, then uses this same set to test the induced tree. The classification accuracy is defined as follows:

$$R^{re}(d) = \frac{1}{N} \sum_{i=1}^{N} I(d(\mathbf{x}_i) = c_i)$$

   where $(\mathbf{x}_i, c_i) \in$ Ł.

2. The second method is *test sample* estimation. Here the cases in set Ł are divided into two subsets $Ł_1$ and $Ł_2$. Only the cases in $Ł_1$ are used as the training set to construct the classification tree *d*. Then the cases in $Ł_2$ are used to estimate the classification accuracy. If N_2 is the number of cases in $Ł_2$, then the test sample estimate $R^{ts}(d)$ is:

$$R^{ts}(d) = \frac{1}{N_2} \sum_{i=1}^{N2} I(d(\mathbf{x}_i) = c_i)$$

   where $(\mathbf{x}_i, c_i) \in Ł_2$.

3. The third method is well known as the *cross-validation* estimate. Here the original set Ł is randomly divided into V subsets, $Ł_1, Ł_2, \ldots, Ł_V$, of as nearly equal size as possible. For every v, $v = 1, 2, \ldots, V$, construct a classification tree using the set $(Ł - Ł_v)$ as a training set and use it to classify those cases in $Ł_v$. A test sample estimate can be obtained for each v using the method described above:

$$R^{ts}(d^{(v)}) = \frac{1}{N_v} \sum_{i=1}^{N_v} I(d^{(v)}(\mathbf{x}_i) = c_i)$$

   where $(\mathbf{x}_i, c_i) \in Ł_v$ and N_v is the number of cases in $Ł_v$. The V-fold cross-validation estimate $R^{cv}(d)$ is defined as the average of all these V test sample estimates:

$$R^{cv}(d) = \frac{1}{V} \sum_{v=1}^{V} R^{ts}(d^{(v)})$$

   One special case of V-fold cross-validation is N-fold cross-validation. For each n, $n = 1, 2, \ldots, N$, the nth case is used to test the classification tree, $d^{(n)}$, constructed using the other $N - 1$ cases. The formula of N-fold cross-validation estimate can be simplified as follows:

$$R_N^{cv}(d) = \frac{1}{N} \sum_{n=1}^{N} I(d^{(n)}(\mathbf{x}_n) = c_n)$$

where $(\mathbf{x}_n, c_n)$ is the nth case in the set Ł.

Comparing the above methods, the resubstitution estimate often gives an overly optimistic picture of the true accuracy of the classification tree d, because it is computed using the same data used to construct d, instead of an independent sample. In order to assess the true accuracy of d, the other two methods are more honest. In case of small sample size, the third method may be the best to use. In particular, the N-fold variant is commonly used and, as Miller (1974) points out, eliminates any arbitrariness in the formation of the subsets. It is important to point out that the difference between estimation by resubstitution and estimation by either of the other two methods may be very large. A pathological example can be found in White and Reed (1986).

It is interesting to note that, if there are no clashes in the training set, classification trees with perfect resubstitution performance can always be generated. However, this does not necessarily mean that they can also perfectly classify cases *outside* the training set. Noise may still be present in the population from which the training set is drawn. If so, classification accuracy should still be estimated by either the test sample or cross-validation method. However, it is worth noting that Efron and Tibshirani (1993) claim (p. 247) that N-fold cross-validation is not really suitable for work with classification trees, because the trees generated are not sufficiently different from the one formed from the entire data set. They claim (from experience) that ten-fold cross-validation works better. Perhaps this is an area that requires further investigation.

The bootstrap is another type of resampling technique. This involves taking random samples, with replacement, from the data set and working with these to derive confidence intervals for the classification error rate. These and other resampling techniques are described in Efron and Tibshirani (1993).

4. SIMULATION EXPERIMENTS IN MACHINE LEARNING

In the machine learning domain it is a common practice to assess the relative merits of different techniques by measuring their performance on a small number of data sets that are known to the researchers working in that area. While such an approach can often be a useful part of the evaluation exercise, relying on these results alone is problematic, because real data sets may well possess particular *ad hoc* features that give an advantage to one particular method, without that method being *generally* better than its competitors.

An alternative approach is to use a number of synthetic data sets, each generated randomly from the same population, designed to investigate discrimination performance on data with particular features. Perhaps 100 such data sets might be used, in order to be able to make comparisons of reasonable power. This allows standard errors of performance to be estimated and, in fact, enables the entire experiment to be designed and analyzed using an appropriate, orthodox experimental design. It should be noted that some technique for providing an unbiased estimate of classification performance (such as cross-validation) is still required. Examples of this approach may be found in Liu and White (1994, 1995).

5. DISCRIMINATION POWER AND CLASSIFICATION THRESHOLDS

It is obvious that an important measure of performance of a classification algorithm operating on a particular data set is the power to discriminate between the classes, as measured under cross-validation (or one of the other techniques of estimating classification performance, mentioned in the previous section). What is perhaps less obvious is that the classification error rate may not always be the most appropriate measure for this purpose. This is particularly true in situations where the class frequencies are not equal. Consider the following confusion matrices from the same classification problem, in Table 1.1. In the first table, the error rate is 23%, whereas it is 30% in the second table. However, the first table does not represent a better discrimination than the

TABLE 1.1. Confusion Matrices for a Classification Problem

	Predicted Class 1	Predicted Class 2	
Actual Class 1	76	4	80
Actual Class 2	19	1	20
	95	5	

	Predicted Class 1	Predicted Class 2	
Actual Class 1	60	20	80
Actual Class 2	10	10	20
	70	30	

TABLE 1.2. Confusion Matrix for the Classification Outcomes in a Two-Class Discrimination Task

		Predicted Class	
		1	0
Actual Class	1	a (hit)	b (miss)
	2	c (FP)	d (CR)

FP represents false positive classification and CR represents correct rejection.

second table. This is obvious if some measure of association between actual and predicted classes is applied, in turn, to each table. Using the odds ratio (defined as ad/bc in Table 1.2), we obtain a value of 1 for the first table but 3 for the second table. Obviously, discrimination is better in the second case. The disparity arises because, in the second case, discrimination is biased more heavily toward the correct detection of cases in the less frequent class, with the result that more errors are made on cases in the more frequent class.

These considerations are obviously of importance when comparisons are being made between different techniques, one of which is biased and the other of which is not. For example, Cestnik (1990) discusses the use of Laplace's law of succession for the estimation of class probabilities at the terminal nodes of the tree. This states that, for a two-class problem, if there are N cases at a terminal node, of which n belong to a particular class, then the corresponding population probability is estimated as $(n + 1)/(N + 2)$. It is a simple matter to show that, in regions of the data space where class probabilities are not equal to 1/2, this is a biased estimator. There will be a tendency for probabilities greater than 1/2 to be underestimated and for probabilities less than 1/2 to be overestimated. The effect of using this estimator in a machine-learning task would therefore be to bias the classification in favor of the less frequent class.

Thus there is a justification for assessing discrimination power by using a measure of association between the actual and predicted classes, rather than the classification error rate. Furthermore, this measure should be calculated using any appropriate resampling technique, such as cross-validation, in order to avoid overestimation of the degree of association.

Of course, this topic is closely related to that of deliberately setting the classification thresholds so as to bias classification toward one class at the expense of the remaining classes. Thus, in classification tasks with two classes, instead of simply classifying each case as belonging to the class with the highest posterior probability, some value of the posterior probability other than 1/2 can be used as the decision threshold. The lower the threshold is set for a given class, the more the classification process will be biased in favor of classification in the direction of that class. This aspect of the matter is described and illustrated in detail by Liu et al. (1994). In that paper, the task of using machine-learning software for the early diagnosis of gastric cancer is described. In the field of medical diagnosis, it is usually the case that a test for a particular disease does not operate with perfect accuracy. This leads to the possibility of two different types of error—false positive error, in which the disease is predicted when it is actually absent, and false negative error (or 'miss'), where the test result is negative when the disease is present. It is clear that, in this application, the second type of error is more serious than the first. Consequently, the usual approach of classifying according to the largest posterior probability found at the terminal node is not satisfactory in such a situation. A more appropriate approach is to adopt classification *threshold* probabilities. With the diagnosis of serious medical conditions, the classification threshold would typically be set at a lower value for predicting the presence of the disease, than for predicting its absence.

More precisely, let us define $C(i\,|\,j)$ as the cost of misclassifying a case from class j as belonging to class i, such that:

$$C(i\,|\,j) \geqslant 0, \qquad i \neq j$$

and

$$C(i\,|\,j) = 0, \qquad i = j$$

Given a terminal node in the classification tree with estimated node probabilities $\hat{p}(j), j = 1, 2$, then if a randomly selected case of unknown class is filtered down to that node and is classified as being of class i, then the estimated expected misclassification cost is given by:

$$\sum_j C(i\,|\,j)\hat{p}(j), \qquad i \neq j$$

An obvious node classification rule is to select the predicted class, i, so as to minimize this expression. If the classification threshold probability for class i is represented by q_i, then as a corollary of an argument by Breiman et al. (1984, pp. 112–115), it can be shown that this is equivalent to setting the ratio of the classification threshold probabilities as follows:

$$\frac{q_2}{q_1} = \frac{C(2\,|\,1)}{C(1\,|\,2)}, \qquad q_1 + q_2 = 1$$

Of course, the ability to vary the classification threshold means that, as the error rate for false negatives is decreased, the error rate for false positives increases. 'Hits' and correct rejections are similarly inversely related. Table 1.2

displays the four possible outcomes for a two-class discrimination task, in the form of a table called a 'confusion matrix.' From this table, some important quantities can be defined. In the terminology of medical diagnosis, *sensitivity* and *specificity* are defined as follows. Sensitivity is the conditional probability of correct classification, given that the disease is present, and specificity is the conditional probability of correct classification, given that the disease is absent. Algebraically, from Table 1.2, sensitivity is given by $a/(a + b)$ and specificity is defined as $d/(c + d)$. As the classification threshold for the disease is reduced, sensitivity will *increase* and specificity will *decrease*, that is, there is a trade-off between the two. If the classification threshold probability is varied from 0 to 1, values for sensitivity and specificity are generated as a series of number-pairs, which can be plotted on a graph.

Sensitivity-specificity curves run from coordinates (0, 1) to (1, 0). A straight line corresponds to an odds ratio of one (i.e. absence of any effective discrimination). In general, a sensitivity-specificity curve will 'bulge' into the upper-right quadrant of the graph. The more extreme the curve, the larger the corresponding odds ratio (i.e., the better the discrimination). Sensitivity-specificity curves are useful because they show the relationship between these two quantities throughout the range of threshold values and, with the use of odds ratio contours, indicate the discrimination power at different threshold settings.

An example demonstrates these various features. Part of the database of dyspeptic patients described by Hallissey et al. (1990) was used for this purpose. Briefly, this consisted of records on 1,401 patients over the age of 40 who were referred to dyspepsia clinics because of symptoms of dyspepsia. The data on each patient comprised a gold-standard diagnosis (made by upper-gastrointestinal endoscopy) and a total of 85 epidemiological and clinical variables. The diagnoses were grouped into two classes. Class 1 (the high-risk group) consisted of those patients diagnosed as either having gastric cancer, or belonging to any of three other diagnostic categories regarded as being at risk for developing the disease because of the mucosal changes typically associated with these conditions. These were gastric ulcer, atrophic gastritis, and gastric polyp. A total of 370 cases fell into the high-risk group. Class 2 (the low-risk group) comprised the remaining cases.

Classification was performed by *Predictor*. Various aspects of *Predictor* have been described elsewhere by White and Liu (1993), Liu (1993), and Liu and White (1994), and will not be described here in detail. Briefly, *Predictor* operates by a recursive binary partitioning of the data space, under a form of statistical control which branches preferentially on the more important variables. A stopping rule based on significance testing principles (White and Liu, 1994) guards against excessive branching and the specification of a minimum terminal node frequency provides additional control against overfitting. Missing values are dealt with by dynamic path generation. In essence, this involves dealing with just one case at a time and constructing the classification path necessary to classify just this case. Only variables which are non-missing for the case under consideration are branched on. Obviously, from a computational point of view, this approach fits in well with *N*-fold cross-validation. Further details are given in the references just cited.

In order to produce the cross-validated sensitivity–specificity curve, the significance level in *Predictor* was set at 0.5 and the minimum terminal node frequency at 5. Cross-validation was used. Another option in *Predictor* produces the posterior probabilities of class membership for each case classified. This feature was employed and the resulting probabilities were post-processed by separate software to produce the data required for the sensitivity–specificity graph. Part of the cross-validated sensitivity–specificity curve is displayed in Figure 1.3. It should be noted that a large proportion of spurious missing values was discovered on some of the variables, due to problems encountered in the data before they were transferred to the computer used for running *Predictor*. Nevertheless, the machine-learning algorithm was able to perform competently in spite of this difficulty.

Some idea of the power of the discrimination may be gained by looking closely at the graph in Figure 1.3. Most of the curve corresponds to criterion settings giving odds ratios (for the cross-validated classification matrix) of 5 or more. Better discrimination is apparent at the high-sensitivity end of the curve, which is beneficial, because of the greater interest in performing discrimination with a high-sensitivity criterion, because of the seriousness of gastric cancer if

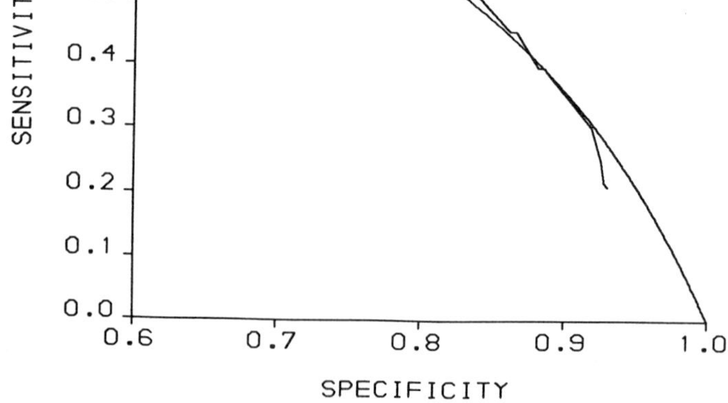

Figure 1.3. *Part of the cross-validated sensitivity-specificity curve for detection of those at high risk for gastric cancer. The smooth curve represents a contour line for an odds ratio of 5. See text for further explanation.*

diagnosis is missed in the early stages. For example, one particular point on this curve (obtained by classifying cases as high-risk if the estimated posterior probability of membership of this class is greater than or equal to 0.2) gives a sensitivity of 0.768 and a specificity of 0.664, with an odds ratio of 6.54 for the cross-validated classification matrix. With this criterion setting, 45.0% of the sample are classified as high-risk.

6. BRANCHING CRITERIA

At each node in a classification tree undergoing construction, the particular attribute (independent variable) having the strongest association with class is selected for branching. Suppose that we are dealing with a problem with k classes and that an attribute, A, with m distinct values is under consideration at a particular node. Table 1.3 represents the cross-classification of classes and attribute values, where C_i $(i = 1 \ldots k)$ and a_j $(j = 1 \ldots m)$ represent class and attribute values, respectively; n_{ij} $(i = 1 \ldots k; j = 1 \ldots m)$ represent the frequency counts of cases with attribute value a_j and class C_i; and

$$n_{i.} = \sum_{j=1}^{m} n_{ij}; \qquad n_{.j} = \sum_{i=1}^{k} n_{ij}; \qquad n_{..} = \sum_{i=1}^{k} \sum_{j=1}^{m} n_{ij} = N$$

Various probabilities can be estimated, as follows:

$$p_{ij} = \frac{n_{ij}}{n_{..}}; \qquad p_{i.} = \frac{n_{i.}}{n_{..}}; \qquad p_{.j} = \frac{n_{.j}}{n_{..}}$$

Given an event with l possible outcomes, each of probability p_i $(i = 1, 2, \ldots, l)$, the information associated with this event is (Edwards, 1964):

$$-\sum_{i=1}^{l} p_i \log_2 p_i$$

Applying this to the contingency table above, we can define the information associated with each possible cell, class and attribute, respectively, as:

$$H_{cell} = -\sum_{i=1}^{k} \sum_{j=1}^{m} p_{ij} \log_2 p_{ij}$$

$$H_C = -\sum_{i=1}^{k} p_{i.} \log_2 p_{i.}$$

$$H_A = -\sum_{j=1}^{m} p_{.j} \log_2 p_{.j}$$

TABLE 1.3. A General Contingency Table

	a_1	a_2		a_m	
C_1	n_{11}	n_{12}	$\cdots$	n_{1m}	$n_{1.}$
C_2	n_{21}	n_{22}	$\cdots$	n_{2m}	$n_{2.}$
$\vdots$	$\vdots$	$\vdots$	$\cdots$	$\vdots$	$\vdots$
C_k	n_{k1}	n_{k2}	$\cdots$	n_{km}	$n_{k.}$
	$n_{.1}$	$n_{.2}$	$\cdots$	$n_{.m}$	$n_{..}$

From these quantities, transmitted information may be defined as:

$$H_T = H_C + H_A - H_{cell}$$

The concept of transmitted information is very useful. In the current context, it means *the information about class membership that is conveyed by attribute value.*

Traditionally, this measure, also called 'information gain' by Quinlan (1986), has been using for the purpose of deciding which of the available attributes should be branched on. However, in recent years, researchers such as Kononenko et al. (1984) have become aware that this measure is liable to favor unfairly attributes with large numbers of values at the expense of those with few.

As a preferred alternative, Quinlan (1986) suggested the use of another information-based measure, which he termed the 'gain ratio.' This was derived from information gain by dividing by attribute information. This acted as a sort of normalizing factor by virtue of the fact that attribute information tends to increase as the number of possible values increases. At the time, it was thought that this would eliminate the bias. However, it was acknowledged that attributes with very *low* information values (i.e, low *attribute* information) then appeared to gain an unfair advantage.

Another measure of interest is the G statistic. This was introduced by Kullback (1959, pp. 158–159) and described by Mingers (1987, 1989), as

$$G = 2NH_T$$

This is somewhat misleading. G has the property that it is distributed as chi-square only if H_T is computed with natural logarithms, whereas convention dictates that information-based measures use logarithms of base 2, in order that the quantity concerned is expressed in bits. Therefore, the definition is better stated as

$$G = 2NH_T \log_e 2$$

We also need to define a statistical measure, χ^2, from the same table:

$$\chi^2 = \sum_i \sum_j \frac{(E_{ij} - O_{ij})^2}{E_{ij}}$$

where O_{ij} is the observed number of cases with value a_j in class C_i, that is, $O_{ij} = n_{ij}$, and E_{ij} is the expected number of cases which should be in cell (C_i, a_j) in the contingency table, if the null hypothesis (of no association between attribute and class) is true:

$$E_{ij} = \frac{n_{.j} n_{i.}}{n_{..}}$$

Both the G statistic and χ^2 are well approximated by the chi-square distribution with v degrees of freedom, where:

$$v = (k - 1)(m - 1)$$

Of course, statisticians are well aware that an easy solution to the problem of deciding between contending attributes with different numbers of values is provided by comparing the *probabilities* from the upper tail of the chi-square distribution, corresponding to the values of G or χ^2 that were obtained. The problem is thus solved by virtue of the fact that the chi-square distribution takes into account the different number of attribute values in the calculation of the number of degrees of freedom. This was pointed out by White and Liu (1994).

However, another problem arises with attributes which have a large number of different values. When dealt with in the manner just described, they lead to heavily branched trees with small numbers of cases in each path. This also means that the contingency tables used to evaluate association with class have too many cells and consequently suffer from unacceptably small expected frequencies, so that the distribution of the test statistic (χ^2 or G) is no longer adequately approximated by the chi-square distribution. This problem is necessarily at its most severe with continuous variables, where each case takes a different value. One approach for dealing with this problem is to recast *all* the attributes as binary variables, so that the resulting tree will be binary. For ordered attributes with more than two values, this is done simply by splitting each attribute at that particular point which gives the strongest association with class. Thus, for an ordered attribute having m distinct values, there are $m - 1$ possible ways of splitting it in order to derive the corresponding binary variable. For categorical attributes, on the other hand, the argument is different. With this type of variable, there are $2^{m-1} - 1$ ways of partitioning m values into two subsets.

At first sight, this approach appears to offer a satisfactory way of overcoming the problem of excessive multiple branching, described earlier. Thus, all the attributes are reduced to the 'common denominator' of a binary form, allowing easy comparisons to be made between different types of attribute in terms of their discrimination power between the classes. Unfortunately, things are not this straightforward. The problem lies in the fact that the larger the number of values over which the maximum is taken, the larger the maximum is likely to be. Thus, this process is biased in favour of attributes with a large number of values. In other words, such attributes are more likely to be branched on than variables that have few distinct values. Attempts are currently being made to use simulation techniques to derive compensation functions to deal with this problem, along the lines suggested by White and Liu (1993). However, it cannot yet be claimed that really satisfactory results have been obtained.

7. CORRESPONDENCE WITH LOGIT MODELS

The formal correspondence of probabilistic classification trees in machine learning with logit models in mathematical statistics is described in detail, with examples, by White and Reed (1989). (Unfortunately, there are a number of typographical errors in that paper. Nevertheless, the basic mathematics are sound.)

The purpose of a logit model is to express a binary dependent variable, Y, in terms of a number of binary independent variables, X_i, thus:

$$\log(O(Y)) = c + c_1X_1 + c_2X_2 + c_3X_3 + \cdots$$

where the various c_i are fitted coefficients and $O(Y)$ is the odds of Y taking one of its values. Higher order models can be derived by taking products of the various independent variables. Any logit model in which no higher order term is included unless corresponding lower-order terms are also present is known as a *hierarchical logit model*. A *saturated* logit model is one in which all possible terms have been included.

Actually, a logit model is a particular type of log-linear model. It is somewhat easier to talk in terms of a log-linear model to begin with and to explain how it is related to the frequency table from which it is derived and then to adapt the discussion to logit models at a later stage. Rather than attempt a completely general explanation, we shall use examples with three variables, which are just about the simplest type of situation containing the required degree of complexity.

Let us suppose that we have a database that consists of a number of cases, taking values on each of three binary variables: A_1, A_2, A_3. Thus there are eight

TABLE 1.4. Frequency Table for All Possible Patterns of Three Binary Attributes

f	A_1	A_2	A_3
f_{000}	0	0	0
f_{001}	0	0	1
f_{010}	0	1	0
f_{011}	0	1	1
f_{100}	1	0	0
f_{101}	1	0	1
f_{110}	1	1	0
f_{111}	1	1	1

types of case in the database and the numbers of each type are given by the frequency counts f_{ijk}, as indicated in Table 1.4. Now, a saturated log-linear model to summarize these data will have the following form:

$$\log(f_{ijk}) = c + c_1 A_1 + c_2 A_2 + c_3 A_3 + c_{12} A_1 A_2 + c_{13} A_1 A_3 + c_{23} A_2 A_3 + c_{123} A_1 A_2 A_3$$

From Table 1.4 it is clear that the coefficients for this model are as follows:

$$c = \log(f_{000})$$

$$c_1 = \log\left(\frac{f_{100}}{f_{000}}\right)$$

$$c_2 = \log\left(\frac{f_{010}}{f_{000}}\right)$$

$$c_3 = \log\left(\frac{f_{001}}{f_{000}}\right)$$

$$c_{12} = \log\left(\frac{f_{000} f_{110}}{f_{100} f_{010}}\right)$$

$$c_{13} = \log\left(\frac{f_{000} f_{101}}{f_{100} f_{001}}\right)$$

$$c_{23} = \log\left(\frac{f_{000} f_{011}}{f_{010} f_{001}}\right)$$

$$c_{123} = \log\left(\frac{f_{100} f_{010} f_{001} f_{111}}{f_{000} f_{110} f_{101} f_{011}}\right)$$

Now let us suppose that we have a similar situation but with a classification variable, Y, added. In the simplest (two-class) case, Y will take either of two possible values (1 or 2) and the frequency table will now need to contain *two* frequencies, g_{ijk} and h_{ijk}, for each possible pattern of attribute values on A_i, A_j, A_k in order to represent the frequencies for each class. The logit model for this type of situation has the same form on the right-hand side of the equation as for the log-linear model described previously. However, the left-hand side of the model takes the form of a *logit*. This is the algorithm of the odds that observations belong to one class rather than the other, that is, $\log(h_{ijk}/g_{ijk})$. (A slight difficulty with this approach is that frequencies of zero constitute a problem. A way out of this difficulty is to add 1/2 to each of the frequencies before taking the logits.)

The basis of the correspondence between a logit model in mathematical statistics and a classification tree describing the same data in an equivalent way is that the process of *conditioning* on a variable in the statistical representation is equivalent to the process of *branching* in the tree representation. The consequence of this is that each path in an induced classification tree corresponds to a specific hierarchical logit model. The logit model that matches the probability on a particular terminal node is a hierarchical one constructed from all the terms corresponding to all the nodes branched on down the path to that particular leaf. In the general case, this means that the data space is fitted by a number of logit models, rather than just one, as is conventionally the case in mathematical statistics. In order to illustrate this point, let us consider an example very similar to the second one discussed by White and Reed (1989). Table 1.5 shows the frequency table for a data set with a binary dependent variable (Y) and three independent variables. (In machine learning terminology, we would call these the class variable and the attributes, respectively). Figure 1.4 shows the induced classification tree for this data set, obtained by using a simple forward branching technique with a stopping rule of $\chi^2 < 3.84$ for the class-attribute association. This was done using *Predictor*, as described by White and Liu (1993) and Liu (1993).

The three corresponding logit models, derived as previously described, are:

$$\begin{aligned}\log(O(Y)) &= 1.723 - 3.381A_1 - 1.569A_2 - 1.650A_3 + 3.359A_1A_2\\ &\quad + 2.520A_1A_3 + 1.531A_2A_3 - 3.541A_1A_2A_3\\ \log(O(Y)) &= 0.5349 - 0.9686A_1 - 0.4813A_3 - 0.03846A_1A_3\\ \log(O(Y)) &= 0.3822 - 1.788A_1 - 0.2988A_2 + 1.434A_1A_2\end{aligned}$$

It is clear that the leaves numbered 1, 2, 5, and 6 are fitted by the saturated hierarchical model in all three attributes. Leaf number 3 is fitted by the hierarchical model in A_1 and A_3, and leaf number 4 by the model in A_1 and A_2.

However, it is possible to go further than this. Clearly, it is unsatisfactory to use one model with eight terms and two models with four terms each in

TABLE 1.5. Frequency Table for an Example with Two Classes and Three Attributes

f	Y	A_1	A_2	A_3
5	1	0	0	0
28	2	0	0	0
53	1	0	0	1
57	2	0	0	1
36	1	0	1	0
42	2	0	1	0
56	1	0	1	1
58	2	0	1	1
84	1	1	0	0
16	2	1	0	0
22	1	1	0	1
10	2	1	0	1
78	1	1	1	0
89	2	1	1	0
74	1	1	1	1
27	2	1	1	1

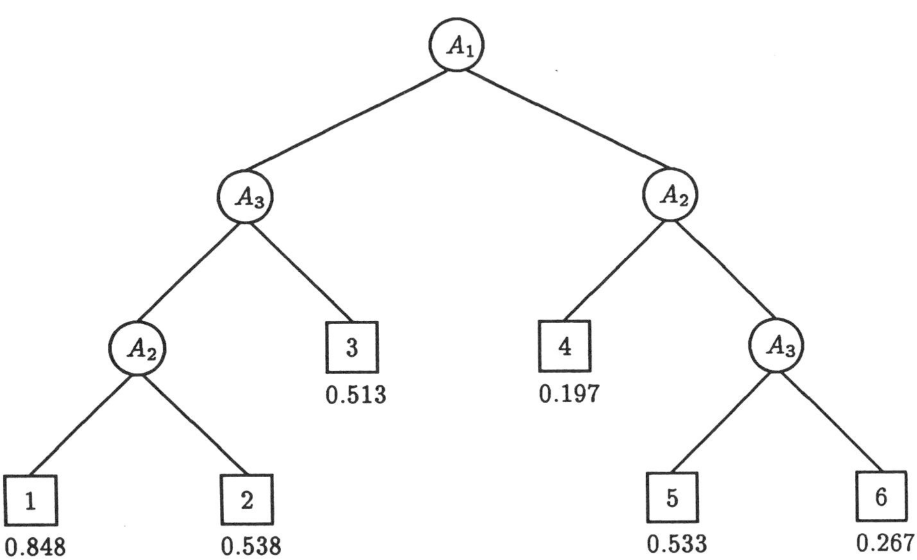

Figure 1.4. *The induced classification tree for the data set shown in Table 1.5. Probabilities of membership of class 2 are given below each terminal node. All attribute values of zero branch to the left and values of one branch to the right.*

TABLE 1.6. Parameter Constraints in the Eight-Term Log-Linear Model Arising from the Combination of Cells in the Corresponding Frequency Table

Cell Combined	Parameter Constraint
f_{000}, f_{100}	$c_1 = 0$
f_{000}, f_{010}	$c_2 = 0$
f_{000}, f_{001}	$c_3 = 0$
f_{010}, f_{110}	$c_{12} = -c_1$
f_{001}, f_{101}	$c_{13} = -c_1$
f_{100}, f_{110}	$c_{12} = -c_2$
f_{001}, f_{011}	$c_{23} = -c_2$
f_{100}, f_{101}	$c_{13} = -c_3$
f_{010}, f_{011}	$c_{23} = -c_3$
f_{011}, f_{111}	$c_{123} = -c_1 - c_{13} - c_{12}$
f_{101}, f_{111}	$c_{123} = -c_2 - c_{23} - c_{12}$
f_{110}, f_{111}	$c_{123} = -c_3 - c_{23} - c_{13}$

order to fit six values (i.e., the class probabilities at each of the six terminal nodes). In fact, it is possible to derive a single six-parameter model which corresponds to the tree shown in Figure 1.4. This is done by combining those cells in the frequency table which correspond to those frequency patterns which are filtered through to the same terminal node in the tree. For the purposes of calculation, cells that have been combined can be assigned the mean of their frequencies. If this is done and the coefficients are calculated, constraints will be observed among them, as indicated in Table 1.6 for the log-linear model. (For the logit model, the only difference is that the averaging process must be conducted separately for the frequencies relating to each class.)

For the example shown in Table 1.5 and Figure 1.4, cells (0, 0, 1) and (0, 1, 1) need to be combined and also cells (1, 0, 0) and (1, 0, 1). If this is done, the second and fourth rows in Table 1.5 will both be changed to 54.5 for class 1 and 57.5 for class 2. Similarly, the fifth and sixth rows will be given the values 53 for class 1 and 13 for class 2. When the logit model is fitted to these data, it becomes:

$$\log(O(Y)) = 1.723 - 3.128A_1 - 1.569A_2 - 1.669A_3 + 3.106A_1A_2 + 1.669A_1A_3 + 1.569A_2A_3 - 2.709A_1A_2A_3$$

Thus we find that $c_{23} = -c_2$ and $c_{13} = -c_3$, exactly as predicted in Table 1.6. Hence we can rewrite the model with six parameters, as follows:

$$\log(O(Y)) = 1.723 - 3.128A_1 + 1.569A_2(A_3 - 1) + 1.669A_3(A_1 - 1) + 3.106A_1A_2 - 2.709A_1A_2A_3$$

(It is interesting to note that, when a model is formulated by combining cells in this way, the fitting error is confined to the cells which have been merged. The remaining cells are fitted *exactly* by the model. This outcome is different from the more usual approach, where if the highest-order terms are omitted from a conventional hierarchical model, the fitting error is distributed over *all* the cells.)

Finally, we indicate briefly how problems with more than two classes are handled. The usual approach is to construct $k - 1$ logit models for problems with k classes. In that scheme, one class is chosen as a reference base and logits are formed between the frequencies for this class and each of the other classes.

8. CONCLUDING REMARKS

The foregoing sections support the idea that, if a classification problem includes uncertainty, then interesting features emerge which require the attention of the discipline of statistics. This applies to tree-based techniques just as much as to any of the other methods used by mathematical statistics. However, many of the *ideas* that have arisen in the area of machine learning have come from other disciplines, such as computer science. What is required is a fruitful collaboration, in which these ideas are subjected to the rigors of statistical analysis and, where appropriate, developed further.

REFERENCES

Breiman, L., Friedman, J. H., Olshen, R. A., and Stone, C. J. (1984). *Classification and Regression Trees.* Wadsworth, Belmont, CA.

Cestnik, B. (1990). Estimating probabilities: A crucial task in machine learning. In *Proceedings of the Ninth European Conference on Artificial Intelligence*, (L. C. Aicllo, ed.), pp. 147–149. Pitman, London, UK.

Clark, L. A. and Pregibon, D. (1992). Tree-based models. In *Statistical Models in S* (J. M. Chambers and T. J. Hastie, eds.), pp. 377–419. Wadsworth and Brooks/Cole, Pacific Grove, CA.

Edwards, E. (1964). *Information Transmission, an Introductory Guide to the Application of the Theory of Information to the Human Sciences.* Chapman & Hall, London.

Efron, B., and Tibshirani, R. J. (1993). *An Introduction to the Bootstrap.* Chapman & Hall, London.

Hallissey, M. T., Allum, W. H., Jewkes, A. J., Ellis, D. J., and Fielding, J. W. L. (1990). Early detection of gastric cancer. *Br. Med. J.* **301**, 513–515.

Hunt, E. B., Marin, J. and Stone, P. J. (1966). *Experiments in Induction.* Academic Press, New York.

Kass, G. V. (1980). An exploratory technique for investigating large quantities of categorical data. *Appl. Stat.* **29**, 119–127.

Kononenko, I., Bratko, I., and Roskar, E. (1984). *Experiments in Automatic Learning of Medical Diagnostic Rules*, Tech. Rep. Jozef Stefan Institute, Ljubjana, Yugoslavia.

Kullback, S. (1959). *Information Theory and Statistics*. Wiley, New York.

Liu, W. Z. (1993). Aspects of machine learning. Unpublished Ph.D. Thesis, University of Birmingham, UK.

Liu, W. Z., and White, A. P. (1991). A review of inductive learning. In *Research and Development in Expert Systems VIII* (I. M. Graham and R. W. Milne, eds.), pp. 112–126. Cambridge Univ. Press, Cambridge, UK.

Liu, W. Z., and White, A. P. (1994). The importance of attribute selection measures in decision tree induction. *Mach. Learn.* **15**, 25–41.

Liu, W. Z., and White, A. P. (1995). A comparison of nearest neighbour and tree-based methods of non-parametric discriminant analysis. *J. Stat. Comput. Simul. 53*, 41–50.

Liu, W. Z., White, A. P., and Hallissey, M. T. (1994). Early screening for gastric cancer using machine learning techniques. In *Machine Learning: ECML-94* (F. Bergadano and L. DeRaedt, eds.), pp. 391–394. Springer-Verlag, Berlin.

Miller, R. G. (1974). The jacknife—a review. *Biometrika* **61**, 1–15.

Mingers, J. (1987). Expert systems—rule induction with statistical data. *J. Oper. Res. Soc.* **38**, 39–47.

Mingers, J. (1989). An empirical comparison of selection measures for decision tree induction. *Mach. Learn.* **3**, 319–342.

Quinlan, J. R. (1979). Discovering rules by induction from large collections of examples. In *Expert Systems in the Micro-Electronic Age* (D. Michie, ed.), pp. 168–201. Edinburgh Univ. Press, Edinburgh.

Quinlan, J. R. (1986). Induction of decision trees. *Mach. Learn.* **1**, 81–106.

Quinlan, J. R., Compton, P., Horn, K. A., and Lazarus, L. (1986). *Inductive Knowledge Acquisition: A Case Study*, Tech. Rep. 86.4. School of Computing Science, New South Wales Institute of Technology.

White, A. P., and Liu, W. Z. (1993). Fairness of attribute selection in probabilistic induction. In *Research and Development in Expert Systems IX* (M. A. Bramer and R. W. Milne, eds.), pp. 209–224. Cambridge Univ. Press, Cambridge, UK.

White, A. P., and Liu, W. Z. (1994). Bias in information-based measures in decision tree induction. *Mach. Learn.* **15**, 321–329.

White, A. P., and Reed, A. (1986). Some predictive difficulties in automatic induction. In *Proceedings of the International Meeting on Advances in Learning* (*Les Arcs, 1986*) (Y. Kodratoff and R. S. Michalski, eds.), pp. 132–139. Université de Paris Sud, Paris.

White, A. P., and Reed, A. (1989). Probabilistic induction and logit models. In *Machine and Human Learning* (Y. Kodratoff and A. Hutchinson, eds.), pp. 221–225. Kogan Page, London.

CHAPTER 2

The Decision Tree Algorithm CAL5 Based on a Statistical Approach to Its Splitting Algorithm

W. MÜLLER
F. WYSOTZKI

1. INTRODUCTION

Methods of machine learning are, at present, well on the way to be utilized for practical industrial applications. A specially important field is the analysis, diagnosis, and control of such processes in environment and technology which cannot (or can only partly) be modeled mathematically.

This chapter deals with a special method of classification learning, namely, the automatic construction of decision trees mainly for object descriptions by real valued attributes, which are transformed into discrete valued ones (intervals) using statistical splitting methods. Furthermore, some relations between statistical and other machine learning approaches will be discussed.

For the aim of classification the objects are to be described by (continuous or discrete) values of certain attributes like color, temperature, age, size, and so forth. In the following the value of an attribute will also be called a feature. This way, objects are represented in the simplest case by feature vectors, that is, as points in a multidimensional feature space. For example, in the case of diagnosis and control of dynamic processes the feature space is identical with the state space used in classical control theory, that is, the space defined by the observable input and output parameters of a process. Decision tree construction is a special kind of supervised learning, that is, a "teacher" has to provide a training set of preclassified objects.

The task of the learning algorithm is then to generalize this classification in such a way that *every* object description (not necessarily seen before) can be

Machine Learning and Statistics, The Interface, Edited by G. Nakhaeizadeh and C. C. Taylor.
ISBN 0-471-14890-3

classified with minimum error or costs. In terms of the feature space this means that the "cluster" of points corresponding to objects of the training set and belonging to the same class has to be extrapolated to a region within the feature space for that class, which is distinct from the regions of other classes.

Seen from another point of view, the learning algorithm has to construct a classifier (classification function) which accepts any unseen object description as input and delivers a class membership as output. The training set of preclassified examples acts as a set of support for the classification function to be learned.

Some results of this chapter are based on the ESPRIT-project *StatLog*, where about 20 of the recent learning algorithms were tested on more than 20 real-world data sets (Michie et al., 1994). In general, and this was also done in *StatLog*, one can try to classify the learning algorithms into three different main categories: statistical, decision tree, and neural net algorithms. However, it can be shown that some of them contain elements of several categories. For example, DIPOL92 (Michie et al., 1994; Schulmeister and Wysotzki, 1994) could be defined as both a statistical and neural network approach and the decision tree algorithm C4.5 (Michie et al., 1994) uses a statistical method (χ^2 test) as termination function.

As mentioned above, the decision tree algorithm CAL5 is specifically designed to convert real-valued attributes into intervals using statistical considerations. The trees are automatically pruned during their construction with the help of a given threshold for the estimated class probabilities within an interval. By means of this threshold the user can control the complexity of the resulting tree, that is, the degree of approximation of class regions in the feature space. The hypothesis that the estimated class probability is greater than, or equal to, the threshold is tested by the determination of a confidence interval computed from a user-defined confidence level for estimation of the probability.

Real-valued attributes are transformed by CAL5 into discrete-valued (ordered) ones, where the number of values are determined by the algorithm. There is no restriction to binary ones as, for example, in CART (Michie et al., 1994) or in modifications of ID3. This feature and the complexity control (pruning) guided by statistical considerations during the tree-construction phase are the most significant differences to other decision tree algorithms. (e.g. C4.5, CART in Michie et al., 1994). The confidence interval approach described below was firstly published by Unger and Wysotzki in 1981 and improved in our laboratory in recent years. A similar method is also described by Sagalnicoff (1993) for supporting splitting decisions in k-D trees.

2. THE SPLITTING ALGORITHM OF CAL5

As in other decision tree algorithms the tree will be constructed sequentially, starting with one attribute and branching out with other attributes recursively if no sufficent discrimination of classes can be achieved. That is, if no decision for a class c_i can be made at a node, a branch (subtree) formed with a new attribute is appended. If this attribute is continuous, a discretization, that is, a creation of intervals corresponding to qualitative values has to be established.

The sequence of the attributes used for tree construction essentially affects the quality of the resulting decision tree according to their classification performances. Therefore, in a first step the attribute with the best local discrimination property is chosen using some evaluation function. Since relations to statistics are the main topic of the current chapter, it is useful to concentrate on the algorithm of interval construction (splitting algorithm).

Let E be a set of n preclassified examples reaching a current terminal node x in the decision tree, constructed so far, where no unique class decision can be made (see below) and c_i, $i = 1, \ldots, l$ are the different classes. All examples have to be ordered along the axis of the attribute a according to increasing values. Now intervals I, which contain an ordered set of values of the attribute, are formed recursively collecting examples from left to right until a class decision can be made or it is sure (due to some given confidence level) that no class decision can be made.

The class decision can be obtained by means of an estimate of the conditional probability $p(c_i \mid x)$ in the interval I and by testing the hypothesis

H1: "There exists a class c_i in I with

$$p(c_i \mid x) \geqslant S.$$

"against

H2: "For all classes c_i occurring in I the inequality

$$p(c_i \mid x) < S$$

holds."

where S is a user-tunable threshold.

Now the following "metadecision" on the dominance of a class in I can be defined:

(a) If a class c_i exists, for which **H1** is true then c_i dominates in I, the interval I will be closed. The corresponding path of the tree is terminated by the class symbol c_i.

(b) If the hypothesis **H2** is true for all classes appearing in I, then no class dominates in I. In this case the interval will also be closed and it will be labeled as an indication for further branching.

(c) If neither (a) nor (b) can be decided due to lack of sufficient data (see below), the interval I will be extended by collecting the next observation of the order of values of the current attribute. This corresponds to an increase of the amount of data to decide for **H1** and **H2**, respectively. If there are no more samples for a further extension of I, a majority decision, that is, a decision for the most probable class, is made.

2.1. Merging of Intervals

Consequently a lot of small intervals I_l with attached class labels (unique class label or an indication for further branching) are obtained by automatic determination of their bounds. After that, adjacent intervals I_l, I_{l+1} with the same class labels are merged. Those merged intervals with attached unique class labels yield the terminal nodes (leaves) of the decision tree. Merged intervals, where no class dominates, yield the intermediate nodes of the decision tree, for which further branching will be performed. As a result the continuous attributes become discretized automatically in a statistically optimal sense. The above described procedure has to be used recursively for each intermediate node.

3. TESTING THE HYPOTHESES H1 and H2

We use a "confidence interval" approach (Unger and Wysotzki, 1981) for testing the above introduced hypotheses. Therefore a confidence level $1 - \alpha$ with the significance level α (to be given by the user) has to be assumed. Let $n_i \leqslant n$ be the number of examples with the class label c_i reaching the current node x handled by the learning algorithm. Under the assumption that the set of examples is created by an independent and stationary source in terms of its classes, n_i has a Bernoulli distribution. Then

$$q = \frac{n_i}{n}$$

is the estimate of $p(c_i \mid x)$. A confidence interval can be derived from the Chebyshev inequality (Unger and Wysotzki, 1981)

$$P(|n_i - E(n_i)| \leqslant \varepsilon) \geqslant 1 - \frac{\sigma^2}{\varepsilon^2}$$

where σ^2 is the standard deviation of n_i. Since n_i is Bernoulli distributed one obtains

$$P(|n_i - np| \leqslant \varepsilon) \geqslant 1 - \frac{np(1 - p)}{\varepsilon^2} = 1 - \alpha$$

and, furthermore, after some transformations one gets the formula for the bounds of the confidence interval

$$d_{1,2} = \frac{2\alpha n_i + 1}{2\alpha n + 2} \mp \frac{1}{2\alpha n + 2} \sqrt{4\alpha n_i \left(1 - \frac{n_i}{n}\right) + 1} \tag{1}$$

Now our hypotheses can be tested as follows:

H1 is accepted, if for one class

$$d_1 \geqslant S$$

holds, d_1 being the lower bound of the confidence interval.

H2 is accepted, if, for all classes

$$d_2 < S$$

holds, d_2 is the upper bound of the confidence interval.

Note that, in contrast to the usual probabilistic approaches, there is no need to assume any particular class distribution on the current attribute axis.

Figure 2.1 shows the graphical representation of formula (1). The horizontal axis represents the number of examples n in the current interval and the vertical axis the lower (upper) bounds $d_1(d_2)$ of the confidence intervals, respectively. For this illustration a constant ratio $n_i/n = 0.7$ is assumed and the curves are computed for two different cases of α (0.2 and 0.5).[1] The ratio means that 70 percent of the learning examples belong to class c_i and 30 percent to other classes, respectively. The d1-curves rise and the d2-curves fall steadily and tend to the given ratio for a large value of n. The areas above (below) d_2 (d_1) indicate that a decision for a unique class cannot (can) be made, respectively. The area between both curves d_1 and d_2 represents an uncertainty, that is, neither hypothesis **H1** nor **H2** can be decided. The higher the number of n and the greater the significance level α the smaller is the difference $d_2(n, \alpha) - d_1(n, \alpha)$.

The influence of the threshold S on the splitting procedure can be illustrated in the graph (Figure 2.1) as well. S is to be imagined as a straight line parallel to the n-axis and its intersection, for example, with $d_1(n, \alpha_i)$, delivers the minimal number of examples n_{min} for which the decision for a unique class can be made under the condition that the ratio $n_i/n = 0.7$ is the current estimate of the class probability. A splitting result and a decision for a unique class can be reached only if $S < 0.7$ is chosen in this case.

[1]Note that, in general, the number of examples assigned to a node to be further branched is not very large (some tens of examples). Therefore, α cannot be chosen as small (e.g., 0.05) as it might be used traditionally.

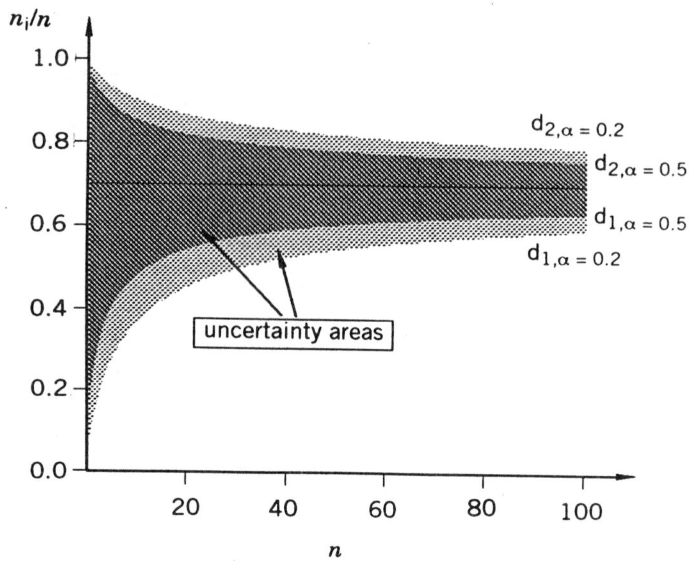

Figure 2.1. *Graph of confidence bounds.*

If neither the hypothesis **H1** nor **H2** can be decided due to lack of sufficient data in the current interval, then the number of examples n has to be increased by extending the interval I and applying formula (1) again. In the case that during increasing n the ratio n_i/n is always exactly equal to S, the algorithm is not able to decide either for a class domination or for no class domination. That is similar to the human's behavior, if a decision has to be made without a well-determined tendency to a unique class. This fact can be proved by substitution of the threshold S for the ratio n_i/n in the formula:

$$d_{1,2} = \frac{2\alpha S + (1/n)}{2\alpha + (2/n)} \mp \frac{1}{2\alpha + (2/n)} \sqrt{\frac{4\alpha S(1 - S) + (1/n)}{n}}$$

and if n tends to infinity it leads to

$$\lim_{n \to \infty} d_{1,2} = \frac{2\alpha S}{2\alpha} = S.$$

But this is only a theoretical case, because in reality a decision will always be caused by a dispersion of the ratio n_i/n if n is sufficiently large. However, a large number of examples might be needed for decision making if that ratio nearly becomes equal to the threshold S.

3.1. Interpretation of the Threshold *S*

Using statistical decision theory it can be shown (Unger and Wysotzki, 1981) that

$$S = \frac{C - C_0}{C},$$

if C is the cost for misclassification of any class and C_0 is the cost of rejection of a decision at all. This corresponds to the case of hypothesis **H2** being true, that is, CAL5 is branching out with the next attribute. In general C, C_0 are not known, therefore S is a parameter to be defined by the user. In fact the optimal S depends on the size of the training set and noise, for example, separability of the classes. A default value in case of a small training set ($n \approx 100$) may be $S = 0.8$. If the costs are class dependent, that is, C_k is the cost of not recognizing class c_k than one can introduce class dependent thresholds S_k (Unger and Wysotzki, 1981) with

$$S_k \approx \frac{const}{C_k},$$

that is, S_k is approximately proportional to $1/C_k$. That means classes with high risk are detected "more carefully". In general the error rate will increase in this case but the mean cost is minimized.

4. COMPARISON OF CAL5 WITH CLASSICAL STATISTICAL METHODS

In the following CAL5 will be compared (Unger and Wysotzki, 1981) with two classical statistical methods: discriminant analysis and Wald's sequential probability ratio test (SPRT). Note that, in both cases, assumptions on the probability distributions of the classes must be made, at least implicitly.

4.1. Discriminant Analysis

For the purposes of comparison the following simple classification problem is considered: There are only two classes c_1 and c_2, which have normal distributions in the k-dimensional feature space with mean vectors μ_1 and μ_2 and equal covariance matrices

$$\Sigma_1 = \Sigma_2 = \Sigma$$

respectively, that is,

$$p(x \mid c_i) = \frac{1}{\sqrt{(2\pi)^k |\Sigma|}} e^{-1/2(x-\mu_i)\Sigma^{-1}(x-\mu_i)} \qquad i = 1, 2. \tag{2}$$

Furthermore, the availability of a sufficiently large training set is assumed in order to obtain very small elementary intervals, constructed by CAL5 by means of the given confidence level $1 - \alpha$. In this case the class decision regions can be approximated by merging these elementary intervals with sufficient accuracy.

It is also assumed that α is sufficiently small so that the conditional class probabilities can be replaced by the corresponding relative frequencies. Then the decision rule of CAL5 is at the end of some path $\tilde{x}$ in the decision tree constructed so far:

1. Decide class c_i if $p(c_i \mid \tilde{x}) \geqslant S$, $0.5 < S < 1$.
2. Reject decision (that is, further branching) if $p(c_j \mid \tilde{x}) < S$ for all classes c_j, $j = 1, 2$. (In fact $\tilde{x}$ corresponds to a hyperinterval, see below)

In discriminant analysis (DA) the whole feature vector $x = (x_1, \ldots, x_k)$ is always used for decision. Now CAL5's decision behavior is "simulated" for DA, that is, a decision threshold S is introduced, which leads to a rejection area as in CAL5, as well. With $f_i(x) = p(x \mid c_i)$ and formula (2) Bayes' theorem gives:

$$p(c_1 \mid x) = \frac{p_1 f_1(x)}{p_1 f_1(x) + p_2 f_2(x)} = \frac{1}{1 + \dfrac{p_2 f_2(x)}{p_1 f_1(x)}}$$

where p_i, $i = 1, 2$ is the a priori probability of class c_i. According to the decision rule class c_1 is decided if

$$\log \frac{p_2 f_2}{p_1 f_1} = (\mu_2 - \mu_1)\Sigma^{-1}x + \frac{1}{2}(\mu_1\Sigma^{-1}\mu_1 - \mu_2\Sigma^{-1}\mu_2) + \log \frac{p_2}{p_1} \leqslant \log \frac{1 - S}{S},$$

(class c_2 similarly), where p_1, p_2, >0 is supposed. The function

$$v' = (\mu_2 - \mu_1)\Sigma^{-1}x$$

is the well-known Fisher discriminant, which corresponds to a linear transformation of $x = (x_1, \ldots, x_k)$ and defines (after normalization) the direction of a new feature derived from $x_1, \ldots, x_k$. With the new feature

$$v = \frac{(\mu_2 - \mu_1)\Sigma^{-1}x}{|(\mu_2 - \mu_1)\Sigma^{-1}|} + \frac{1}{2}\frac{\mu_1\Sigma^{-1}\mu_1 - \mu_2\Sigma^{-1}\mu_2}{|(\mu_2 - \mu_1)\Sigma^{-1}|}.$$

and

$$-\lambda_k = \frac{1}{|(\mu_2 - \mu_1)\Sigma^{-1}|} \log\left(\frac{p_1}{p_2} \cdot \frac{1 - S}{S}\right)$$

the decision rule becomes

1. Decide class c_1 if $v \leqslant -\lambda_k$
2. Decide class c_2 if $v \geqslant \lambda_k$
3. Rejection of decision if $-\lambda_k < v < \lambda_k$

Since the maximal admissible error is fixed a priori by the threshold S, the rejection rate as a criterion for comparison of the decision tree and discriminant method is used, if k attributes $x_1, \ldots, x_k$ are supplied. The rejection rate of the discriminant algorithm is

$$RR_D = \int_{-\lambda_k}^{\lambda_k} (p_1 f_{1v} + p_2 f_{2v})\, dv$$

where f_{1v} and f_{2v} are the probability densities of c_1 and c_2, respectively, projected onto the direction

$$\frac{(\mu_2 - \mu_1)\Sigma^{-1}}{|(\mu_2 - \mu_1)\Sigma^{-1}|}$$

of the new feature v.

To be able to calculate RR_D the following special case (in which *all* attributes $x_1, \ldots, x_k$ must be used for class discrimination) is considered:

$$\mu_1 = \underbrace{(-m, \ldots, -m)}_{n\ times} = -\mu, \qquad m > 0$$

$$\mu_2 = \underbrace{(+m, \ldots, +m)}_{n\ times} = \mu$$

$$\Sigma = I\,(\text{unit matrix}),\ p_1 = p_2 = \tfrac{1}{2}$$

$$f_{1,2}(x) = \frac{1}{\sqrt{(2\pi)^k}}\, e^{-1/2(x \pm \mu)^2}$$

The $x_v (v = 1, \ldots, k)$ are therefore independently and identically distributed variables with probability densities $f_{1,2}(x_v) \propto N(\pm m, 1)$. The case k = 2 and $\lambda_1 = m$ (threshold for a single feature x_1) is illustrated in Figure 2.2. (The probability densities are indicated by level lines; the rejection area is shaded.)

For k dimensions $x_1, \ldots, x_k$ the projected densities are

$$f_{1,2v} = \frac{1}{\sqrt{2\pi}}\, e^{-1/2(v \pm \sqrt{k}m)^2}$$

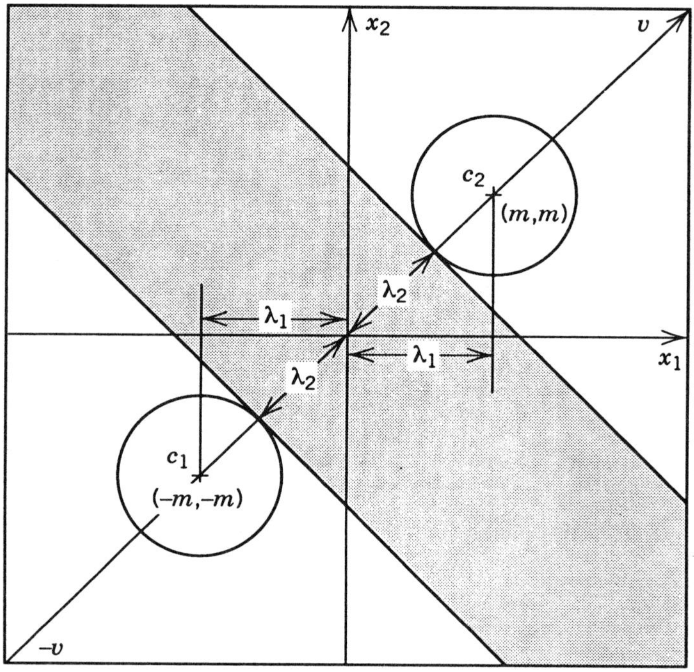

Figure 2.2. *Rejection region using discriminant analysis (see text).*

with

$$v = \frac{\mu}{|\mu|} x = \frac{1}{\sqrt{k}} \Sigma_{\nu=1}^{k} x_\nu .$$

The decision threshold on the new dimension v is

$$\lambda_k = \frac{1}{2\sqrt{km}} \log \frac{S}{1-S}.$$

Thus the rejection rate of the discriminant method becomes

$$RR_D = \frac{1}{2} \int_{-\lambda_k}^{\lambda_k} (f_{1v} + f_{2v}) dv = \Phi(\lambda_k + \sqrt{km}) - \Phi(-\lambda_k + \sqrt{km}),$$

where Φ is the distribution function of the one-dimensional normal (Gaussian) distribution $N(0, 1)$.

$$\Phi(x) = \frac{1}{\sqrt{2\pi}} \int_{-\infty}^{x} e^{-1/2t2}\, dt.$$

We then have

$$\lim_{k \to \infty} RR_D = 0,$$

that is, by the measuring of sufficiently many features (if provided) the rejection rate can be made arbitrarily small.

Now a CAL5-tree is considered, supposing the same class distributions and parameters. Suppose CAL5 is just handling the attribute $x_v (v \geqslant 1)$ and has constructed an interval I_v on the dimension x_v. For a class decision it would implicitly use (since they are not known to CAL5) the one-dimensional probability density functions

$$f_{1,2}(x_v) \propto e^{-1/2(x_v \pm m)^2}$$

The decision rule is: Decide class c_i if

$$p(c_i \mid x_v \in I_v, -\lambda_1 < x_{v-1} < \lambda_1, \ldots, -\lambda_1 < x_1 < \lambda_1) \geqslant S$$

with

$$\lambda_1 = \frac{1}{2m} \log \frac{S}{1-S}.$$

$-\lambda_1 < x_\varrho < \lambda_1$ $(\varrho = 1, \ldots, v-1)$ is the rejection interval constructed by CAL5 for x_ϱ in a previous step by merging elementary intervals. (Note that the availability of a "sufficiently large" training set has been assumed, i.e., that "thinning out" of statistics by successive adding of attributes x_ϱ and partition of the training examples by interval construction can be neglected. Also, each x_ϱ is assumed to be used only once.)

For comparison with discriminant analysis the rejection rate for the CAL5-tree is computed using all available attributes $x_1, \ldots, x_k$. The rejection area for the special case $k = 2$ and $\lambda_1 = m$ is illustrated in Figure 2.3. The rejection rate is

$$RR_{\text{CAL5}} = (\Phi(\lambda_1 + m) - \Phi(\lambda_1 + m))^n$$

and

$$\lim_{k \to \infty} RR_{\text{CAL5}} = 0, \text{ too.}$$

In Figure 2.4 the comparison of RR_d and RR_{CAL5} as a function of m for the special case $k = 2$ and $S = 0.8$ is shown. There is no essential difference between both rates. Note that in the case of using the tree as a classifier fewer

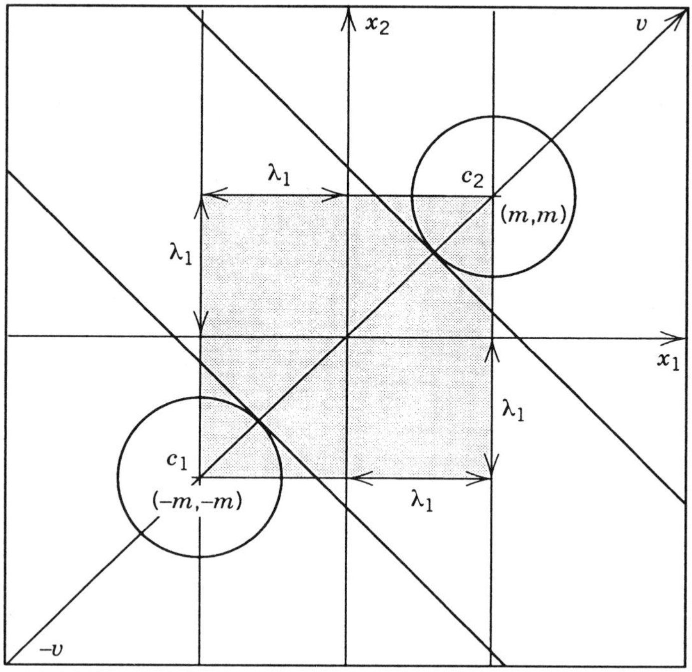

Figure 2.3. *Rejection region using CAL5 (see text).*

measurements have to be performed on the average. Of course, in practical applications with finite or even small training sets, discriminant analysis may perform better because of the "thinning effect" of sequential classification methods, that is, discriminant analysis uses the training set more effectively.

4.2. Sequential Probability Ratio Test (SPRT)

Sequential analysis is already applied by Fu (1968) to problems of pattern recognition. In the case of the SPRT (Wald, 1947) the likelihood ratio

$$R_v = \frac{p(x_1, \ldots, x_v \mid c_1)}{p(x_1, \ldots, x_v \mid c_2)} = \frac{f_1(x_1, \ldots, x_v)}{f_2(x_1, \ldots, x_v)}, \qquad v = 1, 2, \ldots$$

is used for decision, whereby the probability densities f_1 and f_2 must be known a priori. The variables $x_1, x_2, \ldots$ will be measured sequentially and after each step R_v is computed. There are two constants A and B, $0 < B < 1 < A < \infty$ to be defined by the user. If $R_v \leqslant B$ class c_1 and if $R_v \geqslant A$ class c_2 is decided. If $B < R_v < A$ a decision is deferred and x_{v+1} will be measured. The CAL5-

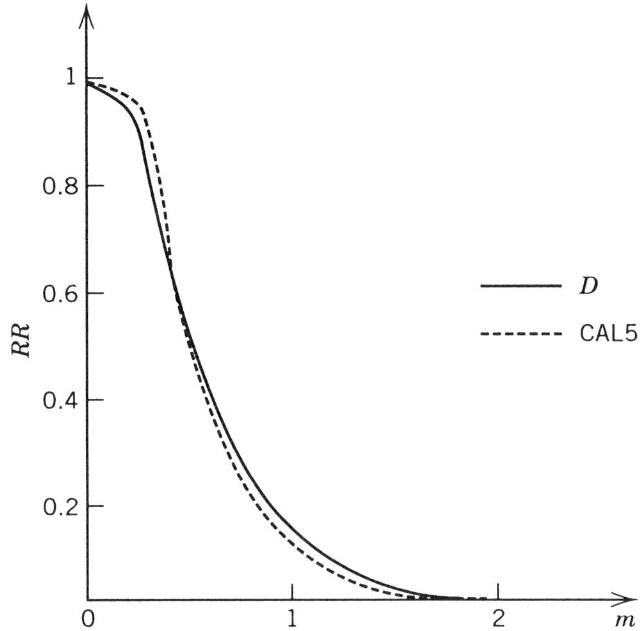

Figure 2.4. *Comparison of theoretical rejection rates (RR) for discriminant analysis and CAL5 for a special problem (two attributes x_1, x_2, $\lambda_1 = m$, $S = 0.8$).*

decision rule

$$p(c_1 \mid x_1, \ldots, x_v) \geqslant S \to c_1$$
$$p(c_2 \mid x_1, \ldots, x_v) \geqslant S \to c_2$$

corresponds to choosing

$$A = \frac{p_2}{p_1} \frac{S}{1 - S}, \qquad B = \frac{p_2}{p_1} \frac{1 - S}{S}$$

The transformation of the decision rule into the format of the SPRT leads to: Decide class c_1 if

$$\tilde{R}_{v\,\mathrm{CAL5}} = \frac{p(x_1 \in I_1 \wedge \cdots \wedge x_v \in I_v \mid c_1)}{p(x_1 \in I_1 \wedge \cdots \wedge x_v \in I_v \mid c_2)} \geqslant \frac{p_2}{p_1} \frac{S}{1 - S} = A, \qquad v = 1, 2, \ldots, k$$

The I_ϱ, $\varrho = 1, \ldots, v - 1$ are the rejection intervals for the variables x_ϱ, respectively, I_v is the current elementary interval constructed for x_v by CAL5. For our

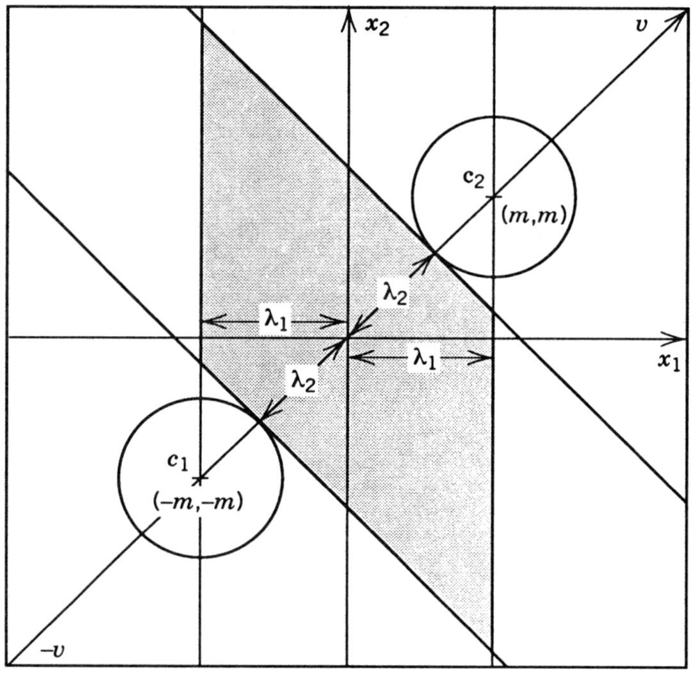

Figure 2.5. *Rejection region using SPRT (see text).*

special case with normally distributed classes and

$$\mu_1 = (-m, \ldots, -m) = -\mu_2, \qquad \Sigma_{1,2} = I, \qquad p_1 = p_2$$

the SPRT provides the rule:

1. Decide c_1 if

$$\log R_v = -2\mu^{(v)}x^{(v)} \geqslant \log A = \log \frac{S}{1-S}$$

$$\text{or } -2\sqrt{vm}v^{(v)} \geqslant \log \frac{S}{1-S}$$

$$\text{or } v^{(v)} \leqslant -\frac{1}{2\sqrt{vm}} \log \frac{S}{1-S} = -\lambda_v$$

with $\mu^{(v)} = (\underbrace{m, \; m, \ldots, m}_{v \text{ times}}), \; x^{(v)} = (x_1, \ldots, x_v)$

$$v^{(\nu)} = \frac{\mu^{(\nu)}}{|\mu^{(\nu)}|} x^{(\nu)} = \frac{1}{\sqrt{\nu}} \Sigma_{r=1}^{\nu} x_r,$$

2. Decide c_2 if

$$v^{\nu} \geqslant \lambda_{\nu}$$

3. Reject if

$$-\lambda_{\nu} < v^{(\nu)} < \lambda_{\nu}.$$

That is, for all $\nu\,(\nu = 2, \ldots, k)$ the decision is made by the Fisher discriminant but within the rejection area of $x_1, \ldots, x_{\nu-1}$.

The rejection rate RR_{SPRT} cannot be computed in a straightforward way. In Figure 2.5 (rejection area shaded) it can be seen, that (in the special case considered) the SPRT resides somewhere between discriminant analysis and the CAL5 classifier, as far as the rejection rate is concerned.

5. RESULTS

CAL5 was tested on about 20 real-world classification tasks within the ESPRIT-Project *StatLog* and compared with the results of other learning algorithms.

The statistical splitting procedure automatically converts real-valued attributes into discrete ones. In contrast to CART or extensions to ID3 (Buntine and Caruana, 1991; and Quinlan, 1986) the resulting decision trees may contain nodes with multiple (not necessarily binary) branches.

In Figure 2.6 the result of comparing one step of the splitting procedures of CAL5 and CART,[2] respectively, is shown. For this an artificial training data set of a 2-class problem containing 30 examples that are specified by two attributes (a,b) is used. Unfortunately, for detecting the attribute with the best discrimination measure, different evaluation functions (GINI-index for CART, entropy measure for CAL5) were applied, so that different splitting attributes were obtained. However, the most significant differences between both approaches can be shown.[3]

Both algorithms try to divide the feature space (attribute space) into regions with unique class labels. CART delivers one bound S_{b1} in the first recursion step and the next one S_{b2} in a second step, that is, in the first step one yields two new nodes after branching (see Figure 2.7). A leaf will be obtained only after the second splitting step. CAL5 directly divides the attribute space into three subregions, which produces three new nodes after branching out. Two of

[2]The INDCART-Version from NASA Ames Research Center (Buntine and Caruana, 1991) was used.

[3]These investigations were provided by M. Bleich (1993).

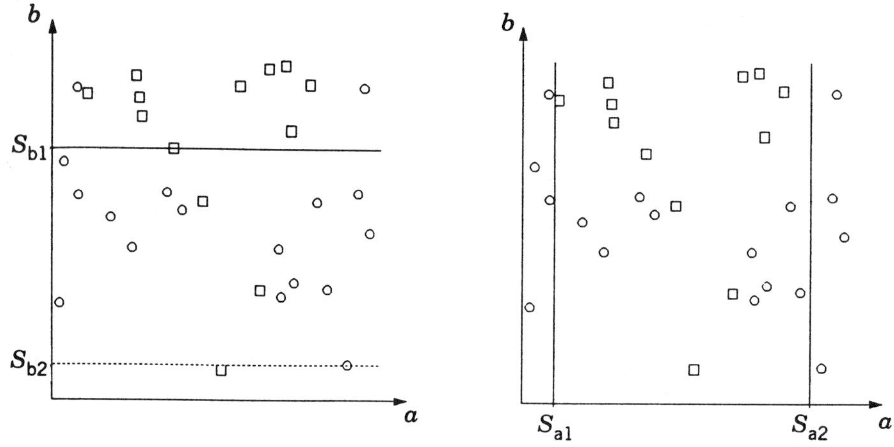

CART CAL5

Figure 2.6. *Feature space split by CART and CAL5.*

these nodes are leaves, that is, the hypothesis **H1** holds. Note that the decision tree produced by CART has to be pruned in an additional final step, but the CAL5-tree does not need any further pruning.

This way CAL5 has a better generalization feature than other decision tree algorithms [e.g. CART, C4.5, AC^2 (Michie et al., 1994)] and its trees are fairly simple in most cases. This can also be illustrated by comparison of the decision trees produced by CART (Fig. 2.8) and by CAL5 (Fig. 2.9) on the basis of one of the *StatLog* real-world data sets (Michie et al., 1994). The problem was the detection of emergency voltage conditions in electrical power stations (see section 6 of this chapter). The error rates obtained by testing the resulting

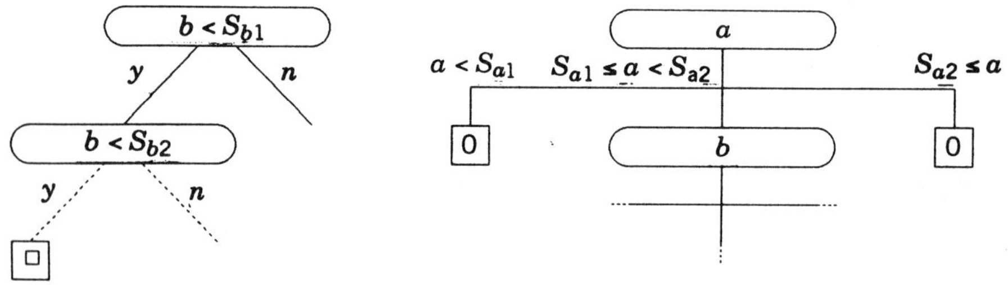

CART CAL5

Figure 2.7. *Parts of decision trees constructed by CART and CAL5.*

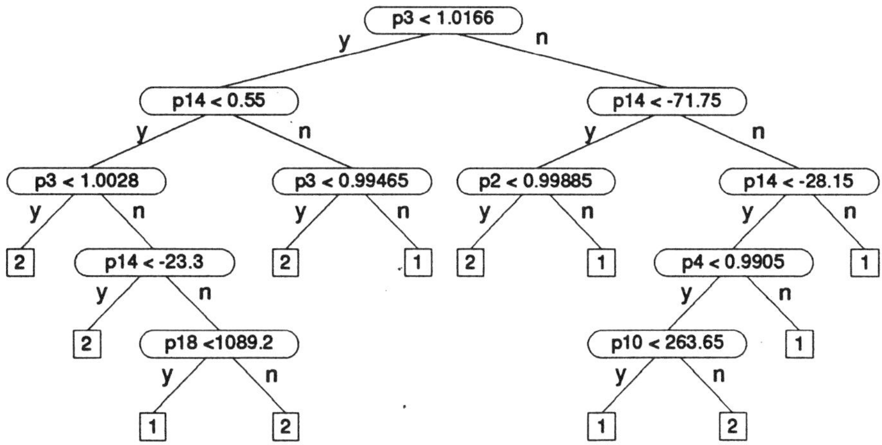

Figure 2.8. *Decision tree constructed by CART.*

decision trees on a test data set independent of the learning data set were nearly similar (3.36% for CART and 2.88% for CAL5, respectively). But the CAL5-tree is much more compact; it contains only eight leaves, in comparison with the 12-leaf tree produced by CART.

CAL5 delivers not only a decision tree, which is able to classify an unseen example, but also a lower bound of probability of its class decisions and a measure of its statistical relevance. If the class distributions in the attribute space of the used training and test data sets (unseen examples) are nearly similar, the resulting decision tree and decision rules derived from it can be assessed in terms of statistical considerations. By "similar" we mean that the examples of the training and test data sets have the same source, that is, have been taken from the same probability distribution. In Figure 2.9 the parameter combination $S = 0.7$ and $\alpha = 0.2$ was used to construct the tree. That means, using this tree as the classifier of unseen examples the probability of correct classification will be greater than or equal to 0.7, on a level of 0.2 as significance. This assumption is valid for all leaves of the tree that are reached by the unseen examples apart from those labeled by a star. For star-labeled leaves only a majority decision could be obtained due to a lack of sufficient data to test the hypotheses **H1** and **H2**. Thus a first assessment of the classification performance of a CAL5-tree without applying the usual verification procedure is set up.

However, the parameters S and α have to be tuned before starting the splitting procedure. This could be a disadvantage, because finding an optimal parameter combination is strongly dependent on the given training data set and requires a search for an optimal solution in the S, α-parameter space, and this may consume a lot of CPU-time (e.g., cross-validation).

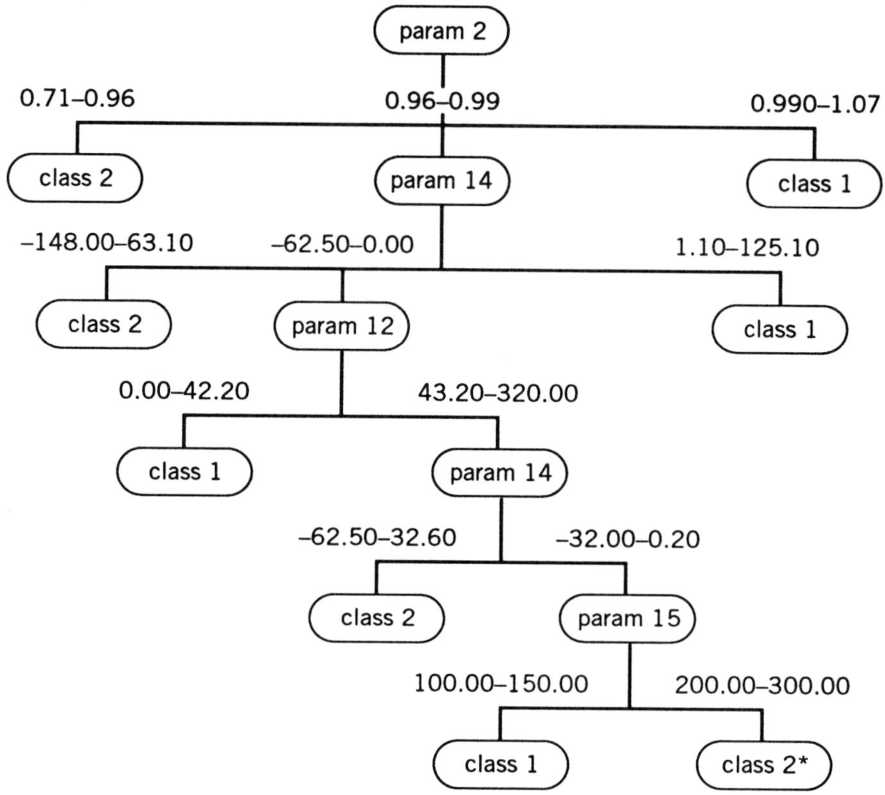

Figure 2.9. *Decision tree constructed by CAL5.*

Different decision trees can be constructed by varying the parameters S and α. In this case it is possible to get rules that approximate the class distribution with different degrees of accuracy or trees with different orders of attributes. This fact might be an advantage in the case of applications of such an algorithm for the structural analysis of processes (e.g. finding cause–effect relationships), which are specified only by their input and output behavior.

6. APPLICATIONS

6.1. Classification

CAL5 was successfully applied to different classification and prediction tasks within the *StatLog* Project, ranging from technical and medical diagnosis and satellite data interpretation to economical and financial forecasting. A former version of CAL5 was applied to problems of quality control [detecting faults in the production of magnetic tapes (Unger and Wysotzki, 1981)] and the diagnosis of complicated processes that could not be modeled mathematically. In the latter case the task was to minimize the coal dust emission in gas production. An optimal region in the parameter space of the process had to be found which corresponds to minimal dust emission. The results obtained by CAL5 (together with other algorithms) led to the construction of an automatic on-line control system of the gas generator by the factory.

Within the ESPRIT-Project *StatLog* more than 20 applications were handled using CAL5. One example of a technical application is the detection of emergency voltage conditions in electrical power stations. (The technical problem is elaborated in Wehenkel et al., (1993)).

The training and test data set were produced by a simulator. Each of them consisted of 1,250 samples. One sample contained 28 features (values of measurable process parameters). There were two classes (class 1 for stable state and class 2 for instable state). The resulting decision tree shown in Figure 2.9 has 13 nodes, and eight of them are terminal nodes with a unique class decision using threshold S. CAL5 has found that there are only four attributes of the given 28 that are relevant for the classification task at hand. Using this tree for decision making, an average number of 1.14 tests is needed. Therefore it has a good real-time performance, for example, a fast runtime and a high degree of compactness. The test run resulted in a success rate of $R = 97.12\%$ (error rate 2.98%).

Another example is the application of CAL5 in medical diagnosis of heart disease. Two-hundred-ninety samples with 13 attributes (e.g. age, sex, resting blood pressure, resting electrocardiographic results, etc.) were available. There were two classes: class 1 (heart disease) and class 2 (normal). In this case the following cost matrix was given:

class	1	2
1	0	1
2	5	0

The costs for not detecting the disease are five times higher than considering a normal person an ill one. This represents the costs to the patient of unnecessary treatment versus failure to treat one who is ill. However, the CAL5 decision tree constructed by using class-dependent thresholds (see section 3.1) may represent the behavior of a doctor, who is more sensitive to detect a special disease in order to minimize the risk of overlooking it. This task was tackled via nine-fold cross-validation. The following results are obtained optimizing Q (minimal risk):

average success rate $R = 78.15\%$
average cost factor $Q = 0.56$

If the cost is used as a quality measure, then a bias is introduced for the estimation of the success rate (which will be decreased in general). The example illustrates the possibility to structure a given statistical data set by using a machine learning algorithm. This way a suitable tool for supporting diagnosis in different fields of application is obtained.

6.2. Forecasting

In the following a special application in a prediction task is described. Within the National Project WISCON supported by the German Ministry of Science and Technology, CAL5 is used for high water flow forecasting in the Werra Valley of Thuringia (Germany). The high water level was to be predicted at some time at the town of Meiningen. The prediction horizon was nine hours. The training data set includes seven separate data sets, each of them describing one flood event as a time sequence. Every sample of the data set contained eight attributes, representing the water levels and water flows, measured at Meiningen and at several locations upstream at appropriate time steps before. Therefore the prediction problem was transformed into a classification problem by definition of intervals of water levels at Meiningen that correspond to discrete classes. This was five classes were formed, which had to be attached on each sample of the data sets. The resulting decision tree was tested by a data set derived from the 8. high water event. A success rate of $R = 83\%$ was obtained. This comes out to be a good result for a nine-hour forecasting of high water events. The tree contains only a subset consisting of three test attributes, that is, those which are relevant to the prediction problem at hand. Note that, by decomposition of a decision tree, a set of production rules, which may serve as a knowledge base for an expert system, can easily be derived. The problem of high water prediction was originally tackled by using a sophisticated mathematical (hydrological) model developed by the Technische Hochschule Ilmenau in Germany. The accuracy of the results of CAL5 was in the order of magnitude of predictions of the hydrological model (depending on the granularity of discretization and the available training data). The effort for the development of that mathematical model, however, was very high, while the decision tree model requires only a training data set and some qualitative knowledge to define the classes and to select the most efficient attributes.

7. CONCLUSIONS

In this Chapter the algorithm CAL5 for learning decision trees is presented, which is based on the determination of confidence bounds for the estimation of class probabilities in a constructed interval. The resulting decision tree is automatically pruned during tree construction and the tree may contain nodes with multiple branches. One of the results of the *StatLog* project (Michie et al., 1994) is that CAL5 constructs the most compact decision trees in comparison with other decision tree algorithms in most cases. This is affected by the performance of the above described splitting procedure.

The classification performance of the CAL5-trees can be approximately assessed by the parameters α and S without applying the sometimes time-consuming evaluation procedure for learning algorithms. Prediction tasks can be successfully transformed into classification problems by including the time history of the parameters (features).

PC- and SUN-versions of CAL5 are available from W. Mueller Fraunhofer-Institute IITB/EPO, Kurstrasse 33, D-19117 Berlin, Germany, and must be licensed for commercial application.

REFERENCES

Bleich, M. (1993). Erweiterung des Klassifikationsalgorithmus CAL5 und sein Einsatz für Probleme der Kommunikationstechnik. Diplomarbeit, TU Dresden, Fakultät Elektrotechnik, Dresden.

Buntine, W., and Caruana, R. (1991). *Introduction to IND and Recursive Partitioning*, User's Manual. NASA Ames Research Center, Moffet Field, CA.

Fu, K. S. (1968). *Sequential Methods in Pattern Recognition and Machine Learning*. Academic Press, New York and London.

Michie, D., Spiegelhalter, D. J., and Taylor, C. C. (1994). *Machine Learning, Neural and Statistical Classification*. Ellis Horwood, Hertfordshire, UK.

Müller, W., and Wysotzki, F. (1994). Automatic construction of decision trees for classification. *Ann. Oper. Res.* **52**, 231–247.

Quinlan, J. R. (1986). Induction of decision trees. *Mach. Learn.* **1**, pp. 81–106.

Sagalnicoff, M. (1993). Density- Adaptive learning and forgetting. In *Proceedings of the Tenth International Conference on Machine Learning, Amherst, MA*, pp. 276–284. Morgan Kaufmann, San Mateo, CA.

Schulmeister, B., and Wysotzki, F. (1994). Piecewise linear classifier DIPOL92. In *Proceedings of the Seventh European Conference on Machine Learning, Catania, Sicily*, 1994.

Unger, S., and Wysotzki, F. (1981). *Lernfähige Klassifizierungsysteme*. Akademie-Verlag, Berlin.

Wald, A. (1947). *Sequential Analysis*. Wiley, New York/Chapman & Hall, London.

Wehenkel, L., Pavella, M., Euxibie, E., and Heilbronn, B. (1993). Decision tree based transient stability assessment—A case study. *IEEE Trans. Power Syst.*

CHAPTER 3

Probabilistic Symbolic Classifiers: An Empirical Comparison from a Statistical Perspective

M. CHIOGNA

1. INTRODUCTION

Inferring a classification system can be considered a general problem of induction. In these types of problems, often referred to as problems of *inductive generalization*, we seek the best predictive rules relating all of, or any set of, the feature variables to a set of classes. In other words, a set of relationships among variables has to become apparent.

Different professional communities have worked on the problem of inductive generalization. Most of the results have been obtained within the 'machine learning' and the statistical communities, but it is only recently that comparisons of symbolic induction with the more traditional statistical solutions have begun to appear in the literature and joint applications have begun to take place (see, for example, *StatLog* Michie et al., 1994). The main difference between the inductive approaches developed by the two communities are related to the nature of the results offered by the techniques and to the emphasis with which the results are interpreted. The symbolic techniques developed in the machine-learning field tend to give a deterministic answer to the problem of classifying new observations; the statistical techniques focus on the estimation of probabilities of the possible results. Moreover, the machine-learning approach tends to emphasize the transparency and the interpretability of the results, whereas the statistical techniques aim to guarantee good numerical performances for the derived techniques.

In this case study, the problem of inferring a classification system for the diagnosis of congenital heart diseases in babies has been tackled with the use

Machine Learning and Statistics, The Interface, Edited by G. Nakhaeizadeh and C. C. Taylor.
ISBN 0-471-14890-3

of two different symbolic network-based strategies, namely, class-probability trees and belief networks. A statistical perspective is adopted in analyzing the construction of the classifiers; in considering the criteria on which to make comparisons, the predictive accuracy and the transparency of the derived systems are equally emphasized. The aim, therefore, is to analyze the effectiveness of the techniques, not only in providing good classification tools, but also in gaining sufficient insight into the process so that understanding of the problem is increased.

The structure of the chapter is as follows. After a general overview on classification problems contained in section 2, a brief description of the problem of congenital heart diseases is given in section 3. The following three sections describe the aim of the study, the performance measures chosen to evaluate the performance of the compared techniques, and the organization of the validatory assessment. The construction of the classifiers is then analyzed. Section 6 deals with the learning process of class-probability trees. Section 7 sketches the derivation of belief networks. The results of the structural and numerical comparisons are collected in section 8 and section 9, and a discussion of the results is presented in section 10.

Standard software has been used to perform the analysis. To build the class-probability trees we have used S-plus implementation of the tree-growing algorithm developed by Clark and Pregibon (1992); to build the belief networks some experimental software (*CoCo.* Badsberg, 1991), BIFROST (Højsgaard et al., 1992) and the package HUGIN (*Hugin Expert A/S*) (Andersen et al., 1989) have been utilized.

2. THE CLASSIFICATION PROBLEM: AN OVERVIEW

The problem of classification is mainly concerned with differentiating between k alternative classes on the basis of some multivariate observations on a set of features thought to be related to the classes. The features, which can be continuous or categorical variables, are usually structured as vectors X, allowed to take values in a space $\mathscr{X}$. The aim is to construct a rule, often referred to as *classifier*, which can be used to classify a new observation on the basis of the values for some of, or all of, the explanatory features.

Mathematically, the problem can be solved by partitioning the feature space $\mathscr{X}$ into k regions $\mathscr{R}_1, \mathscr{R}_2, \ldots, \mathscr{R}_k$, so that if an observation falls in $\mathscr{R}_j$ it is said to belong to the class j. The regions are identified by analyzing the properties of a set of data, the *training sample* $\mathscr{T}$.

Along with the feature vector, the observations in $\mathscr{T}$ may include information on the correct classification for the observations. If this is the case, the problem of inferring the decision rule is said to be *supervised*. In the following, only supervised learning tasks will be considered.

Let $\mathscr{C} = \{c_1, c_2, \ldots, c_k\}$ be the finite ordered set of classes' labels. The true class c_j of observation i can be represented by $y_i = (y_{11}, \ldots, y_{ij}, \ldots, y_{ik}) =$

$(0,\ldots,1,\ldots,0)$ with the 1 in the position corresponding to c_j. Let Y represent the class-vector for a generic observation in $\mathcal{T}$, and $\mathcal{Y}$ the set of possible values for Y. Inducing a classifier from the training set requires to identify a mapping from the feature space $\mathcal{X}$ to $\mathcal{Y}$ that associates to each possible feature vector a class of $\mathcal{C}$.

When the construction of a classifier is considered from a statistical point of view, the focus is on deriving for each of the possible classes in $\mathcal{C}$ a probability of being the correct class for the observations. In this case, the decision rule simply associates the observation to the most probable class.

From this perspective, the quantity of interest is $\Pr(Y|X)$, that is, the conditional probability of the class vector Y given that the feature vector X has been observed. Statistical theory offers two possible routes to estimate $\Pr(Y|X)$. The traditional solution requires the assumption for the distribution of the conditional random variable $Y|X$ of a model with parameter λ, which is assumed to depend on the feature vector X through a specified function f,

$$\lambda = f(\beta X).$$

Here, the features in X are considered to be simple explanatory variables and the randomness is related only to the class-variable Y. The classification task requires to estimate the vector β for the various possible values which can be assumed by X.

Alternatively, it is possible to estimate a joint probability distribution $\Pr(Y, X)$ for the vector (Y, X) and obtain the $\Pr(Y|X)$ simply by conditioning the joint probability over the feature vector X. In this second case, all the variables are considered to be random and all the possible relationships between any sets of variables need to be taken into account to estimate the joint probability distribution $\Pr(Y, X)$.

The relative merits and disadvantages of the two approaches are evident. The flexibility given by the knowledge of $\Pr(Y, X)$ is penalized by the difficulty of its estimation, which requires the evaluation of complex interactions among variables, some of which not totally relevant to the classification task. On the other hand, the simplification obtained by the assumption of a fixed relation f between λ and X does not allow one to deal easily with unseen patterns between features and classes, and with partial information caused by missing values.

In the following, two statistical techniques, each representing one of the two possible approaches, will be employed to identify classification systems for a diagnostic medical problem. Specifically, *class-probability trees* will be applied as an example of the traditional solution and *belief networks* will be utilized to estimate $\Pr(Y, X)$. These techniques are often referred to as *graphical*, because they make use of graphs to represent the sets of induced relations between features and classes. Network representations are also common in machine learning and artificial intelligence, mainly because many patterns of human reasoning follow the pathways laid out by such networks. Therefore graphical techniques seem to offer the most natural bridge between statistical and machine learning solutions to a classification task. A survey of the modern statistical usage of network methods in statistics can be found in Ripley (1994).

3. THE DIAGNOSTIC PROBLEM

Survival of newborn babies suffering from a critical congenital heart disease crucially depends on urgent medical and possibly surgical treatment. Considering the specialized nature of the treatments, these are normally carried out at regional specialist centers, to which are referred the children suspected to suffer from a congenital heart disease. Since the general conditions of the patients on arrival influence the success of the treatment, accurate transportation management is essential. In this view, an accurate early diagnosis is a crucial step.

The current routine followed at the Great Ormond Street Hospital for Sick Children (GOS) in London is that a provisional diagnosis is agreed between the local hospital and the referral center through a telephone conversation and an appropriate treatment decided on the basis of the chosen diagnosis. The items required to perform the diagnosis are related to the findings of a clinical examination: an electrocardiogram, a chest X-ray and, optionally, an arterial blood gas sample. Some of the items are represented by categorical variables, some others by continuous ones and in most cases only limited information is available.

The level of expert knowledge in the domain of congenital heart disease is undoubtedly advanced. In fact, two expert-derived structures have been elicited to help the doctors in the diagnostic task.

Franklin and colleagues (1991) developed an algorithm structured in a binary tree form. The algorithm identifies 27 possible diseases on the basis of up to 21 questions asked over the telephone. The paper contains an evaluation of the algorithm based on a three-phase study, which is seen to improve the performances of both the cardiologists of the referral center and the pediatricians of the local hospitals.

For dealing with a particular subgroup of congenital heart diseases, the so called "blue baby" diseases, a belief network has also been elicited (Spiegelhalter et al., 1993). The network, in the form of a directed acyclic graph, has been designed by the experts by following a sequence of scrutinies and refinements of an initial provisional structure and it follows a first attempt at modeling the problem with the use of a naive statistical scoring system (Franklin et al., 1989).

4. AIM OF THE STUDY

In this case study data-derived class-probability trees and belief networks have been built with respect to the problem of diagnosis of the "blue baby" diseases. The resulting classifiers have been compared to analyze their numerical performance and to evaluate the structure of relationships that they identify for the variable involved in the diagnostic task.

As anticipated, these techniques represent two statistical graphical alternatives of tackling a classification problem. The existence of expert-derived models offers an uncommon and ideal benchmark to accompany the numerical comparison with an evaluation of the potential of the techniques of imitating the equivalent human-provided models.

In this particular case, the comparison is not immediate, as the expert-derived models appear to be more general than the data-derived ones. The expert-derived classification tree deals with 27 possible diseases. However, the "blue baby" diseases are chosen after all the remaining diagnoses have been ruled out and can therefore be easily isolated on the algorithm as a separate diagnostic group. The expert-derived belief network utilizes a set of exploratory variables, which includes additional explanatory variables to the ones that have been used in this case study, and performing the comparison by simply extracting the set of nodes corresponding to the variables relevant to this study could be misleading. However, the structure of the flow of causal relationships among variables in the network identified by the experts gives some insight into the role played by the variables used in this study in determining the global topology of the network. Therefore, the use of the expert-derived structures as a loose term of comparison is seen as highly informative to help grasp the ability of the data-driven techniques to discover relationships in complex domains.

5. THE EXPERIMENT

Data related to the subset of diseases of interest have been extracted for the three phases of the study described in Franklin et al. (1991) and the distribution of the diseases on the set of the 238 patients suffering from them is shown in Table 3.1. They are: obstructed Total Anomalous Pulmonary Venous connection Disorder (TAPVD), Persistent Fetal Circulation (PFC), lung disease (Lung), simple Transposition of the Great Arteries (TGA), pulmonary outflow obstruction—dominant right ventricle (Fallot), Pulmonary Atresia and Intact Ventricular System (PAIVS). To unify the notation, the phase labeling used in the above mentioned study has been maintained.

Ten explanatory variables relevant to the diagnosis of the six diseases have been used, of which five are categorical—sick (Yes, No), birth asphyxia (Yes, No), grunting (Yes, No), X-ray lung fields (normal, oligemic, plethoric, ground glass, asy/patchy), left ventricular hypertrophy (LVH) (Yes, No)—and five are continuous—age, room air blood gases: pH, pCO_2, pO_2 in 100% FiO_2 right-upper quadrant (RUQ), pO_2 in 100% FiO_2 lower body (Lower B.). Since the compared techniques deal only with categorical variables, a discretization of the continuous variables has been performed on the basis of medical advice.

To assess the numerical performance of the techniques, a cross-validation experiment (Stone, 1974) was organized. Using the existing temporal stratifica-

TABLE 3.1. Distribution of the Six Diseases, with Their Frequency of Presentation within the Three Phases

	Phase 0		Phase 2		Phase 3	
Disease (Label)	*n*	%	*n*	%	*n*	%
TAPVD	3	(4.3%)	5	(6%)	7	(8%)
PFC	5	(7.1%)	17	(21%)	11	(12.7%)
Lung	11	(15.7%)	6	(7.5%)	10	(11.5%)
TGA	21	(30%)	30	(37%)	28	(32.2%)
Fallot	18	(25.7%)	17	(21%)	23	(26.5%)
PAIVS	12	(17.2%)	6	(7.5%)	8	(9.1%)
Total	70	(100%)	81	(100%)	87	(100%)

tion of the data into three phases, three different training sets have been created by merging in turn data coming from two different phases. The test set for each of the training sets has been identified with the remaining third phase. In the following, we will indicate as experiment '1', '2', '3' the experiment based on the training set 'phase 0 + phase 2', 'phase 0 + phase 3', 'phase 2 + phase 3', respectively. We note that the obtained threefold cross-validation is not random and that few differences exist in the distribution of the variables in the three training sets. In particular we note a greater proportion of missing values in the training sets 'phase 0 + phase 2' and 'phase 0 + phase 3' due to less frequent information on the continuous variables in phase 0. Data-derived classifiers have been built on each training sample, and then tested on the corresponding test samples.

6. PERFORMANCE MEASURES FOR THE EXPERIMENT

When considering the construction of a classifier from a statistical point of view, it is common to assume that y_i is a multinomial observation vector with parameter $p_i = (p_{i1}, \ldots, p_{ik})$ expressing the probability for each of the possible classes in $\mathscr{C}$. In the experiment, the techniques have been built to create for each classified observation a vector of estimated class probabilities $\hat{p}_i = (\hat{p}_{i1}, \ldots, \hat{p}_{ik})$. This vector has been used to classify the observation, according to the highest estimated probability, and to study the quality of the induced classifiers. Standard *error rates* have been used to analyze accuracy of the classifiers; *proper scoring rules* (Murphy and Winkler, 1984) such as

- *logarithmic score*: $LS_i = -\log \hat{p}_{ij}$ if d_j is the true class observed on y_i;
- *Brier score*: $BS_i = (1 - \hat{p}_{ij})^2 + \Sigma_{k \neq j}\, \hat{p}_{ik}^2$ if d_j is the true class observed on y_i;

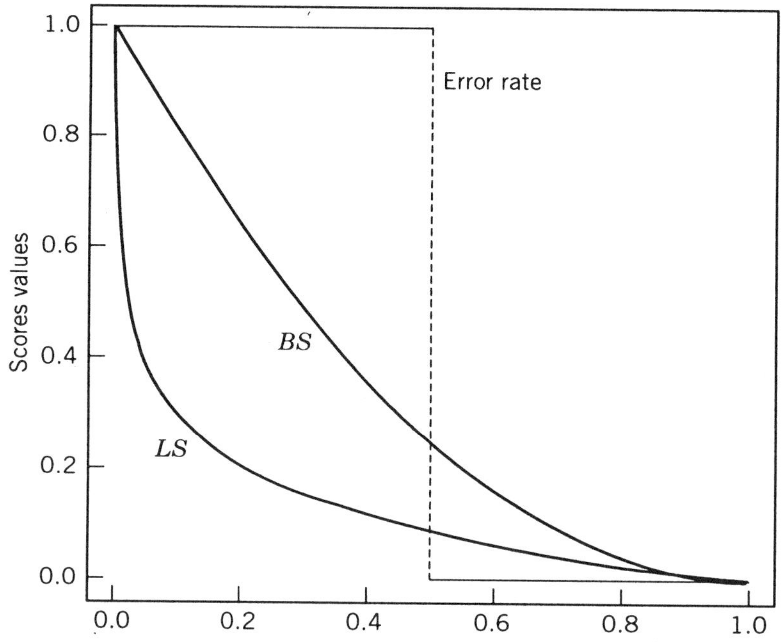

Figure 3.1. *Performance measures.*

have been used to assess the lack of accuracy of the probabilistic predictions. In a single prediction, both the scores assume value 0 if the classifier correctly identifies the true disease and a positive value otherwise. It can be noted that the logarithmic score is simply minus the logarithm of the probability given to the true class. Since a single estimated probability $\hat{p}_{ij} = 0$ will cause the total logarithmic score to be infinity, a minimum probability of 0.01 has been given to each class. Over a set of n cases, they define total penalities $LS = \Sigma_{i=1}^{n} LS_i$ and $BS = \Sigma_{i=1}^{n} BS_i$.

For a two-class problem, a plot of the normalized values assumed by the three performance measures is shown in Figure 3.1. The plot highlights the different weights given by the scores to the estimated values for the probability of the true disease. The logarithmic score heavily penalizes estimates which are very close to zero. The Brier score, by taking into account the estimated probabilities of each class, introduces information on the entropy of the results. In general, for a k-class problem, the normalized score assumes the value 1 when all the probability is concentrated on one wrong class, the value $(k - 1)/2k$ when the probability maximizes the entropy among the classes and the value 0 when the probability is concentrated on the correct class.

7. CLASS-PROBABILITY TREES FROM DATA

Despite the long tradition within the machine-learning community, the statistical applications of class-probability trees are still related to the seminal work of the CART collaboration (Breiman et al., 1984). Recent applications can be found in Segal (1988) and Ripley (1993).

Class-probability trees perform a recursive hierarchical partition of the multivariate sample space into class (diagnosis) regions on the basis of some optimality criterion, which guarantees that the identified class regions are as homogeneous as possible with respect to the diagnoses. The partition is organized in a binary tree structure, where each node of the tree defines a split of the observations into two groups with respect to one variable. We shall adopt the extensions of CART developed by Clark and Pregibon (1992). This approach assumes that the vector y_i for each individual i can be modeled as a multinomial distribution. At each step of the partition procedure, the class probabilities are estimated from the subset of observations reaching the node to be partitioned. This is equivalent to assuming that the class probabilities depend on the subset of explanatory variables used by the tree to identify the node. As the type of link f that relates the class probabilities to the explanatory variables is left unspecified, the construction of class-probability trees can be seen as a non-parametric application of the traditional statistical approach to the construction of a classifier. On each node, the estimated class probabilities are used to measure homogeneity of the classes within the node and within the potential offspring, and to select the next best partition. Over the leaves of the trees, the probabilities are used as empirical estimates of the probability of each diagnosis. Future observations are assigned to a region (leaf) of the partition according to their feature vector values with respect to the nodes conditions.

Optimization criterion. We adopt the deviance-based optimization criterion. The probabilities ${}_Np_i$ of each class on one node N containing n individuals are estimated with their empirical frequencies ${}_N\hat{p}_i = n_i/n$. In statistical terms, this corresponds to performing a maximum likelihood estimation of the unknown parameters of the multinomial distribution when the individuals in N are assumed to be identically distributed. The deviance of the observations can be defined to be

$$D({}_N\hat{p}) = -2 \sum_k n_k \log({}_N\hat{p}_k).$$

We note that the criterion differs from the more traditional entropy-based criteria for a constant factor. The partition proceeds by selecting as best binary split of a node the split that maximizes the quantity $D({}_N\hat{p}) - D({}_N\hat{p}_L) - D({}_N\hat{p}_R)$, with $D({}_N\hat{p}_L)$ and $D({}_N\hat{p}_R)$ representing the deviances of the left-hand and the right-hand extremes of the split, respectively.

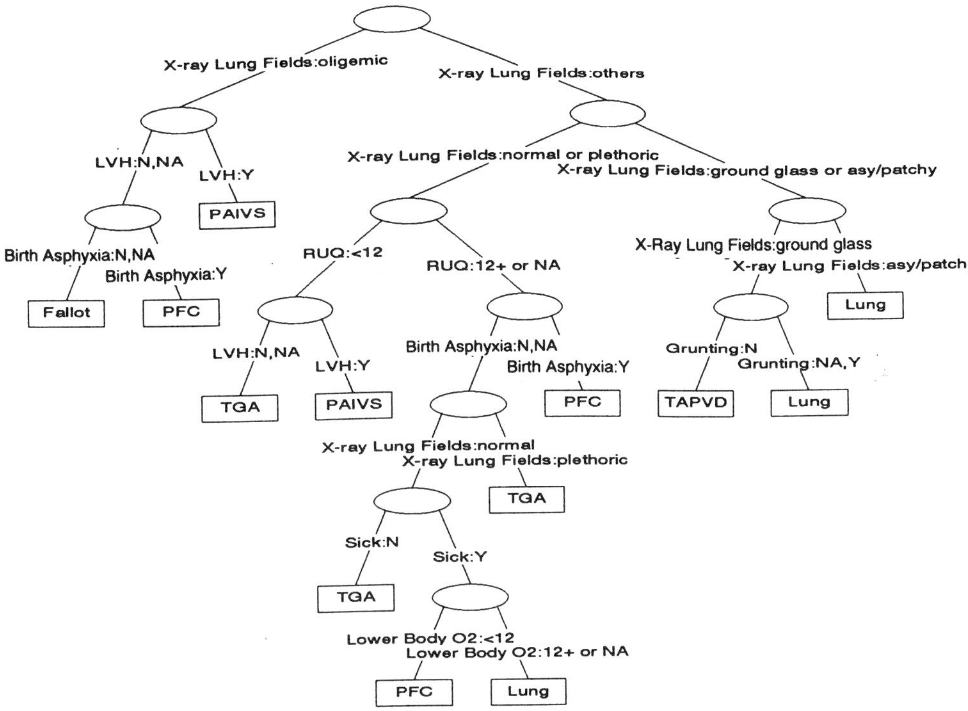

Figure 3.2. *Class-probability tree for 'phase 0 + phase 2' training set (the highest probable diagnoses label the leaves of the tree). Missing values are coded as 'NA'.*

Training algorithm. An exhaustive search for the best split is performed at each node of tree: all the possible one-step look-ahead splits are generated and the criterion evaluated to select the best choice. The partitioning process is stopped when the node deviance is less than one percent of the deviance calculated at the root node, or if the number of observations at the node is less than 10. A refinement of the structure has then been achieved by manual pruning, to avoid the removal of leaves associated with rare diseases performed by some tested automatic pruning procedures.

Missing values. In the training phase, missing values have been treated by adding a new class to the variables showing incomplete information. However this management can create difficulties in the test phase, when a new case, showing a missing value on a variable with only complete training information, needs to be classified. In this case, it has been decided to perform the classification according to the distribution of the classes on the deepest reachable node of the tree.

The structure of the tree for the 'phase 0 + phase 2' training set is reproduced in Figure 3.2.

8. BELIEF NETWORKS FROM DATA

Belief networks are traditionally defined as probabilistic expert systems tools used to represent and deal with expert knowledge. Considering the evident connection of the machine-learning applications with the field of expert systems and applied artificial intellegence, it seems plausible to define a belief network as a probabilistic symbolic technique.

In the probabilistic approach to the construction of expert systems, each item of knowledge affected by uncertainty is represented by a random variable. Typically, random variables representing knowledge take values in a finite set and the interest relies in the joint distribution of the whole set of variables. Indeed, most of the questions usually asked in an expert system consultation can be answered by managing the joint distribution. The joint distribution is defined on the space given by the Cartesian product of the finite domains of each variable. Therefore, the numerical assessment required to completely specify the joint distribution should associate a probability to each of the possible combination of values of the variables in the problem. It is easy to guess that storage and management of this information becomes rapidly computationally untractable as soon as the number of variables increases.

Belief networks offer a formalism to represent concisely and deal efficiently with a complex distribution. They comprise two parts: a qualitative part and a numerical part. The qualitative part is represented by a graph. In the graph, nodes represent the variables involved, and edges (directed or undirected) represent certain dependencies between variables. Graphs can be completely directed, completely undirected, or mixed. Intuitively, a directed edge expresses the notion of causal relevance between variables, whereas an undirected edge represents a symmetric dependence. Instead, a missing edge represents a missing causal or symmetric interaction between two variables. Therefore, graphs are used to give a lucid representation of the key dependences and irrelevancies among the random variables involved.

Associated with the qualitative structure of the model, there are some numerical functions representing the numerical specification for the distribution under consideration. In probabilistic terms, the notion of "irrelevance" embedded in the topology of the graph can be expressed through the device of conditional independence, which allows, in computing a probability, to select only what is of real relevance. By exploting all the conditional independence properties portrayed in the graph, the initial set of variables can be decomposed into subsets, often called universes, which group together variables of valuable relevance for each other. The joint distribution can therefore be decomposed into the product of functions defined over the universes.

The decomposition obtained through the identification of the universes permits to simplify storage and management of the joint distribution. Numerical specification of the distribution only requires numerical specification of the functions defined over the universes, which usually comprise fewer variables than those on which the joint distribution is defined. Efficient algorithms can also be constructed for managing the distribution by exploting local computations on the universes. In the following, the theoretical framework developed by Lauritzen and Spiegelhalter (1988) has been used to construct and use the network.

Normally, human experts identify the properties of the grapical structure; a data-derived perspective of such theory puts the construction of a belief network in the context of the selection of the best set of properties for the probability distribution, that is, of the selection of the best network topology. However, the human expertise can been utilized to group the variables in blocks such that a 'symmetric' dependence can be assumed to exist among the variables within each block and a causal relationship is believed to exist among variables in different blocks. This allows us to organize a stepwise model selection procedure by performing the search for the best model on increasingly larger sets of variables obtained by merging successive blocks. For the diagnostic task at hand, two blocks of variables have been identified (age, birth asphyxia, sick, final disease and grunting, X-ray lung fields, LVH, pH, pCO_2, RUQ, Lower B.).

Future observations are assigned to the class (diagnosis) that shows the highest probability on the conditional distribution obtained by conditioning $\Pr(Y, X)$ to the feature values. For example, the most probable diagnosis for a case showing complete information on all the concomitant variables is found by selecting the most probable class in the conditional distribution obtained by fixing each concomitant variable to be equal to the value observed on the case, like in

$$\Pr(\text{final diagnosis} \mid \text{sick} = \text{Yes, birth asphyxia} = \text{Yes, grunting} = \text{No}, \ldots)$$

Optimization criterion. We use a deviance-based test statistic to select the relevant interactions (edges) between pairs of variables. The selection is based on comparison of pairs of models of which one eliminates the edge under test. The models are chosen within the class of graphical decomposable log-linear models and their parameters, modeling the probabilities of the possible combinations of levels of the variables in (Y, X) are estimated via maximum likelihood procedures. Indicating with *base* the model including the edge and with *current* the competitive model that eliminates the edge and with ${}_B\hat{p}$ and ${}_C\hat{p}$ their estimated vectors of parameters, the test statistic

$$D({}_B\hat{p}, {}_C\hat{p}) = 2 \sum_{i \in \mathscr{I}} n_i \log \frac{{}_B\hat{p}_i}{{}_C\hat{p}_i}$$

measures the difference in goodness of fit between the two models. Here, $\mathscr{I}$ identifies the set of possible combinations of levels for variables in the two models; for the combination i, ${}_B\hat{p}_i$ and ${}_C\hat{p}_i$ represent the maximum likelihood estimates of the parameters under the base and current model, respectively, and

n_i the observed absolute frequency of the cases showing combination i.

Training algorithm. An automatic model selection has been performed based on the backward removal of irrelevant edges through significance tests based on the previous criterion. Starting from a fully interconnected network, the relevance of the possible edges is tested. If an edge does not appear to be relevant, it is removed from the structure of the network. The procedure is recursive with respect to the block structure of the variables. The structure of the network for the variables in the first block is chosen first. The edges which have been considered relevant in this phase of the topology's selection are permanently fixed in the qualitative model of the network and they are not reconsidered in the successive steps of the algorithm. Then, the set of variables is enlarged by including the variables in the second block and a new selection process is performed.

Missing values. In the training phase, only the complete cases have been used to perform the model selection at each step of the edges selection. In the test phase, only the non-missing values of each observation are used in conditioning the joint probability distribution and therefore in determining the

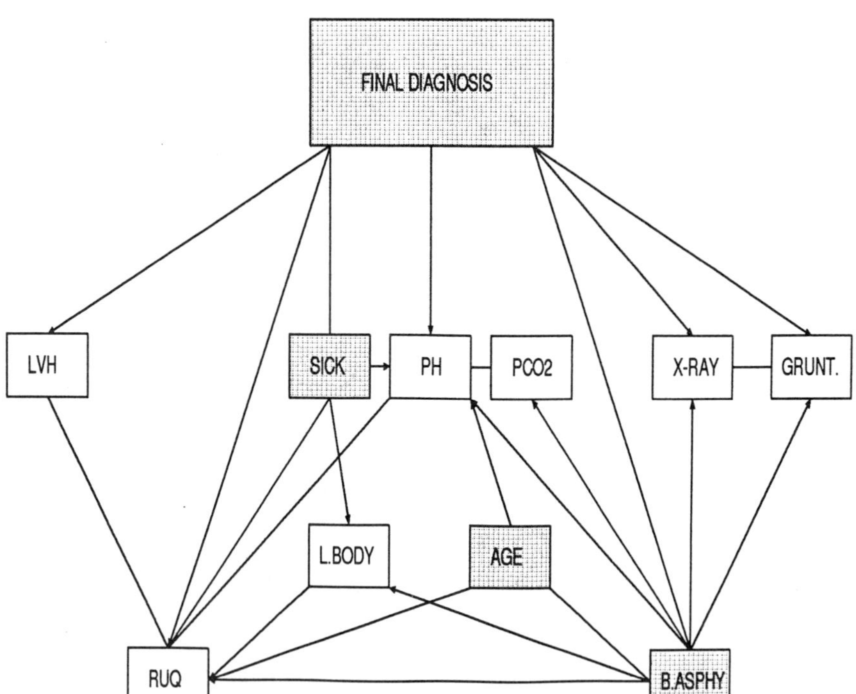

Figure 3.3. *Structure of the belief network for 'phase 0 + phase 2' training sample. The shaded nodes correspond to variables in the first block.*

most probable class for the observation.

We note that, for the same problem, other authors have built belief networks in the same manner (Lauritzen et al., 1994). Although the models selection techniques are very similar, the final structures of the networks obtained by the authors differ from the results presented in this case study. The main differences between the results appear to be related to the discretization of the continuous variables and to the missing values management.

The structure of our network for the 'phase 0 + phase 2' training set is reproduced in Figure 3.3.

9. STRUCTURAL COMPARISON

Insight. To evaluate the ability of the techniques to gain insight into the problem, the structures derived in the experiment can be compared with the equivalent expert-derived structures described in Franklin et al. (1991) and Spiegelhalter et al. (1993), although these give a more general model for the diagnostic problem.

The portion of the expert-derived tree that deals with the six diseases considered in this case-study shows strong similarities with the trees derived in the experiment, in the sense that they reproduce almost exactly the set of conditions leading to some of the diseases (compare, for example, the diagnosis of Fallot). In most of the cases, the order in which the conditions lead to the final diagnoses differs between the expert-derived and data-derived trees, this being also due to the impossibility of identifying multiple features splits in the data-derived tree.

The expert-derived belief network, as already noted, is built on an enlarged set of concomitant variables. However, it is possible to capture a substantially simpler structure of the interrelationships among variables and, in particular, among those that we have included in the second block of the recursive structure. This confirms that the data-derived networks suffer from the sparseness of the data and suggest that, when limited data are available, a more intensive use of the expert knowledge is required to obtain models closer to the human-derived competitors.

Models' Structure. Within each technique, similarities occur in the three structures derived. Class-probability trees identify the same hierarchical ordering for the variables X-ray Lung Fields, LVH, RUQ, birth asphyxia, sick, lower body. As expected, the main differences among the three structures are related to the lower levels of the trees where different variables or the same variables in different order are used. This can be related to peculiarities of the training sets, as well as to the small sample sizes of the nodes in the lower levels of the trees. Belief networks, on the other hand, identify the same topology for the variables in the first block of the recursive structure and same constant structures are identified in the global topology (for example, final diagnosis is always related to LVH, RUQ, X-ray lung field, grunting).

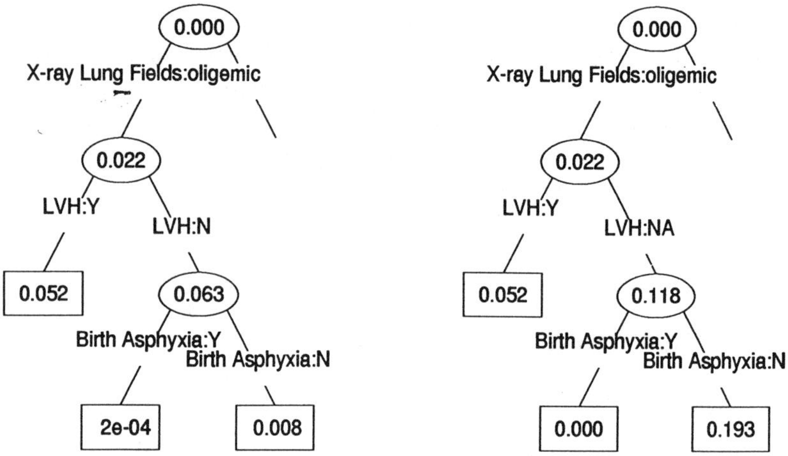

Figure 3.4. *Experiment 1. Two possible sets of rules which can be exploited on the tree choosing the branch 'X-ray Lung Field: oligemic'. Numbers on the nodes represent the distance between the estimates of the class probabilities obtained from the tree and the network.*

Taking into account the differences between the two techniques in building the predictive models, it must be noted that some groups of interacting variables have been equally highlighted by the trees and the networks. For example, LVH and birth asphyxia both interact through RUQ in the structures in Figure 3.2 and Figure 3.3. On the same graphs, Final diagnosis, X-ray lung fields, and grunting are also connected.

When the set of condition on the branches of the trees leading to the diseases are reproduced on the network, 92.5 percent of them identify the same classification with comparable probability.

Figure 3.4 shows the level of agreement between the two techniques in Experiment 1. The figure represents two of the four possible sets of rules exploited by the tree on choosing 'X-ray Lung Field: oligemic'. On each node, the value of the distance

$$\sum_{j=1}^{k} ({}_{tree}\hat{p}_j - {}_{net}\hat{p}_j)^2$$

is printed, where ${}_{tree}\hat{p}_j$ and ${}_{net}\hat{p}_j$ are the estimates of the class-probability vector as estimated from the tree and the network, respectively, in the conditional distributions identified by the conditions on the branches of the tree. The distance assumes value zero if the two estimated class-probability vectors are identical, and a maximum of 2 if the two class-probability vectors concentrate a probability equal to 1 on two different diseases. The range of values assumed by the distance measure suggests an almost perfect agreement between the two structures. It seems, therefore, that the two techniques share the same ability to capture the main characteristics of the partition of the sample space induced by the diseases.

Models' complexity. Identifying the complexity of a class-probability tree with the number of nodes on the trees and the complexity of a belief network with the number of edges in the networks, the techniques show comparable differences in the degree of complexity for the three structures among the training phases. Both the techniques highlight the lowest complexity for the structure of the classifiers in Experiment 1 (12 nodes for the trees and 23 edges for the networks) and the highest complexity in Experiment 3 (14 nodes for the trees and 32 edges for the networks). In general, belief networks result in being highly interconnected, this being related to the sparseness of the data on which the models are estimated more than to existing relationships among the variables. This reflects a high number of parameters for the estimated probability densities and, consequently, a structure that appears to be more complex than the structure of class-probability trees.

Feature selection. As a side effect of the tree construction, a variable selection is also performed. Referring again to Figure 3.2, it can be seen that the variable pH never appears on the tree, whereas in the belief network it plays the role of interconnecting variables in the first block of the recursive structure to variables in the second block. We note that the continuous variables are almost exclusively used in 'phase 2 + phase 3' tree, that is, on the tree built on the training set with fewest missing information on them, as a confirmation of their discriminant power. Feature selection, by contrast, is not performed by belief networks.

10. COMPARATIVE PERFORMANCE

The results of the empirical accuracy comparison are contained in Table 3.2.

Generally, we noted better results for class-probability trees. The trees seemed to combine a better accuracy with estimates of the probabilities that, according to the proper scores, appeared to be less extreme than those obtained from the corresponding networks.

However, the second experiment showed comparable results. From the error rate it appeared that the test set used in Experiment 2, which corresponds to 'phase 2' of the study of Franklin et al. (1991), was for the trees the most difficult to predict. Taking into account the homogeneity criterion employed by the trees in constructing the classifiers, the results of Experiment 2 could suggest that the actual patterns of the data in 'phase 2' slightly differed from those of 'phase 0' and 'phase 3'. This could explain the improved predictive capabilities of the trees built on the two training sets that contain 'phase 2' and the difficulties of the networks in grasping on the same training sets the structure of relationships among variables. A random cross-validatory assessment might control this discrepancy across experiments.

TABLE 3.2. Results of Cross-Validatory Assessment for the Compared Techniques

Training set	phase 0 + 2	phase 0 + 3	phase 2 + 3
(Size of the training set)	(151)	(157)	(168)
Test set	**phase 3**	**phase 2**	**phase 0**
(Size of the test set)	(87)	(81)	(70)
Class-probability trees			
Error rate	20/87	26/81	19/70
Brier score	32.08	44.62	22.36
Log score	76.27	106.50	52.06
Belief networks			
Error rate	27/87	27/81	26/70
Brier score	47.62	50.66	47.08
Log score	107.13	116.25	108.28

The values of the probabilistic scores obtained from the belief networks in Experiment 1 and Experiment 3 seemed surprisingly high. It is expected that models as complex as the belief networks do not perform well as simple classifiers, but normally they are expected to produce better estimates for the probability of the results. On the contrary, the high value for the logarithmic score was an indication that, in the misclassified observations, very small probabilities had been associated with the true diagnoses.

A closer scrutiny of the performed classifications revealed that, in the experiments, most of the misclassifications operated by the networks and not by the trees were due to the assignment of a probability 1 to one particular diagnosis. In fact, the removal from the values of the performance measures of the contribution deriving from this subgroup of misclassifications identified scores' values almost comparable with those obtained from the trees. A monitoring process of the change in the probability distributions on the nodes of the network induced by the conditioning process allowed the identification of the combination of levels causing the wrong classifications to be taken. Some of these combinations corresponded to the set of rules exploited on the branches of the trees which were not reproducible on the network. Other combinations were only partially identifiable on the trees, as they involved features not utilized by the trees. However, the monitoring process seemed to suggest that classification trees can help identify possible lacks-of-fit of belief networks built on the same data.

Although informative, the performance measures say little about the correct classifications operated by the techniques. Restricting the attention to the cases correctly classified by both the techniques, Figure 3.5 shows a comparison of the probabilities of the classifications computed by the two techniques. The numbers at the bottom line of the diagnoses' boxes identify the experiment to which the classifications belong.

To some diseases, one of the two techniques attributed a probability that was almost always systematically bigger (smaller) than the probability

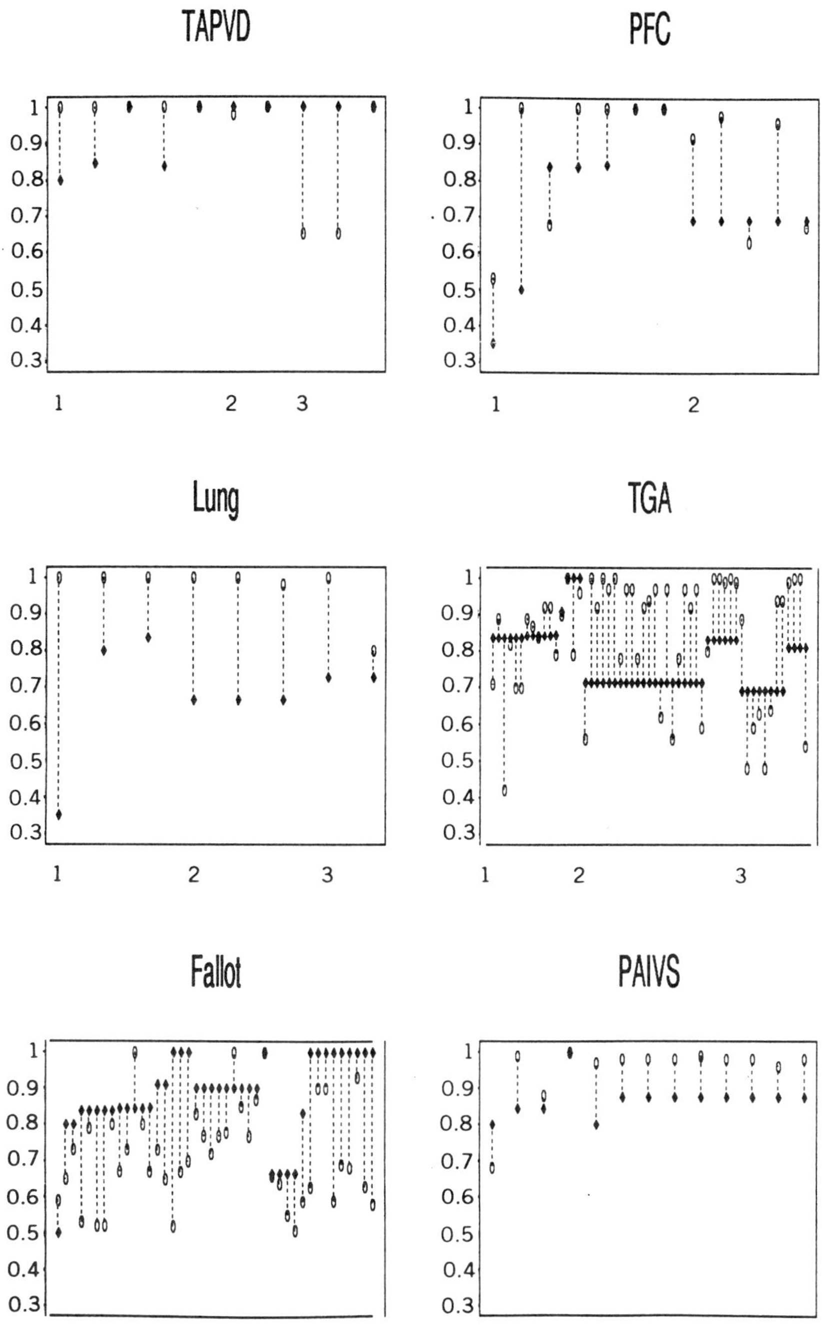

Figure 3.5. *Comparison of the probabilities of the classifications correctly classified by both the techniques. The estimates obtained from the network are marked with an "o". Labels underneath the x-axis highlight the phases to which the observations belong.*

TABLE 3.3. Distribution of the Errors Operated in the Cross-Validatory Assessment

Experiment	1		2		3	
	n	%	*n*	%	*n*	%
Tree and Network	11 + 6	(19.54%)	11 + 8	(23.45%)	2 + 9	(15.71%)
Tree Only	3	(3.44%)	7	(8.64%)	5	(7.14%)
Network Only	10	(11.49%)	8	(9.87%)	15	(21.42%)

Percentages in brackets are computed on the appropriate test sets.

obtained by the other technique. For example, 'Lung' was given by the networks a probability systematically higher than the probability estimated by the trees; on the contrary, 'Fallot' was identified with a generally higher probability by the trees. These remarks seem to contrast with the agreement of the structures found in the structural comparison, according to which the diagnostic rules identified by the trees could be reproduced with comparable probabilities of the final results on the networks. However, it must be noted that only a subset of explanatory variables is selected and used by the trees to perform the classifications. The plot in Figure 3.5 seems therefore to suggest that, in diagnosing some diseases, like 'Lung' or 'TGA', the potential use of other explanatory variables might improve the precision of the estimates, whereas in other cases, like in 'Fallot', the use of more variables could reduce the degree of certainty of the results.

Table 3.3 contains the distribution of the errors operated by the techniques. The first row represents the number of observations misclassified by both the techniques; the left-hand side numbers in the sums represent the number of cases with which the techniques have associated the same wrong diagnosis.

Considering the subset of observations misclassified by both the techniques, we note that in 51 percent of the cases the techniques produced the same misclassification. Therefore there were areas of the sample space on which the techniques seemed to highlight the same wrong structure.

11. DISCUSSION

The results of the study emphasize the difficulties in finding a right balance between good numerical performances of the classifiers and transparent understanding of the true underlying model generating the diagnostic classes.

Potentially, the techniques seem capable of reaching comparable numerical performances, although in relatively small and complex data sets the networks are likely to necessitate guided refinements of the structures automatically derived from the data. Both the techniques seem to benefit from guided assistance either in the derivation or refinement of the classifiers. In pruning the trees, a manual selection of the nodes to be pruned was necessary to avoid the removal of leaves representing rare diseases; a block recursive structure was adopted to control and organize the topology's selection in the networks.

When the transparency of the models is considered, a completely different viewpoint is taken up by the two techniques in highlighting the set of relationships existing among the variables. The construction of the class-probability trees is focused on homogeneity properties of the classes of the response variable in conditional distributions. This explicitly reflects the interaction properties of the classes' distribution with the features, but only implicitly reflects the possible relationships among the feature variables. Belief networks, on the other hand, explicitly interpret the set of relationships among variables by defining a qualitative model for their joint probability distribution. From a strictly explanatory perspective, this should provide considerably more insight into the true structure of the problem and therefore should result in more easily interpretable final models. Although the study shows that belief networks are able to capture the main relationships among the variables involved in the diagnostic task, the transparency of the results is penalized by the complexity of the relationships that the networks need to take into account. Comparatively, the set of rules identified by the trees appears to be more informative and close to the experts' understanding of the process.

The comparison performed in this study seems to suggest that the techniques can complement each other. Class-probability trees can indicate possible lacks-of-fit of the belief network; in this respect, they are likely to indicate how to improve their accuracy. Belief networks, on the other hand, can indicate possible explanatory variables that are worth considering to improve the estimation capabilities of the trees. The extent to which the suggested improvements can be implemented in practice remains to be explored.

Nonetheless, the evidence at hand seems to suggest that, in real and complex situations with limited data available, data-derived class-probability trees outperform data-derived belief networks in supporting a diagnostic task.

ACKNOWLEDGMENTS

The chapter's author expresses thanks to Dr. David Spiegelhalter, at the MRC Biostatistics Unit of Cambridge, who suggested this application, and to Dr. Kate Bull at the Great Ormond Street Hospital for Sick Children of London, who gave insight into the medical problem, and the data on which this experiment is based.

REFERENCES

Andersen, S. K., Olesen, K. G., Jensen, F. V., and Jensen, F. (1989). HUGIN—A shell for building Bayesian belief universes for expert systems. In *Proceedings of the 11th International Joint Conference on Artificial Intelligence, Detroit*, pp. 1080–1085. Morgan Kaufmann, San Mateo, CA.

Badsberg, J. H. (1991). *A Guide to CoCo*, Tech. Rep. R-91-43. Department of Mathematics and Computer Science, Aalborg University, Aalborg, Denmark.

Breiman, L., Friedman, J. H., Olshena, R. A., and Stone, C. J. (1984). *Classification and Regression Trees.* Wadsworth, Belmont, CA.

Clark, L. A., and Pregibon, D. (1992). Tree-based models. In *Statistical Models in S* (J. M. Chambers and T. J. Hastie, eds.), pp. 377–420. Wadsworth & Brooks/Cole, Pacific-Grove, CA.

Franklin, R. C. G., Spiegelhalter, D. J., Macartney, F., and Bull, K. (1989). Combining clinical judgments and statistical data in expert systems: Over the telephone management decisions for critical congenital heart disease in the first month of life. *Int. J. Clin. Monit. Comput.* **6**, 157–166.

Franklin, R. C. G., Spiegelhalter, D. J., Macartney, F., and Bull, K. (1991). Evaluation of a diagnostic algorithm for heart disease in neonates. *Br. Med. J.* **302**, 935–939.

Højsgaard, S., Skjøth, F., and Thiesson, B. (1992). *User's Guide to BIFROST*, Tech. Rep. R-92-2001. Institute for Electronic Systems, Aalborg University, Aalborg, Denmark.

Lauritzen, S. L., and Spiegelhalter, D. J. (1988). Local computation with probabilities on graphical structures and their applications to expert systems. *J. R. Stat. Soc., Ser. B* **50**, 157–224.

Lauritzen, S. L., Thiesson, B., and Spiegelhalter, D. J. (1994). Diagnostic systems created by model selection methods—a case study. In *Selecting Models from Data: Artificial Intelligence and Statistics* (P. Cheesman and R. W. Olford, eds.), pp. 143–152. Springer-Verlag, New York.

Michie, D., Spiegelhalter, D. J., and Taylor, C. C. (1994). *Machine Learning, Neural and Statistical Classification.* Ellis Horwood, Chichester, UK.

Murphy, A. H., and Winkler, R. L. (1984). Probability forecasting in meteorology. *J. Am. Stat. Assoc.* **79**, 489–500.

Ripley, B. D. (1993). Statistical aspects of neural networks. In *Networks and Chaos: Statistical and Probabilistic Aspects* (O. Barndorf-Nielsen, J. L. Jensen, and W. S. Kendall, eds.), pp. 40–123. Chapman & Hall, London.

Ripley, B. D. (1994). Network methods in statistics. In *Probability, Statistics, Optimisation; A Tribute to Peter Whittle* (F. P. Kelly, ed.), pp. 241–253. Wiley, New York.

Segal, M. R. (1988). Regression trees for censored data. *Biometrics* **44**, 35–47.

Spiegelhalter, D. J., Dawid, A. P., Lauritzen, S. L., and Cowell, R. G. (1993). Bayesian analysis in expert systems. *Stat. Sci.* **8**, 219–247.

Stone, M. (1974). Cross-validation choice and assessment of statistical predictions. *J. R. Stat. Soc. Ser. B* **36**, 111–147.

CHAPTER 4

A Multistrategy Approach to Learning Multiple Dependent Concepts

D. MALERBA
G. SEMERARO
F. ESPOSITO

1. INTRODUCTION

In concept learning from examples, it is instructive to distinguish the situation in which only one concept has to be induced (*single-concept learning*) from that in which there are several concepts to be learned (*multiple-concept learning*). In the former case, the learner is provided with examples (positive instances) and optionally counterexamples (negative instances) of the single concept and produces a generalization that is complete (all positive instances are explained by means of the generalization) and possibly consistent (negative instances, if any, cannot be explained). Many learning systems presented in the literature can actually learn only one concept at a time. Some of the most renowned are Vere's Thoth (Vere, 1975), Mitchell's candidate-elimination algorithm (Mitchell, 1977), Stepp's AQ7UNI (Stepp, 1979), and the recent FOIL system developed by Quinlan (1990a).

Few researchers have taken an interest in the problem of learning many concepts. Indeed, it is commonly believed that the task of learning m distinct concepts can simply be cast as m single concept learning tasks. No attention is paid to the sequential order in which concepts have to be considered, and sometimes it is claimed that all concepts could be learned in parallel by assigning each single-concept learning task to a distinct processor. Actually, this view of the matter is adequate only if we assume that concepts are mutually independent (*independence assumption*). Concept independence entails the independence of the m single-concept learning problems, in which case

Machine Learning and Statistics, The Interface, Edited by G. Nakhaeizadeh and C. C. Taylor.
ISBN 0-471-14890-3

there would be no reason to worry about the order in which concepts are learned.

Even though in many applications such an assumption is reasonable, it cannot be accepted as a general rule. In many other cases, it is possible to facilitate the learning task by exploiting information about the mutual relationships between concepts. In fact, when the independence assumption is dropped, the learner is required to hypothesize the description of each single concept, as well as the dependencies between concepts.

In this chapter we present problems that occur when the independence assumption is made, such as a decrease in predictive accuracy and an increase in the computational complexity of the learning task. We also point out some new issues that the consideration of concept dependencies can raise, namely, the definition of the correct order in which concepts have to be learned, the adoption of the notion of extensional coverage, and the control of mutual recursion. For the first of them, we propose a multistrategy approach in which statistical techniques are profitably exploited in order to infer concept dependencies before starting the learning process of each single concept. Knowledge on dependencies can subsequently be used by the learner that can perform the opportune shift of language before trying to learn each single concept. Since the appropriateness of the statistical techniques depends on the power of the representation language adopted, we have organized our presentation into two main sections: the next is devoted to attribute-based domains, and the third section is dedicated to structural domains. Some positive results obtained in a real-world problem with a first-order learning system, called INDUBI/CSL, are also illustrated.

2. MULTIPLE-CONCEPT LEARNING IN ATTRIBUTE-BASED DOMAINS

Traces of the independence assumption can already be found in some of the early studies that represented milestones for work on learning from examples in attribute-based domains. For instance, given a set of hypotheses, $V = \{V_i\}$, $i = 1, \ldots, m$, and a family of facts $F = \{F_i\}$, which are only partially described by the hypotheses, Michalski and Larson (1978, p. 2) define the learning problem as the production of a new set of hypotheses, $V^1 = \{V_i^1\}$, where each V_i^1 describes all the facts in the set F_i, and does *not* describe facts in any other set F_j, $j \neq i$. In this statement of the problem, there is an implicit assumption that each example must belong to only one class, that is, that classes are mutually exclusive. Since concepts are intensional descriptions of classes, this means that the possibility of dependence among concepts is excluded a priori. Quinlan (1986, p. 86) is more explicit when he establishes that each object in the universe belongs to one of a set of mutually exclusive classes. For Rendell (1986, p. 181), a hypothesis could be an assertion involving any number of classes, but it is always possible to consider one dichotomy at a time. For example, a universe of visual grids could be categorized into 26 letters plus a nonsense class, but instead each letter can be learned *separately* by performing induction 26 times. Once again, there is no space for expressing dependencies among concepts.

Generally, the learning problem for attribute-based domains is stated as follows:

Given a set of examples (or observations) described by means of a feature vector:

$$\langle a_1, a_2, \ldots, a_n, c \rangle$$

Find a hypothesis for predicting the value of c from the values of $a_1, a_2, \ldots, a_n$.

The attributes a_i are *known* attributes, while c is the only attribute to be predicted (*target* attribute) and represents the class of each example. In a more general case, however, there are several target attributes and each observation is described as a feature vector:

$$\langle a_1, a_2, \ldots, a_n, c_1, c_2, \ldots, c_m \rangle,$$

where $c_1, c_2, \ldots, c_m$ are attributes to be predicted. This is the case in which we need to learn several concepts of the same domain (Datta and Kibler, 1993). For instance, the horse-colic database, which is available in the machine learning repository at the University of California, Irvine, collects data concerning horses hospitalized at Guelph (Canada) that suffer from "colicky" symptoms (McLeish and Cecile, 1990). For each case, there are at least four variables to be predicted, namely:

1. $c_1 \equiv V23$: outcome (lived, died, was euthanized);
2. $c_2 \equiv V24$: surgical lesion? (yes, no);
3. $c_3 \equiv V25$: site of lesion (gastric, sm intestine, lg colon, ...); and
4. $c_4 \equiv V26$: type of lesion (simple, strangulation, inflammation, other).

Note that $c_1, c_2, \ldots, c_m$ do not represent mutually exclusive classes in which the universe of observations is partitioned, but conversely, each of them defines a distinct partitioning. Therefore, it is important to take into account possible dependencies among target attributes, since they can help to produce simpler and more accurate and comprehensible hypotheses. In the example above, the prediction of the type and site of lesion can help to establish the outcome and whether the problem is surgical or not. Another instance is that given in Figure 4.1. In this case there are only two known attributes and three target attributes, each of which defines a distinct partitioning of the feature space. If the learner had no possibility of defining piecewise region boundaries in its language of

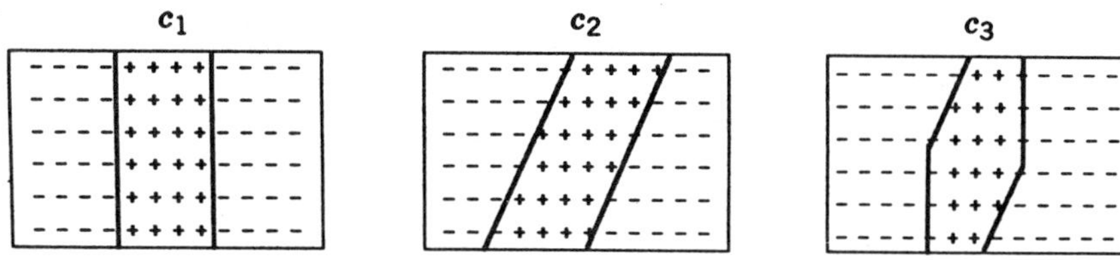

Figure 4.1. *Concept c_3 can easily be recognized when concepts c_1 and c_2 have already been learned.*

hypotheses, then it would not be able to find any generalization for c_3. On the contrary, by taking concept dependencies into consideration, the system could easily learn the concepts c_1 and c_2, and could then express c_3 as a logical *and* of two tests on c_1 and c_2: $c_3 = dc_1 \wedge c_2$.

2.1. The Problem of the Order

The main issue raised by the abolition of the independence assumption is the need to define the order in which concepts should be learned, since the definition of a concept may depend on the definition of other previously learned concepts. In particular, if a target attribute c_i is relevant for the target attribute c_j, whereas all known attributes are not, then it is extremely important to learn c_i first, and then to use it as a known attribute while learning c_j. Wrong ordering may also affect the learnability of some target concepts. Basically we can say that the definition of an order in which concepts are to be learned is the main aspect of learning several dependent concepts. But how is it possible to find the right order?

In some application domains, the user already knows which are the possible dependencies among concepts to be learned. For instance, in the case of the horse-colic database, it is not difficult to establish that c_1 and c_2 depend on c_3 and c_4. When concept dependencies are acyclic, it is possible to use directed acyclic graphs, called *dependency graphs*, to represent them (see Figure 4.2). In dependency graphs a node represents a concept to be learned, while an arrow from c_i to c_j indicates that c_j depends on c_i. For each concept c_i, it is possible to associate a level according to the following criterion: if c_i depends on $\{c_{i_1}, c_{i_2}, \ldots, c_{i_m}\}$, then $level(c_i) = 1 + max\{level(c_{i_1}), level(c_{i_2}), \ldots, level(c_{i_m})\}$. By definition, if c_i does not depend on any concept, then $level(c_i) = 0$. The order in which concepts should be learned is fully defined by the dependency graph. In particular, concepts at the zeroth level, called *minimally dependent concepts* (Malerba, 1993) or *golden point* (Baroglio et al., 1992), have to be learned first, since they do not depend on any other concept. Concepts at the i-th level ($i > 0$) will be learned only after all concepts at the j-th level, with $j < i$, have been learned, since concepts at a lower level must be considered as known attributes/predicates while learning concepts at the i-th level. In other words,

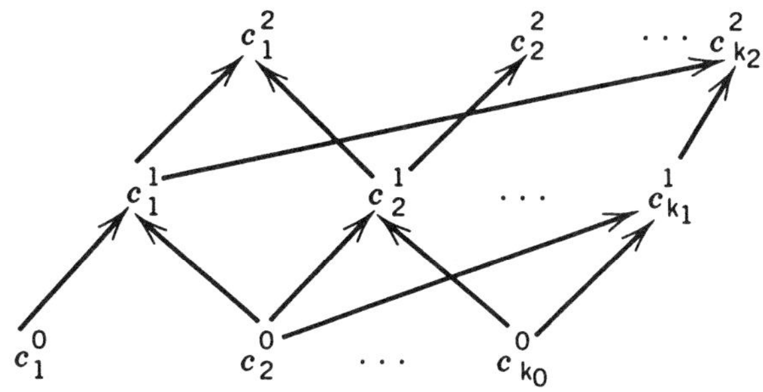

Figure 4.2. *An example of a dependency graph. The upper index denotes the level of each concept.*

the order in which concepts have to be learned defines the appropriate shift of language that is necessary while learning multiple concepts.

When the user does not know the dependencies among concepts, it is possible to infer the graph before learning the single concepts. The approach followed to discover the graph is inherently multistrategic. Statistics provide several tools for studying probabilistic (in)dependencies between variables, the best known of them being the χ^2 test for contingency tables. Given two variables, say c_i and c_j, then the independence hypothesis:

$$H_0: P(c_i \cap c_j) = P(c_i) \cdot P(c_j)$$

can be tested by comparing the expected cell frequencies to the actual values according to the following formula:

$$\chi^2 = \Sigma_k[(f_o^k - f_e^k)^2/f_e^k]$$

where f_o^k equals the observed frequency in each cell, and f_e^k equals the expected frequency (Hogg and Tanis, 1977). By rejecting the null hypothesis, however, we can only conclude that there is a systematic relationship between c_i and c_j, without being able to indicate a *direction* of dependence. For this purpose, several asymmetric *measures of association* have been proposed in statistics.

Lambda (λ) is a measure of association for *nominal* variables and gives the percentage of improvement in our ability to predict the value of the dependent variable once we know the value of the independent variable (Mueller et al., 1970). This is based on the assumption that the best strategy for prediction is to select the value of the dependent variable that covers most cases, since this choice will minimize the number of wrong guesses. Given two concepts, say c_i and c_j, we have:

$$0 \leqslant \lambda(c_i, c_j) \neq \lambda(c_j, c_i) \leqslant 1$$

A value of zero means no improvement in predicting one concept given the other one, while the value 1.0 occurs when one concept can be deterministically predicted from the other one. Thus we can choose the dependency direction $c_i \rightarrow c_j$ if $\lambda(c_i, c_j) > \lambda(c_j, c_i)$.

Piatetsky-Shapiro (1991) proposed a different measure for the dependency-analysis of nominal values, called *probabilistic dependency* (*pdep*). If $c_i^{(k)}$ denotes the value taken by the target attribute c_i in the k-th observation, then $pdep(c_i, c_j)$ is the conditional probability that c_j will take on the same value in two observations in which c_i is the same, $P(c_j^{(k)} = c_j^{(l)} \mid c_i^{(k)} = c_i^{(l)})$. Normalization of *pdep* using a proportional reduction in variation results in Goodman and Kruskal's τ measure of association (1972), which is always between 0 and 1. Once again, we choose the dependence direction $c_i \rightarrow c_j$ if $\tau(c_i, c_j) > \tau(c_j, c_i)$. Finally, Somer's D is an asymmetric measure of association suitable for ordinal variables (Gibbons, 1993). It is based on the computation of the number of concordant and discordant pairs. Given two pairs of observations, $(c_i^{(k)}, c_i^{(l)})$ and $(c_j^{(k)}, c_j^{(l)})$, they are *concordant* if the relative ordering of $c_i^{(k)}$ and $c_i^{(l)}$ is the same as the relative ordering of $c_j^{(k)}$ and $c_j^{(l)}$; otherwise they are discordant. Somer's D assumes a value in the interval $[-1, 1]$, where the absolute value is proportional to the strength of the correlation between the two variables, while the sign indicates whether the correlation is positive (the dependent variable grows with the independent variable), or negative.

It is worthwhile noting that probabilistic measures of association discover simple (in)dependencies between *pairs* of variables. Nevertheless, the dependency graph can be simplified when *conditional* independencies between variables are considered. For instance, if $c_1 \rightarrow c_2 \rightarrow c_3$ is the true underlying dependency graph of three concepts, then non-conditional tests of independence for some probabilistic measure of association might also find a direct dependency between c_1 and c_3. Several studies have been performed on the problem of inferring probabilistic dependency graphs from data. In particular, Cooper and Herskovits (1992) proposed a Bayesian method for constructing *Bayesian belief-networks* (Neapolitan, 1990; Pearl, 1988) from a database. A Bayesian belief-network is a directed acyclic graph in which nodes represent domain variables (i.e., concepts, in our case) and arcs between nodes represent probabilistic dependencies.

When concept dependencies express causal influence, it is possible to exploit other statistical methods for discovering causal models from empirical data, such as Pearl and Verma's Inductive-Causation Algorithm (Malerba et al., 1994; Pearl and Verma, 1991). Basically, a *causal model* is a directed acyclic graph whose nodes are variables and whose edges denote direct causal reationships between pairs of variables. Causal relationships can be *potential* or *genuine*, according to whether we have sufficient (control) knowledge to be able to reject the hypothesis of a *spurious* association. In the latter case, correlation between two variables can only be explained by postulating the existence of a common latent cause. Finally, it is possible to have *indefinite* associations, for which it is not even possible to hypothesize the existence of a

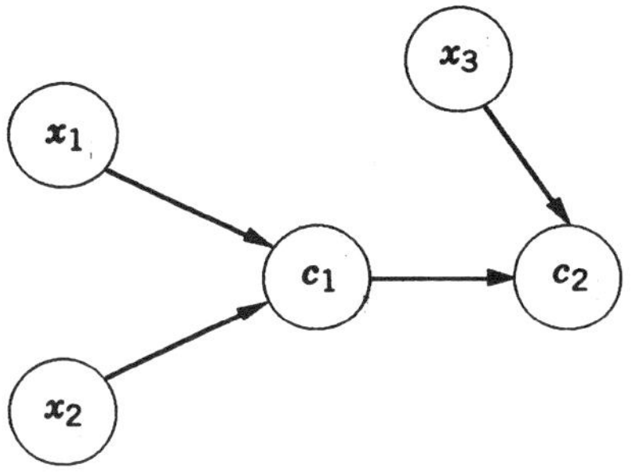

Figure 4.3. *An example of causal model.*

common latent cause. Pearl and Verma provide a theory of causal inference in which they show how it is possible to discriminate among these kinds of relationships on the basis of the conditional independencies between variables. Henceforth, the independence of two variables, say x_1 and x_2, given a set S of variables, will be denoted as $I(x_1, x_2 \mid S)$. The set S is said to be the *context* in which the independence holds. To show how causal models discovered by means of the Inductive-Causation Algorithm can be exploited in multiple-concept learning, we generated a sample of 1,000 observations concerning the model in Figure 4.3.

Each attribute has an ordinal domain with three possible values: low (L), medium (M), and high (H). Two of them are target attributes, while the other three are known. We built a contingency table for each pair of variables to test the independence in the empty context, and we used a χ^2 test on each table. For testing conditional independence with non-empty contexts we had to generate as many subtables as the number of possible values that variables in the context could take. At the significance level $\alpha = 0.05$ we detected the following independencies:

$$I(x_1, x_2 \mid \emptyset) \qquad I(x_1, x_3 \mid \emptyset) \qquad I(x_1, c_2 \mid \{c_1\})$$
$$I(x_2, x_3 \mid \emptyset) \qquad I(x_2, c_2 \mid \{c_1\}) \qquad I(x_3, c_1 \mid \emptyset)$$

The Inductive-Causation Algorithm builds the original model but with the difference that all the causal relationships except $c_1 \rightarrow c_2$ are potential and not genuine.

Having discovered the causal dependencies between the variables, we used a learning system to induce the causal rules from the data. In particular, two learning problems were defined: the first for learning the causal law that relates the study level and the intelligence of a student with his/her preparation, and the second for learning how the preparation and leniency of the examiner affect the final result of the examination. In order to solve both problems we used

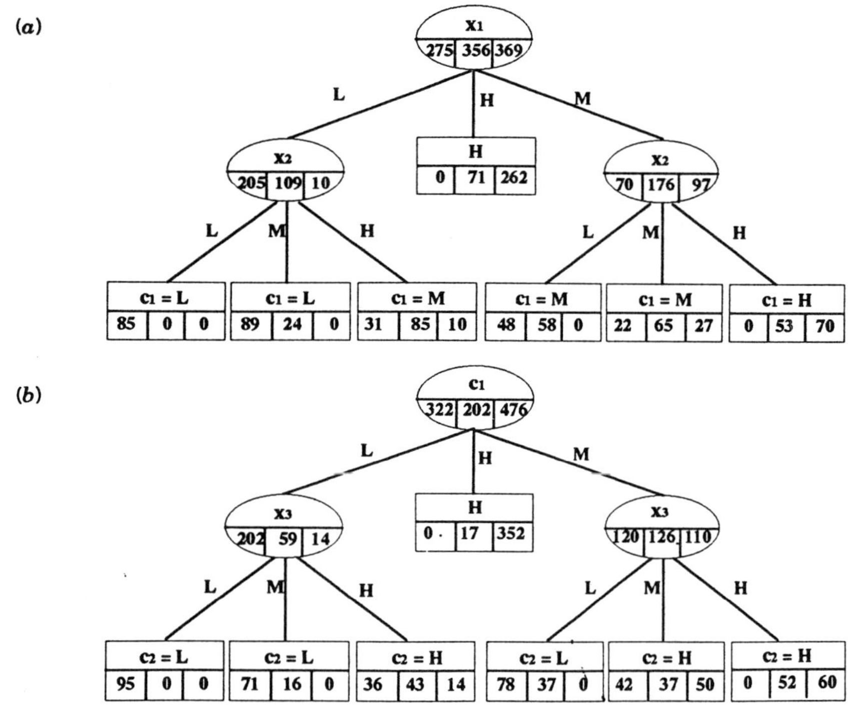

Figure 4.4. *The two decision trees generated by C4.5 for the causal model in Figure 4.3. Here, leaves are represented by boxes. Triplets of numbers in each node represent the distribution of examples reaching the node with respect to the values taken by the target attribute, namely L, M and H.*

C4.5 (Quinlan, 1993), a learning system that induces decision trees from examples. In particular, in the first problem, examples are described by means of x_1, x_2, and the class (target) attribute c_1, while, in the second problem, examples are described by means of c_1, x_3, and the class (target) attribute c_2. The results are shown in Figure 4.4. The composition of the training cases at a leaf F gives a probability $P(K \mid F)$ that a case at F belongs to class K, where the probability can be estimated as a relative frequency (Quinlan, 1990b). For instance, in the case of a moderately lenient examiner and an insufficiently prepared student, the probability of having a low score is $71/(71 + 16) = 0.82$, while the probability of having a medium score is $16/(71 + 16) = 0.18$. From tree b) in Figure 4.4, we can also draw the conclusion that even students with an insufficient or moderately good preparation can get a high score if the examiner is particularly lenient. These and other rules can be directly derived in an explicit form by transforming each decision tree into a set of production rules. Classification rules whose condition parts contain target concepts are called *contextual rules* (Malerba, 1993).

3. MULTIPLE-CONCEPT LEARNING IN STRUCTURAL DOMAINS

Issues raised by the independence assumption are even more evident when we consider a more powerful representation language, such as first-order predicate logic, which allows us to represent structural descriptions. In this case, each concept can be represented by means of a single predicate, hence the name *mutiple predicate learning* (DeRaedt et al., 1993). In addition to the problem of the order observed for attribute-based representations, there are other difficulties, caused by the adoption of a more powerful language, that concern the particular model of generalization employed by the learning system and the generation of mutually recursive rules.

3.1. The Issues of Extensional Coverage and Mutual Recursion

Several systems that learn first-order logic theories expressed as sets of Horn clauses resort to the *syntactic* notion of *extensional coverage*, in order to define the properties of completeness and consistency for a hypothesis H. Let E be the set of positive (E^+) and negative (E^-) instances of one or more target predicates $p_1, p_2, \ldots, p_n$, that is, $E = E^+ \cup E^-$. Then H *extensionally covers* an example $e \in E$, if and only if there exist a clause $c \in H$ and a substitution θ, such that the head of c matches the head of e ($head(c)\theta = head(e)$) and the body of c θ-subsumes the body of e ($body(c)\theta \subseteq$ body(e)) (Plotkin, 1970). Such a syntactic notion of coverage is adopted in many learning systems, since it is more mechanizable than the *semantic* notion of intensional coverage: Given a background knowledge B, which possibly includes previously learned hypotheses, we say that H and B intensionally cover an example $e \in E$, if and only if $B \cup H \cup body(e)$ logically entails $head(e)$ ($B \cup H \cup body(e) \models head(e)$). In fact, the implementation of the notion of intensional coverage requires a theorem prover, while a pattern matcher is enough for the extensional coverage.

The adoption of the notion of extensional coverage to check whether a hypothesis H is complete and consistent can lead to a number of problems, especially when E does not include all positive and negative instances of $p_1, p_2, \ldots, p_n$, for a given set of constant symbols. In fact, we can have hypotheses that are extensionally but not intensionally consistent, as well as hypotheses that are intensionally but not extensionally complete. Note that the notion of extensional coverage causes troubles in the case of single predicates to be learned and there is therefore all the more reason for believing that it could be a source of trouble in the case of learning many predicates. Indeed, when there are several concepts to be learned, it is also possible to have hypotheses that are extensionally but not intensionally complete. This problem occurs when the generated hypotheses are mutually recursive and causes a non-terminating computation.

Mutual recursion is another issue, which is independent from the adopted notion of coverage. In fact, it requires a change in the solutions for the ordering problem, since the definition of a concept c_i depends on the definition of another concept c_j, which, in its turn, depends on the very concept c_i that cannot be completely defined. In this case, the dependency graph is cyclic, and we have to consider the possibility of learning single clauses instead of complete predicate definitions before moving on to a dependent concept. Obviously, the order in which single clauses are learned markedly affects the final result.

3.2. Inferring Concept Dependencies in Structural Domains

Finding the right order in which concepts have to be learned is a more difficult task when we move on to structural domains. This is due to several reasons. First, statistical methods for testing independence are suitable for attribute-based representations and not for structural representations. Second, first-order learning systems adopt stronger language/search biases than zeroth-order learners, thus the right choice of the order in which concepts should be learned is much more critical in the former systems than in the latter. Third, it is possible to represent mutually recursive definitions, which makes dependency graphs no longer acyclic. Here, we do not present any general solution to the problem of inferring the right order while learning multiple concepts in structural domains, but we illustrate an *ad hoc* solution that seems to work well on those images analysis and interpretation problems in which both symbolic and numeric descriptions are available. We have successfully tried this approach in the domains of cartographic map interpretation (Barbanente et al., 1992) and document understanding. The latter is described below.

3.2.1. The Problem of Document Understanding. According to international standards (Horak, 1985), any document is characterized by two different structures representing its spatial organization and its content: the *layout* and the *logical* structure. The former concerns the *organization* of the document on the presentation medium. Each page layout can be decomposed into a set of layout objects that are rectangular areas, each of which is associated with a portion of content. Ideally, each layout component should contain only text or graphics, but sometimes it can contain both, owing to the difficulties of automatically separating text from graphics in a document bitmap image. Conversely, the logical structure is concerned with "meaning" of some layout objects, such as sender or receiver in a business letter (see Figure 4.5a). The set of possible "meanings" associated with the layout objects of a document is named *logical classes*, while the layout objects associated with a logical class are called *logical objects*. The term *document understanding* denotes the process of identification of logical objects in a document image.

In our previous works (Esposito et al., 1993a, b), we investigated the possibility of using machine learning techniques for document understanding. More precisely, given a set of page layouts of single-page documents belonging to the same class (e.g., business letters, scientific papers, or magazine indexes) and assuming that the user-trainer has already labeled some layout objects

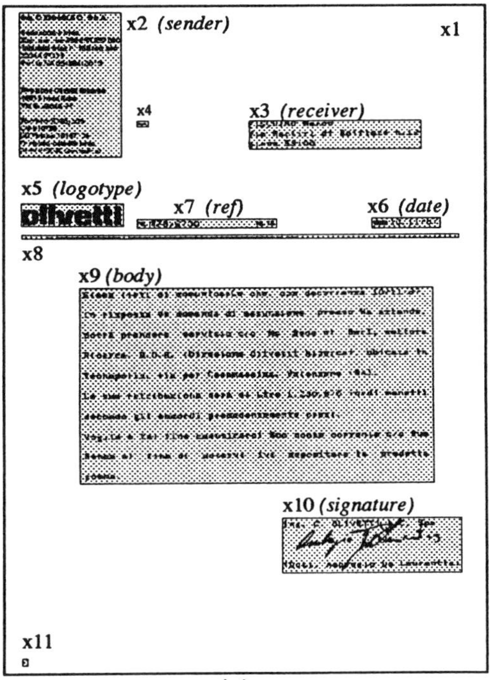

(a)

[logic_type(x6)=date] ←

[part_of(x1,x2)]
[part_of(x1,x3)]
[part_of(x1,x4)]
[part_of(x1,x5)]
[part_of(x1,x6)]
[part_of(x1,x7)]
[part_of(x1,x8)]
[part_of(x1,x9)]
[part_of(x1,x10)]
[part_of(x1,x11)]
[position(x2)=north_west]
[position(x3)=north]
[position(x4)=north_west]
[position(x5)=north_west]
[position(x6)=north_east]
[position(x7)=north]
[position(x8)=center]
[position(x9)=center]
[position(x10)=south_east]
[position(x11)=south_west]
[width(x2)=medium]
[width(x3)=medium_large]
[width(x4)=smallest]
[width(x5)=medium]
[width(x6)=medium_small]
[width(x7)=medium_large]
[width(x8)=very_very_large]
[width(x9)=very_very_large]
[width(x10)=medium_large]
[width(x11)=smallest]
[height(x2)=medium_large]
[height(x3)=small]
[height(x4)=smallest]
[height(x5)=very_small]
[height(x6)=very_very_small]
[height(x7)=very_very_small]
[height(x8)=smallest]
[height(x9)=large]
[height(x10)=medium_small]
[height(x11)=smallest]
[type(x2)=text]
[type(x3)=text]
[type(x4)=text]
[type(x5)=picture]
[type(x6)=text]
[type(x7)=text]
[type(x8)=text]
[type(x9)=text]
[type(x10)=mixture]
[type(x11)=text]
[on_top(x5,x8)]
[on_top(x6,x8)]
[on_top(x7,x8)]
[on_top(x9,x10)]
[to_right(x2,x4)]
[to_right(x5,x7)]
[aligned(x2,x5)=both_columns]
[aligned(x5,x7)=ending_row]
[aligned(x4,x7)=starting_col]
[aligned(x7,x6)=both_rows]
[aligned(x8,x9)=ending_col]
[aligned(x4,x3)=starting_row]
[aligned(x8,x11)=starting_col]

(b)

Figure 4.5. *a) Page layout of an Olivetti business letter. b) A VL_{21} definite clause describing an instance of the logical class* date.

according to their meanings, the problem is that of learning rules that allow logical components to be correctly identified on the basis only of layout information. The training instances are symbolic descriptions of the page layout of each document, together with information on the logical class of some layout objects. Note that concepts to be learned may be related to each other in some way (e.g., the receiver is reported above the date), thus we should learn contextual rules in order to benefit from concept dependencies. For this purpose, we adopted a first-order learning system, called INDUBI/CSL (Contextual Supervised Learner) developed at the University of Bari, Italy.

3.2.2. INDUBI/CSL. INDUBI/CSL, an extension of INDUBI and INDUBI/H (Esposito et al., 1994a), can learn contextual rules expressed as VL_{21} linked clauses, which have the same expressive power as Horn clauses. In VL_{21} (Michalski, 1980), the equivalent of an atom in predicate logic is the notion of *selector* or *relational statement*, which is written as:

$$[L = R]$$

where L, named *referee*, is a function symbol with its *arguments*, and R, named *reference*, is a set of values in the range of the function. A range can be an unordered (nominal), totally ordered (linear), or partially-ordered set. The semantics of a selector is the following: it is true if L takes one of the values in R. A VL_{21} *definite clause* is an expression of the kind:

$$[f(t_1, t_2, \ldots, t_n) = \lambda] \leftarrow \phi(t_1, t_2, \ldots, t_m)$$

where f is a function symbol, t_i's may be constants or variables, λ is a value in the range of f, and $\phi(t_1, t_2, \ldots, t_m)$ is a conjunction of selectors whose arguments contain the terms $t_1, t_2, \ldots, t_m$. An example of VL_{21} definite clause is given in Figure 4.5b.

The body of the clause describes the page layout of the document in Figure 4.5a, while the head defines the logical class of the layout object denoted by $x6$.

Definite clauses generated by INDUBI/CSL satisfy two constraints:

1. All variables in the head must occur in the body of the clause (that is, the clause is *range-restricted*).
2. Clauses must be linked. A VL_{21} definite clause is *linked* if all its selectors are. A selector is linked if at least one of its arguments is. An argument of a selector is linked if either the selector is the head of the clause or another argument of the same selector is linked.

Both training/testing examples and hypotheses are expressed as VL_{21} clauses, but there are some differences:

1. Each training/testing example is represented by a single ground VL_{21} clause.
2. Each hypothesis is expressed as a set of constant-free VL_{21} clauses having the same head.
3. The reference of each selector in any training/testing example has one value only.
4. In the examples, false predicates are omitted, while all the attributes are specified.
5. In the hypotheses, each omitted selector is assumed to be a function, which takes on any value of its range.
6. Variables with different names must be bound to distinct constants (Esposito et al., 1994b), since they are distinctly existentially quantified (Michalski, 1980).

Note that all these constraints on the languge of hypotheses are also valid for the languge of background knowledge. In addition to the training examples and the background knowledge. INDUBI/CSL requires a dependency graph in order to perform the appropriate shift of languge (procedure update_example_description) after all concepts at the same level have been learned (see Algorithm 1).

Algorithm 1. *Learning Contextual Rules and Shift of Language*

---

procedure learn_multiple_concepts(Examples, BK, $\{\{C_1^0, \ldots, C_{n1}^0\}, \ldots, \{C_1^l, \ldots, C_{nl}^l\}\}$)
  update_example_description(Examples, BK)
  AllLearnedRules: $= \emptyset$
  for $i := 0$ **to** l **do**
    LearnedRulesSameLevel: $= \emptyset$
    foreach C_j^i in $\{C_1^i, C_2^i, \ldots, C_{ni}^i\}$ **do**
      LearnedRules: = Separate_and_Conquer(Examples, C_j^i)
      LearnedRulesSameLevel: = LearnedRulesSameLevel∪LearnedRules
    endforeach
  update_example_description(Examples, LearnedRulesSameLevel)
  AllLearnedRules: = AllLearnedRules∪LearnedRulesSameLevel
  endfor
  return AllLearnedRules

INDUBI/CSL implements a separate-and-conquer search strategy at the high level, in order to construct a set of clauses (see Algorithm 2), while it adopts a beam search at the low level, that is, for the construction of a single clause. More precisely, INDUBI/CSL starts with a positive example e^+ and generates a set *Cons* of at least *m* distinct generalizations, which are consistent and cover e^+. Such generalizations are extended-against and stored in *MQ* [see Esposito et al. (1994a) for a detailed description of the extension-against generalization rule]. Then, the best generalization is selected from *MQ* according to a preference criterion, which takes into acount the number of positive examples covered and the complexity of the generalization expressed by the number of its selectors, as well as the total cost of each generalization. Then, positive examples covered by the best generalization are removed from the set

Algorithm 2. *High-level Separate-and-conquer Strategy for Learning a Set of* VL_{21} *Definite Clauses*

---

procedure separate_and_conquer(Examples, Class)
  E^+: = set of positive Examples belonging to Class
  E^-: = set of negative Examples belonging to Class
  LearnedRules: $= \emptyset$
  while $E^+ \neq \emptyset$ **do**
    $MQ := \emptyset$
    randomly select e^+ from E^+
    Cons: = Beam_Search_for_consistent_hypotheses(e^+, E^+, E^-, m)
    foreach generalization G in *Cons* **do**
      $MQ := MQ$∪extension_against(G, E^+, E^-)
    endforeach
    Best: = FindBest(MQ)
    LearnedRules: = LearnedRules∪{Best}
    $E^+ := E^+$ − Covers(Best, E^+)
  endwhile
  return LearnedRules

---

of positive examples and a new clause is generated, if the set of remaining positive examples is not empty.

At the low level, INDUBI/CSL proceeds top-down, specializing the unit clause:

$$[f(t_1, t_2, \ldots, t_n) = \lambda] \leftarrow$$

by adding one of the selectors in the positive example e^+ and turning constants into variables. Thanks to the constraint on the binding of variables, it is possible to turn distinct constants in the example description into distinct variables. Only a subset of n selectors among all selectors in the example are considered: they are chosen according to the cost associated to each function symbol in the referee and according to the arity of the function symbol, so that the greater the arity, the better. The former criterion offers the user a way to express any preference for some literals, while the latter criterion guarantees that relations are not treated unfairly. Obviously, selectors that cause the partial clause to become unlinked are not considered at all. All specialized VL_{21} definite clauses, which cover e^+ and possibly other positive examples, are ranked according to another preference criterion, which takes into account the number of positive and negative examples covered and the complexity of the clause, as well as the total cost. The first p generalizations are selected and stored in a set *PS*. Consistent generalizations are removed from *PS* and stored in *Cons* (see Algorithm 3).

To sum up, INDUBI/CSL adopts two generalization rules during the inductive learning process, the turning-constants-into-variables rule and the extension-against rule, and one specialization rule called adding-a-selector. Moreover, when background knowledge expressed in the form of VL_{21} definite clauses is available, INDUBI/CSL applies the *elementary saturation* operator

Algorithm 3. *Low-level Beam Search Strategy for Learning One VL_{21} Definite Clause*

```
procedure Beam_Search_for_consistent_hypotheses(e+, E+, E-, m)
  PS:={[f(t1,t2,...,tp)=λ]←}
  OldPS:=∅
  Cons:=∅
  while PS≠OldPS and |Cons|≤m do
  OldPS:=PS
  PS:=∅
  foreach VL21 generalization G in OldPS do
    S:=choose_best_linked_selectors(e+, G, n)
    S:=turn_constants_into_variables(S)
    PS:=PS∪specialize_G_by_adding_a_selector_in_S(G,S)
  endforeach
  Cons:=Cons∪(consistent(PS)∩range-restricted(PS))
  PS:=select_best_p_generalizations(PS)
  foreach VL21 generalization G in CONS do
    MQ:=MQ∪extension_against(G, E+, E-)
  endforeach
endwhile
return Cons
```

TABLE 4.1. Experimental Results with INDUBI/CSL: Independence Assumption

Rule/Run	1	2	3	4	5	6	Av. Error
logo	0/0	0/0	0/0	0/0	0/0	0/0	0.0%
sender	0/0	0/1	0/1	0/1	0/9	0/2	16.9%
ref	1/0	1/3	1/0	1/1	1/4	0/2	14.3%
date	0/4	1/4	0/2	0/4	0/2	0/4	28.8%
receiver	2/1	0/0	0/0	0/3	0/5	0/2	17.9%
signature	0/1	0/2	0/2	0/0	0/0	0/0	8.9%
body	1/0	2/0	2/1	0/2	0/1	1/2	20%
Total	4/6	4/10	3/6	1/11	1/21	1/12	11%

described by Rouveirol (1992), before starting the learning process of all concepts. In particular, all clauses in the background knowledge are repeatedly matched against the training examples and all their conclusions are added to the examples in a forward-chaining way. The process ends when no new conclusion expressed as a selector can be added to any training instance. The added selectors are then considered by the procedure choose_best_linked_selectors in the low-level beam-search strategy, so that they can be selected only if they appear useful.

It is worthwhile noting the INDUBI/CSL adopts the extensional notion of coverage, but provides an answer to some of the related problems we discussed in Section 3.1. Indeed, the problem of the generation of extensionally, but not intensionally, consistent hypotheses is faced by updating the example descriptions every time all concepts at the same level have been learned. Moreover, there is no possibility of generating extensionally, but not intensionally, complete hypotheses, since this could occur only with recursive definitions, which cannot be generated by INDUBI/CSL. Finally, the combination of this latter limit with the updating of the example descriptions prevents the system from generating intensionally, but not extensionally, complete hypotheses.

3.2.3. Experimental Results and Inference of Dependencies. INDUBI/CSL has been applied to the problem of document understanding. Several

TABLE 4.2. Experimental Results with INDUBI/CSL: Contextual Rules with Predefined Order

Rule/Run	1	2	3	4	5	6	Av. Error
logo	0/0	0/0	0/0	0/0	0/0	0/0	0.0%
sender	0/0	0/1	0/1	0/1	0/9	0/1	15.6%
ref	2/0	0/1	0/1	0/1	1/1	0/2	10.6%
date	0/3	0/0	1/0	0/5	0/6	1/2	23.5%
receiver	0/1	0/0	0/0	0/2	0/5	0/2	13.5%
signature	1/0	0/2	0/2	0/0	0/0	0/0	8.9%
body	0/1	2/0	2/1	0/2	0/1	1/1	18.3%
Total	3/5	2/4	3/5	0/11	1/22	2/8	9.0%

experiments have been organized to verify whether learning contextual rules leads to better results than learning under the independence assumption. For this purpose, we considered a set of 30 business letters,[1] with 354 layout objects. In each experiment, six different runs were made by randomly splitting the set of documents into two subsets according to the following criteria:

- 20 documents for the training set; and
- 10 documents for the test set.

Seven concepts have to be learned, namely, *sender* of the letter, *receiver*, *logotype*, *reference number (ref)*, *date*, *body*, and *signature*. Obviously, not all layout objects are instances of one of these concepts: some layout components have no logical class associated with them, while in some cases more than one layout object in a document is associated with the same logical class.

INDUBI/CSL has been used for learning both contextual and non-contextual rules. In the former case, we considered a linear dependency order, namely:

$$log \rightarrow signature \rightarrow body \rightarrow sender \rightarrow receiver \rightarrow ref \rightarrow date$$

Such an order is user-defined and can be partly explained in terms of spatial reasoning. Indeed, when the *logotype* and the *signature* have been recognized, the understanding of the geometrically contiguous logical objects, namely, *sender* for the *logotype* and *body* for the *signature*, should be easier. Following the same line of reasoning, we expect recognition of the *sender* to help to identify the *receiver*, which, together with the identification of *logo*, should help to locate the *ref* and, finally, the *date*. Comparison of Tables 4.1 and 4.2 shows that there is a uniform decrease in error rate for all the logical classes when the independence assumption is not made. Globally, the total average error is 2.0% lower for contextual rules than for noncontextual rules. The learning time was the same in both experiments.

Since the dependency order is linear, we have also tried to discover the order by means of statistics. The idea is the following: the accuracy of a classifier that takes into consideration only the characteristics of each layout object can provide information on how easily a logical object can be recognized without taking into account other logical objects in the context. We are aware that the dependency order defined in this way may be just a rule of thumb, but, as we will show later, it can help to identify at least the minimally dependent concepts. The classifiers that we adopted for this third experimentation are Fisher's linear discriminant functions (Hand, 1981), which take the following form:

$$q_i(x) = \mathbf{v} \cdot \mathbf{x} - v_0$$

where:

- $\mathbf{x}$ is an M-dimensional feature vector representing an observation;
- $\mathbf{v}$ is a vector of M coefficients; and
- v_0 is a constant term.

[1]Data are available via anonymous ftp at the UCI machine learning repository.

TABLE 4.3. Experimental Results with INDUBI/CSL: Contextual Rules with Order Defined by the Discriminant Analysis

Class/Run	1	2	3	4	5	6	Av. Error
logo	0/0	0/0	0/0	0/0	0/0	0/0	0.00%
sender	0/0	0/1	0/1	0/1	0/9	0/2	16.9%
ref	1/0	0/1	0/1	0/2	1/0	0/2	9.2%
date	0/2	0/0	0/1	0/5	0/2	0/4	16.4%
receiver	0/1	0/0	0/0	0/3	0/4	0/2	13.5%
signature	1/0	0/2	0/2	0/0	0/0	0/0	8.9%
body	0/1	2/0	2/1	1/2	0/1	0/2	20.0%
Total	2/4	2/4	2/6	1/13	1/16	0/12	8.7%

In order to find those logical components that can more easily be recognized by means of layout information, the discriminant functions have been applied directly on a subset of numerical features describing each single layout object. Such features result from previous phases of document processing, and concern the coordinates of the centroid of an object, the height, width, and eccentricity of a layout component, the number of black pixels in a layout object, and so on. For each logical class c_i we computed the following coefficient:

$$\kappa_i = (\text{number of positive instances covered} + \text{number of negative instances not covered})$$

and we ordered the logical classes according to the decreasing value of κ_i. Ties are broken by preferring the class with the greater number of positive instances. In the following, we list the different orderings obtained in the six experiments:

1. *logo→ref→date→body→signature→sender→receiver*
2. *logo→ref→signature→date→sender→body→receiver*
3. *logo→signature→ref→body→date→sender→receiver*
4. *logo→ref→date→signature→sender→body→receiver*
5. *logo→sender→date→signature→ref→body→receiver*
6. *logo→sender→ref→date→signature→body→receiver*

Table 4.3 shows the experimental results obtained by exploiting the ordering defined by the discriminant analysis. The idea of using statistical methods in order to define the dependency graph seems to work well, since the average error rate was slighty lower even than that obtained with the predefined order. Only in two experiments were the results worse, while globally, the number of commission errors decreased.

Finally, it is worthwhile noting that the results of the first two experiments agree with those obtained with a different first-order learning system, called

FOCL (Pazzani and Kibler, 1992). Even in that case we have observed an improvement in accuracy when the concept dependencies are considerd (Esposito et al., 1993a; Malerba, 1993): this result corroborates our opinion that taking concept dependencies into account is important.

4. CONCLUSIONS

In this chapter we have criticized the assumption of concept independence made by most of the well-known learning systems. Problems caused by the independence assumption are particularly evident in at least three situations: learning multiple attributes in attribute-based domains, learning multiple predicates in inductive logic programming, and learning classification rules for labeling problems. However, dropping the independence assumption raises several issues in its turn, the most important of which is the definition of an order in which concepts have to be learned.

When a dependency graph is given, it is possible to learn contextual rules by performing a shift of language according to the level assigned to each single concept in the dependency graph. On the contrary, when the user does not know which concepts depend on which, statistical methods can help to discover concept dependencies. In particular, statistical tests for independence, such as χ^2, can help to find significant correlations between concepts, while asymmetric measures of association, such as λ, Goodman and Kruskal's τ, and Somer's D can help to define a direction of dependence between two variables. Such a direction, however, simply establishes which variable can be more easily predicted given the other, but does not say anything about the possible causal relationships between them. In order to infer causal information from data, different methods have to be used, such as the one implemented in Pearl and Verma's Inductive-Causation Algorithm.

All these approaches are appropriate for attribute-based domains, but no easy solution is yet available for the case of structural domains. In this chapter we have presented an ad hoc solution for a real-world problem: document understanding. Experimental results in this domain confirm that by taking into account concept dependencies, it is possible to improve the classification accuracy. These encouraging results were obtained both when the order was defined by the user and when it was found by using linear dicriminant functions. In the latter case, the classification accuracy of a classifier that takes into consideration only the numerical characteristics of each layout object was exploited in order to understand how easily a logical component can be recognized without taking into account other logical components in the context. This simple rule proved effective not only for identifying the minimally dependent concepts, but also for suggesting the whole sequence in which concepts should be learned.

REFERENCES

Barbanente, A., Borri, D., Esposito, F., Leo, P., Maciocco, G., and Selicato, F. (1992). Automatically acquiring knowledge by digital maps in artificial intelligence planning. In *Theories and Methods of Spatio-Temporal Reasoning in Geographic Space.* Lecture Notes in Computer Science (A. Frank, I. Campari and U. Formentini, eds.), **639**, pp. 379–401. Springer-Verlag, Berlin.

Baroglio, C., Giordana, A., and Saitta, L. (1992). Learning mutually dependent relations. *J. Intell. Inf. Syst.* **1**, 159–176.

Cooper, G., and Herskovits, E. (1992). A Bayesian method for the induction of probabilistic networks from data. *Mach. Learn.* **9**, 309–347.

Datta, P., and Kibler, D. (1993). Concept sharing: A means to improve multi-concept learning. In *Proceedings of the Tenth International Conference on Machine Learning, Amherst, MA*, pp. 89–96. Morgan Kaufmann, San Mateo, CA.

DeRaedt, L. D., Lavrač, N., and Dzeroški, S. (1993). Multiple predicate learning. In *Proceedings of the Third International Workshop on Inductive Logic Programming*, pp. 221–240.

Esposito, F., Malerba, D., and Semeraro, G. (1993). Automated acquisition of rules for document understanding. In *Proceedings of the Second International Conference on Document Analysis and Recognition*, pp. 650–654. IEEE Computer Society Press, Los Alamitos, CA.

Esposito, F., Malerba, D., Semeraro, G., and Pazzani, M. (1993b). A machine learning approach to document understanding. In *Proceedings of the Second International Workshop on Multistrategy Learning*, pp. 276–292. George Mason University, Harpers Ferry, WV.

Esposito, F., Malerba, D., and Semeraro, G. (1994a). Multistrategy learning for document recognition. *Appl. Artif. Intell.* **8**, 33–84.

Esposito, F., Malerba, D., Semeraro, G., Brunk, C., and Pazzani, M. (1994b). Traps and pitfalls when learning logical definitions from relations. In *Methodologies for Intelligent Systems, Lecture Notes in Computer Science* (Z. W. Ras and M. Zemankova, eds.), **869**, pp. 376–385. Springer-Verlag, Berlin.

Gibbons, J. (1993). *Nonparametric Measures of Association, Sage University Paper, Series on Quantitative Applications in Social Sciences.* Sage, Newbury Park, CA.

Goodman, L., and Kruskal, W. (1972). Measures of association for cross classification IV: Simplification of asymptotic variances. *J. Am. Stat. Assoc.* **67**, 415–421.

Hand, D. (1981). *Discrimination and Classification.* Wiley, London.

Hogg, R., and Tanis, E. (1977). *Probability and Statistical Inference.* Macmillan, New York.

Horak, W. (1985) Office document architecture and office document interchange formats: Current status of international standardization. *IEEE Comput.* **18**(10), pp. 50–60.

Malerba, D. (1993). *Document Understanding: A Machine Learning Approch*, Tech. Rep., Esprit Proj. 5203 INTREPID.

Malerba, D., Semeraro, G., and Esposito, F. (1994). An analytic and empirical comparison of two methods for discovering probabilistic causal relationships. In *Machine Learning: ECML-94, Lecture Notes in Artificial Intelligence*,, (F. Bergadano and L. DeRaedt, eds.), **784**, 198–216. Springer-Verlag, Berlin.

McLeish, M. M., and Cecile, M. (1990). Enhancing medical expert systems with knowledge obtained from statistical data. *Ann. Math. Artif. Intelli.* **2**, 261–276.

Michalski, R. (1980). Pattern recognition as rule-guided inductive inference. *IEEE Trans. Pattern Anal. Mach. Intell.* **PAMI-2**, 349–361.

Michalski, R., and Larson, J. (1978). *Selection of the Most Representative Training Examples and Incremental Generation of VL_1 Hypotheses: The Underlying Methodology and the Description of Programs Esel and Aq11*, Tech. Rep. UIUCDCS-R-78-867. Dept. of Computer Science, University of Illinois, Urbana.

Mitchell, T. (1977). Version spaces: A candidate elimination approach to rule learning. In *Proceedings of the Fifth International Joint Conference on Artificial Intelligence, Cambridge, MA*, pp. 305–310.

Mueller, J., Schessler, K., and Costner, H. (1970). *Statistical Reasoning in Sociology.* Houghton Mifflin, Boston.

Neapolitan, R. (1990). *Probabilistic Reasoning in Expert Systems.* Wiley, New York.

Pazzani, M., and Kibler, D. (1992). The utility of knowledge in inductive learning. *Mach. Learn.* **9**, 57–94.

Pearl, J. (1988). *Probabilistic Reasoning in Intelligent Systems.* Morgan Kaufmann, San Mateo, CA.

Pearl, J., and Verma, T. (1991). A theory of inferred causation. In *Proceedings of the Second International Conference on Principles of Knowledge Representation and Reasoning*, p. 441–452. Morgan Kaufmann, San Mateo, CA.

Piatetsky-Shapiro, G. (1991). Probabilistic data dependencies. In *Proceedings of the ML-92 Workshop on Machine Discovery*, pp. 441–451, Morgan Kaufmann, San Mateo, CA.

Plotkin, G. (1970). A note on inductive generalization. In *Machine Intelligence 5* (B. Meltzer and D. Michie, eds.), pp. 153–163. Edinburgh Univ. Press, Edinburgh.

Quinlan, J. (1986). Induction of decision trees. *Mach. Learn.* **1**, 81–106.

Quinlan, J. (1990a). Learning logical definitions from relations. *Mach. Learn.* **1**, 177–226.

Quinlan, J. (1990b). Probabilistic decision trees. In *Machine Learning: An Artificial Intelligence Approach* (Y. Kodratoff and R. Michalski, eds.), Vol. 3, pp. 140–152. Morgan Kaufmann, San Mateo, CA.

Quinlan, J. (1993). *C4.5: Programs for Machine Learning.* Morgan Kaufmann, San Mateo, CA.

Rendell, L. (1986). A general framework for induction and a study of selective induction. *Mach. Learn.* **1**, 177–226.

Rouveirol, C. (1992). Extensions of inversion of resolution applied to theory completion. In *Inductive Logic Programming* (S. Muggleton, ed.), pp. 63–90. Academic Press, London.

Stepp, R. (1979). *Learning without Negative Examples via Variable Valued Logic*, Tech. Rep. UIUCDCS-R-79-982. Dept. of Computer Science, University of Illinois, Urbana.

Vere, S. (1975). Induction of concepts in the predicate calculus. In *Proceedings of the Fourth International Joint Conference on Artificial Intelligence, Tblisi, USSR*, pp. 281–287.

CHAPTER 5

Quality of Decision Rules: Definitions and Classification Schemes for Multiple Rules

I. BRUHA

1. INTRODUCTION: PARADIGMS OF LEARNING ALGORITHMS

There exist a few paradigms of learning from examples. One commonly used paradigm is called *divide-and-conquer*. It is widely utilized by the family of TDIDT (*T*op-*D*own *I*nduction of *D*ecision *T*rees) learning algorithms that induce decision trees from examples. One of its first pioneers is the ID3 algorithm (Quinlan, 1986), but currently mostly used and well-known member is the C4.5 algorithm (Quinlan, 1992).

Another widely used stream in symbolic learning uses the *covering* paradigm. The AQ*x* family (Michalski, 1980; Michalski et al., 1986) is one of the first ones utilizing this paradigm. Another family exploiting it is formed by the CN*x* learning algorithms. The most known member of this family is the CN2 algorithm (Clark and Niblett, 1989; Clark and Boswell, 1991). It induces a set of decision rules of the form

if Condition then class is C

It handles noisy data by applying statistical techniques in order to select the most predictive and reliable conditions. Due to the relaxation on consistency and completeness of the decision set inferred, the algorithm exhibits quite promising performance in processing noisy data. CN2 uses for that purpose a heuristic to terminate the rule construction, based on an estimate of the noise present in the training data. The original version uses either entropy (Clark and Niblett, 1989) or a Laplacian error estimate (Clark and Boswell, 1991). It

Machine Learning and Statistics, The Interface, Edited by G. Nakhaeizadeh and C. C. Taylor.
ISBN 0-471-14890-3

should be also pointed out that CN2 induces either *ordered* or *unordered* set of rules.

To include some extensions, we have decided to design and implement our own version of this inductive algorithm which we (with kind permission of the authors of CN2, P. *C*lark and T. *N*iblett) call CN4. First, we have enhanced the algorithm itself by adding more robust stopping conditions when generating the rules, and by improving numerical attribute processing. Then, following in the spirit of the extensions of ID3 that handle attribute cost (Nunez, 1988; Tan and Schlimmer, 1990), we have incorporated routines and heuristics that are able to process attribute cost and thus to economize the classification of unseen objects. Last, but not least, inspired by processing unknown attribute values in ID3 (Quinlan, 1989; Brazdil and Bruha, 1992), we have added several routines for processing unknown attribute values to CN4, too. Detailed descriptions of our version can be found in Bruha and Kočková (1993) and Kočková and Bruha (1993). The entire algorithm has been written in C and runs under the Unix system. The newest version deals with rule quality and various schemes for combining these qualities.

The purpose of this contribution is to exhibit a numerical factor, called rule quality, that could be very useful when classifying unseen objects by an unordered set of decision rules. The CN4 algorithm is used here only as a learning vehicle in order to compare various definitions of rule qualities. The methodology may be, however, applied to any system that derives an unordered set of decision rules. It should be noticed, too, that decision trees generated by, say, ID3, may be easily converted to an unordered set of rules (Quinlan, 1987); therefore the methodology of rule quality may be applied to such sets, too.

Section 2 provides a brief explanation of the covering paradigm and the CNx family of rule-inductive algorithms, including a simple example. Section 3 introduces the terms and definitions required for defining rule quality. Section 4 presents various definitions of rule quality, namely, empirical formulas and statistical ones. Section 5 provides analysis of the formulas from the theoretical viewpoint. Section 6 discusses three schemes for combining rule qualities, and section 7 gives results of experiments.

2. THE COVERING PARADIGM AND THE CN*x* FAMILY: AN OVERVIEW

A learning algorithm using the *covering paradigm* works in sweeps. In each sweep, it tries to find a reliable and powerful rule which covers a subset of training examples;[1] the covered examples are then removed from the training set and the algorithm tries to cover the remaining examples in the next sweeps. The entire process is terminated if all examples are covered or a certain stopping condition is satisfied.

[1]A rule *covers* an example, if the example *satisfies* (*matches*) its condition, that is, the condition is *true* when applied to the example. All these terms are used in literature.

Covering learning algorithms exhibit two modes: ordered and unordered. In the ordered mode, the order of rules generated by a covering learning algorithm is vital, that is, the rules create a 'cascade' for testing that cannot be changed. Ordered-mode covering paradigm may be generally formalized by the procedure *ORDERED_COVER(E)* where E is a given training set:

procedure *ORDERED_COVER(E)*
ListOfRules:= *NIL*
until $E = 0$ **do**
1. Find a rule condition *Cond* according to a certain technique
2. Let E' be the subset of training examples covered by *Cond*
   $E := E \setminus E'$
3. Add the rule

   if Cond then class is C

   to the end of *ListOfRules* where C is the desired class of examples in E'

enduntil
return *ListOfRules*

All the learning algorithms utilizing this paradigm differ, namely, in the step 1 and in the stopping condition, which is here (for simplicity) expressed as $E = 0$, that is, the process stops if the entire training set is covered. Some algorithms induce rule conditions that may cover examples from more than one class; in this case, the class C in the step 3 is the *majority* class of examples in E'.

In the unordered mode, the rules are generated for each class independently, that is, rules can be tested within classification in any fashion. A general procedure for unordered-mode covering paradigm looks as follows:

procedure *UNORDERED_COVER(E)*
ListofRules:= *NIL*
for each class C **do**
  Let E^+ be examples of the class C in E
  until $E^+ = 0$ **do**
  1. Find a rule condition *Cond* with respect to C
  2. Let E' be the subset of training examples covered by *Cond*
     $E^+ := E^+ \setminus E'$
  3. Add the rule

     if Cond then class is C

     to *ListOfRules*

enduntil
endfor
return *ListOfRules*

Again, the covering algorithms with unordered mode differ in the step 1 and the stopping condition. The term 'with respect to C' is mostly interpreted as 'where C is the majority class'.[2]

Covering learning algorithms of the CNx family use the star methodology and the beam search in a way similar to the AQx family (Michalski, 1980; Michalski, et al., 1986), but they handle noisy data by applying statistical techniques, analogous to those used for tree pruning in TDIDT (Cestnik et al., 1988; Quinlan, 1992). They relax the constraint of both consistency and completeness. A rule induced may cover examples from different classes, that is, some training examples need not be classified perfectly. Moreover, some examples need not be covered by any rule induced. Due to this relaxation, the algorithms exhibit quite promising performance in processing noisy data.

A rule condition (see step 1 above) is searched for by a procedure that selects rule conditions according to a certain heuristic; moreover, the condition has to satisfy a so-called significance test, that is, it must be a promising and reliable one. If no such rule condition is found, the inductive process terminates.

To illustrate the performance of CN4 we will at least present its output (set of decision rules) both for the ordered and unordered mode. For simplicity, let us consider the well-known set of the weather problem (Quinlan 1986) with 14 examples, two classes, and 4 symbolic attributes (windy, humidity, outlook, temperature). The order mode of CN4 yields:

```
if outlook=overcast then class is +;
   Kr=[ 4 0]; signif=5.099; quality=0.889; cost=1
else if windy=false && humidity=normal then class is +;
   Kr=[ 3 0]; signif=3.825; quality=0.867; cost=2
else if humidity=high && outlook=sunny then class is -;
   Kr=[ 0 3]; signif=8.913; quality=0.920; cost=0
else if windy=true && outlook=rain then class is -;
   Kr = [ 0 2]; signif=5.942; quality=0.880; cost=0
else if true then class is +;
   Kr=[ 2 0]; signif=2.550; quality=0.844; cost=0
```

Here Kr represents the (absolute) frequency of examples covered by each rule, signif is the rule significance, cost is its cost if we consider economy learning (by default cost of each attribute is 1), and quality depicts the quality of each rule (its default is formula (1)). Note that the structure if – else if --- forms the cascade of the rules whose order cannot be changed.

[2]In literature, one may find an alternative procedure *UNORDERED_COVER_1(E, C)* which induces rules for the explicitly specified class *C*; its body is identical to the core of the **for**-loop above.

The unordered mode of CN4 yields:

```
if humidity=high && outlook=sunny then class is -;
  Kr=[ 0 3]; signif=8.913; quality=0.920; cost=2
if outlook=overcast then class is +;
  Kr=[ 4 0]; signif=5.099; quality=0.889; cost=1
if windy=false && humidity=normal then class is +;
  Kr=[ 4 0]; signif=5.099; quality=0.889; cost=2
if windy=true && outlook=rain then class is -;
  Kr=[ 0 2]; signif=5.942; quality=0.880; cost=2
if windy=false && outlook=rain then class is +;
  Kr=[ 3 0]; signif=3.825; quality=0.867; cost=2
if humidity=normal && temperature=mild then class is +;
  Kr=[ 2 0]; signif=2.550; quality=0.844; cost=2
if true then class is +;
  Kr=[ 9 5]; signif=0.000; quality=0.714; cost=0
```

Notice that the algorithm always generates the default rule with its condition equal to true. It is used only if no rule matches an unseen object to be classified; the object is then classified to the majority class of the given task.

3. CLASSIFICATION BY UNORDERED SET

A set of decision rules derived by an inductive algorithm may be directly exploited by a classifier as its decision scheme. If the decision set is *ordered*, then classifying an unseen object is quite straightforward: the classifier goes through the list of rules and looks for the first rule that matches ('fires') the given object; it is then categorized into the class attached to the rule.

However, the important and natural way seems to be the *unordered* mode that is utilized in various expert systems. However, classification exploiting an unordered set of decision rules exhibits a significant deficiency, not immediately apparent. Three cases are possible:

1. If the unseen object satisfies one or more rules of the same class, then the object is categorized to the class assigned to the rule(s).
2. If the unseen object is not covered by any rule, then either the classifier informs the user about its inability to decide, or the object is assigned by default to the majority class in the training set, or some similar techniques are invoked.
3. Difficulty arises if the object satisfies more rules assigned to different classes. Then some schemes have to be applied to assign the unseen object to the most appropriate class.

One possibility to clarify the conflict (case 3) is to associate each rule in the decision scheme of a classifier with a numerical factor that can express its properties and characterize a measure or belief of the rule, its power, predictability, reliability, likelihood, and so forth. A collection of these properties is symbolized by a function commonly called the *rule quality*.

Then, one classification scheme exploiting rule quality orders the decision rules into an ordered list according to their qualities, and looks for the first rule satisfied by the object. Another scheme also considers rule quality, but goes through the entire set of rules, combines somehow qualities of the matched rules of the same class, and assigns the unseen object to the class for which the *quality combination* reaches the maximum.

If we accept the idea of rule qualities for elucidating the above conflict, then two questions arise: first, how to define rule quality, second, how to combine rule qualities for several rules.

A general characteristic required for the rule quality is its *monotony* toward its arguments. Its common arguments are the *consistency* and *completeness* factors of decision rules. The reason for exploiting the above characteristics is obvious. Any machine-learning algorithm dealing with real-world noisy data is to induce decision rules that cover larger numbers of training examples even with a few negative ones (not belonging to class of the rule). In other words, the decision set induced must be not only reliable but also powerful. Its reliability is characterized by a consistency factor and its power by a completeness factor.

Let a given task to be classified be characterized by a set of training examples that belong to two classes, named C and $\bar{C}$. Let R be a decision rule of the class C, that is, of the form

R: **if** *Condition* **then** class is C

Behavior of the above decision rule can be formally depicted by the 2×2 *contingency table* (Bishop et al., 1991), which is commonly used in machine learning (Quinlan, 1987; Mingers, 1989):

	class C	not class C	
rule R covers	rc	$r\bar{c}$	r
does not cover	$\bar{r}c$	$\bar{r}\bar{c}$	$\bar{r}$
	c	$\bar{c}$	K

where

rc is the number of training examples that are covered by the rule R and belong to the class C;

$r\bar{c}$ is the number of examples covered by the rule R but not belonging to the class C, etc.;
$r = rc + r\bar{c}$ is the number of examples covered by R;
$c = rc + \bar{r}c$ is the number of the training examples of the class C, etc.; and
$K = c + \bar{c} = r + \bar{r}$ is the number of all training examples in the given task.

Using the elements of the contingency table, we may define the *consistency* of a rule R (also called the *sensitivity*; Weiss et al., 1987) by

$$cons(R) = \frac{rc}{r}$$

and its *completeness* (*coverage*, or *positive predictive value*) by

$$compl(R) = \frac{rc}{c}$$

Notice that the terms 'consistency' and 'completeness' (not only in this chapter but also in other publications) have two meanings: a property or a numerical factor, but one may easily find the right meaning.

Some formulas for the rule quality introduced in this chapter require *relative* rather than absolute frequencies in the contingency table:

	class C	not class C	
rule R covers	f_{11}	f_{12}	f_{1+}
does not cover	f_{21}	f_{22}	f_{2+}
	f_{+1}	f_{+2}	1

where

$$f_{11} = \frac{rc}{K}, \quad f_{12} = \frac{r\bar{c}}{K}, \quad f_{21} = \frac{\bar{r}c}{K}, \quad f_{22} = \frac{\overline{rc}}{K}$$

and

$$\sum_{i,j} f_{ij} = 1$$

Note that $f_{+1} = c/K$ denotes the relative frequency of the class C toward the entire training set; $f_{1+} = r/K$ is the relative frequency of examples covered by R toward the entire training set.

4. FORMULAS FOR RULE QUALITY

I. Empirical Formulas

The problem of defining rule quality remains an interesting issue in machine learning. Several attempts to define such a criterion have been made (Bergadano et al., 1988; Brazdil and Torgo, 1990; Pearl, 1978; Torgo, 1993). However, the majority of these formulas represent an empirical, ad hoc approach to the definition of rule quality. We emphasize the term 'empirical' here, since the work is based on intuitive logic and not necessarily backed by statistical or information theories.

The two characteristics of consistency and completeness can be combined to yield a single number, the rule quality. The meaning of the result should be more or less understandable to human experts. The common policy is to use a weighted sum or multiplication of the above components.

(1). Rule quality as a *weighted sum* of the consistency and completeness has the form (named after one of its authors):

$$Quality_{\text{Michalski}}(R) = w_1\, cons(R) + w_2\, compl(R) \tag{1}$$

where $w_1, w_2 \in (0, 1)$ are user-defined weights, usually $w_1 + w_2 = 1$.

Formula (1) is used in Bergadano et al. (1988) as part of a more complex scenario which also comprises comprehensibility and the cost of concept description. However, Bergadano et al. (1988) does not state how to determine the above weights; only an 'experimental' determination of the weights is proposed.

Formula (1) is also used in the CN2 unordered algorithm (Clark and Boswell, 1991), and in CN4 with an unordered mode (Kočková and Bruha, 1993) as one of the selectable formulas. An interesting application of the weighted sum formula for rule quality is used in the incremental learning system YAILS (Torgo, 1993). The weights in YAILS are not user-defined but specified by the algorithm itself:

$$\begin{aligned} w_1 &= 0.5 + \tfrac{1}{4}\, cons(R) \\ w_2 &= 0.5 - \tfrac{1}{4}\, cons(R) \end{aligned} \tag{2}$$

These heuristic formulas follow from the results of experiments in Torgo (1993) and emphasize consistency. Here, the weights are made dependent on consistency, which introduces a flexibility into the procedure: the larger the consistency value, the more influence consistency has on rule quality.

(2). Rule quality as a *product* of consistency and completeness has the form (named after one of the authors):

$$Quality_{\text{Brazdil}}(R) = cons(R) \cdot f(compl(R)) \tag{3}$$

where f is an increasing function. Appropriate determination of f may, similarly to (2), prefer the consistency. The completeness works here as a factor that reflects a *confidence* to the consistency of the given rule. This scenario corresponds to the noisy-data-processing paradigm, mentioned above.

A large number of experimental tests performed in the knowledge integration system *Integ.3* (Brazdil and Torgo, 1990) led its authors to the following form of the function f in the formula (3):

$$f(x) = \exp(x - 1)$$

II. Statistical Formulas

The empirical formulas for rule quality are easy to understand and provide quite comparative results. On the other hand, they are ad hoc ones, without any theoretical support. Statistics and information theory provide large collections of various procedures and formulas, usually not easily understandable. Nevertheless, they may be interpreted and/or modified as reasonable schemes for defining rule quality. This chapter discusses three such sources.

(1). The theory on *contingency tables* seems to be one of these sources, since the performance of any rule can be characterized by them. Generally, there are two groups of measurements (see, e.g., Bishop et al., 1991); association and agreement.

a. A measure of *association* indicates a relationship between rows and columns in a 2×2 contingency table so that the i-th row may be 'associated' with the j-th column. In other words, this measure reflects an 'association' on both diagonals of the contingency table.

Measures of association are described in detail in Bishop et al. (1991). We have found that the following two statistics may serve as reasonable schemes for the rule quality:

- The Pearson χ^2 statistic applied to the 2×2 contingency table:

$$Quality_{\chi^2}(R) = \sum_{i,j} \frac{(f_{ij} - f_{i+}f_{+j})^2}{f_{i+}f_{+j}} = \frac{(rc\,.\,\bar{r}\bar{c} - r\bar{c}\,.\,\bar{r}c)^2}{c\,.\,\bar{c}\,.\,r\,.\,\bar{r}} \tag{4}$$

- The $G2$-likelihood statistic

$$Quality_{G2}(R) = 2\left(rc \ln \frac{rc\,.\,K}{r\,.\,c} + r\bar{c} \ln \frac{r\bar{c}\,.\,K}{r\,.\,\bar{c}}\right) \tag{5}$$

where logarithm here is to the base e.

Both statistics are distributed asymptotically as χ^2 with one degree of freedom.

It should be noted that the *J-measure* introduced in Smyth and Goodman (1990) and often quoted as one of the most promising measures (Kononenko, 1992), in fact, is equal to the $G2$-likelihood statistic divided by $2K$:

$$Quality_J(R) = \frac{Quality_{G2}(R)}{2K}$$

which could be also written as

$$Quality_J(R) = f_{1+}\left(cons(R)\ ln\ \frac{cons(R)}{f_{+1}} + (1\text{-}cons(R))\ \ln\ \frac{1\text{-}cons(R)}{f_{+2}}\right)$$

From the viewpoint of machine learning, the above formulas work properly only if the class C is the majority class of examples covered by the rule. If not, then the rule quality symbolized by any measure of association may yield the largest values for the rule that does not cover any example of the class C. One may easily prove that both statistics reach their maximum for $rc = 0$.

b. A measure of *agreement* is a special case of association that indicates an 'association' of the elements of a contingency table on its main diagonal only. The simplest measure proposed in Bishop et al. (1991) is the sum of the main diagonal:

$$f_{11} + f_{22}$$

Cohen (1960) suggests comparison of the *actual* agreement $Aagree = \Sigma_i f_{ii}$ with the *chance* agreement $Cagree = \Sigma_i f_{i+} f_{+i}$, which occurs if the row variable is independent of the column variable, that is, if the rule's coverage does not yield any information about the class C. The difference $Aagree - Cagree$ is then normalized by its maximum possible value. This leads to the measure of agreement that we interpret as rule quality (named after its author):

$$Quality_{\text{Cohen}}(R) = \frac{\Sigma_i f_{ii} - \Sigma_i f_{i+} f_{+i}}{1 - \Sigma_i f_{i+} f_{+i}}$$

or, if absolute frequencies are used:

$$Quality_{\text{Cohen}}(R) = \frac{K.rc + K.\overline{rc} - r.c - \bar{r}.\bar{c}}{K^2 - r.c - \bar{r}.\bar{c}} \tag{6}$$

or introducing consistency and completeness:

$$Quality_{\text{Cohen}}(R) = \frac{cons(R) - f_{+1}}{\frac{1}{2}\left(1 + \frac{cons(R)}{compl(R)}\right) - f_{+1}} \tag{7}$$

Note that the formulas (6) and (7) depend on four variables: r, c, rc, K, or $cons(R)$, $compl(R)$, rc, K, respectively.

Cohen's statistic provides a formula which responds to the agreement on the main diagonal of the contingency table, that is, when both components rc and $\overline{rc}$ are reasonably large. Coleman (Bishop et al., 1991) defines a measure of agreement that indicates an 'association' between the first column and any particular row in the contingency table. For the purpose of the rule quality definition, the agreement between the first column ('example from the class C') and the first row ('rule R covers an example') is the proper one. The formula follows Cohen's statistic in principle by normalizing the difference between the 'actual' and 'chance' agreement. Hence, Coleman's formula for the rule quality is

$$Quality_{\text{Coleman}}(R) = \frac{f_{11} - f_{1+}f_{+1}}{f_{1+} - f_{1+}f_{+1}}$$

or, using the absolute frequencies,

$$Quality_{\text{Coleman}}(R) = \frac{K\,.\,rc - r\,.\,c}{r\,.\,\bar{c}} \tag{8}$$

or, introducing consistency and completeness,

$$Quality_{\text{Coleman}}(R) = \frac{K\,.\,cons(R)\,.\,compl(R) - rc}{K\,.\,compl(R) - rc} \tag{9}$$

Coleman's statistic intuitively exhibits necessary properties of a rule quality: agreement between 'rule covers' and 'example from class C'. Nevertheless, one can find by simple analysis that Coleman's formula does not properly comprise the completeness of a rule; substituting rc by $c\,.\,compl(R)$ changes (9) to

$$Quality_{\text{Coleman}}(R) = \frac{cons(R) - f_{+1}}{1 - f_{+1}}$$

On the other hand, Cohen's statistic is more completeness-based. Therefore, we have modified Coleman's formula in two ways to obtain these definitions:

$$Quality_{C1}(R) = Quality_{Coleman}(R)\,\frac{2 + Quality_{Cohen}(R)}{3} \tag{10}$$

$$Quality_{C2}(R) = Quality_{Coleman}(R)\,\frac{1 + compl(R)}{2} \tag{11}$$

where the coefficients 2, 3 and 1, 2 have been added to normalize the statistics.

(2). *Information theory* is another source of statistical measurements suitable for defining rule quality. Following this theory, Kononenko and Bratko (1991) define the so-called information score for a classifier when classifying an unseen object; their formula can be also applied to rule quality:

$$Quality_{IKIB}(R) = -\log_2 f_{+1} + \log_2 cons(R) \tag{12}$$

where the logarithm here is of base 2. More precisely, this formula is valid only if

$$cons(R) \geqslant f_{+1}$$

which could be interpreted as a necessary condition for the reliability of a rule.

(3). The vehicle of the *confidence sets* can be also exploited. The unknown conditional probability

$$p = P(\text{example of } C \mid \text{rule } R \text{ fires})$$

has $cons(R) = rc/r$ as its estimate. The random variable

$$X^2 = \frac{(rc - r\,p)^2}{r\,p} + \frac{(r\bar{c} - r(1-p))^2}{r(1-p)}$$

has a χ^2 distribution with one degree of freedom. Hence,

$$P(X^2 \leqslant \chi_1^2(\alpha)) \doteq 1 - \alpha$$

for the selected factor α.

The limits of the confidence interval $\langle p_1, p_2 \rangle$ for the unknown probability p are given by the solution to the following equation:

$$\frac{(rc - r\,p)^2}{r\,p} + \frac{(r\bar{c} - r(1-p))^2}{r(1-p)} = \chi_1^2(\alpha) \tag{13}$$

and rule quality can be characterized by the lower bound p_1 of the confidence interval. Quadratic equation (13) has two solutions p_1, p_2, and the lower result is

$$Quality_{\text{Conf}}(R) = \frac{rc^2 + S - r\bar{c}^2 - \sqrt{(r\bar{c}^2 - rc^2 - S)^2 - 4S + rc^2}}{2S}$$

where

$$S = r(r + \chi_1^2(\alpha))$$

5. ANALYSIS

Figures 5.1 to 5.3 exhibit the values of the rule quality for various formulas. Each formula is depicted by two figures for various values of c, and each figure represents the quality as a function of rc and r. The size K of a training set is for simplicity equal to 100. Figure 5.4 shows the rule qualities for various formulas so as to compare their behavior. Particularly, the rule quality is shown as a function of consistency and completeness with other parameters being fixed.

A. Statistical formulas for the rule quality exhibit the so-called *independence* property: they yield the value zero, if and only if the rule does not have any expressive capability, that is, if its distribution is equal to the distribution of the entire training set. The independence property can also be expressed by the consistency or completeness; it arises if

$$cons(R) = f_{+1}, \qquad compl(R) = f_{1+}$$

The name is derived from the fact that the columns and rows in the corresponding contingency table are independent if and only if the formulas yield the value zero.

B. One may easily prove that the empirical, agreement-based, and information-based qualities are non-decreasing functions of consistency and completeness. Figures 5.4 and 5.5 illustrate the same property.

C. Coleman's statistic does not comprise the completeness of the rule. As we have already mentioned, this was the reason for its modification, see (10), (11). Formula (14) exhibits the same property.

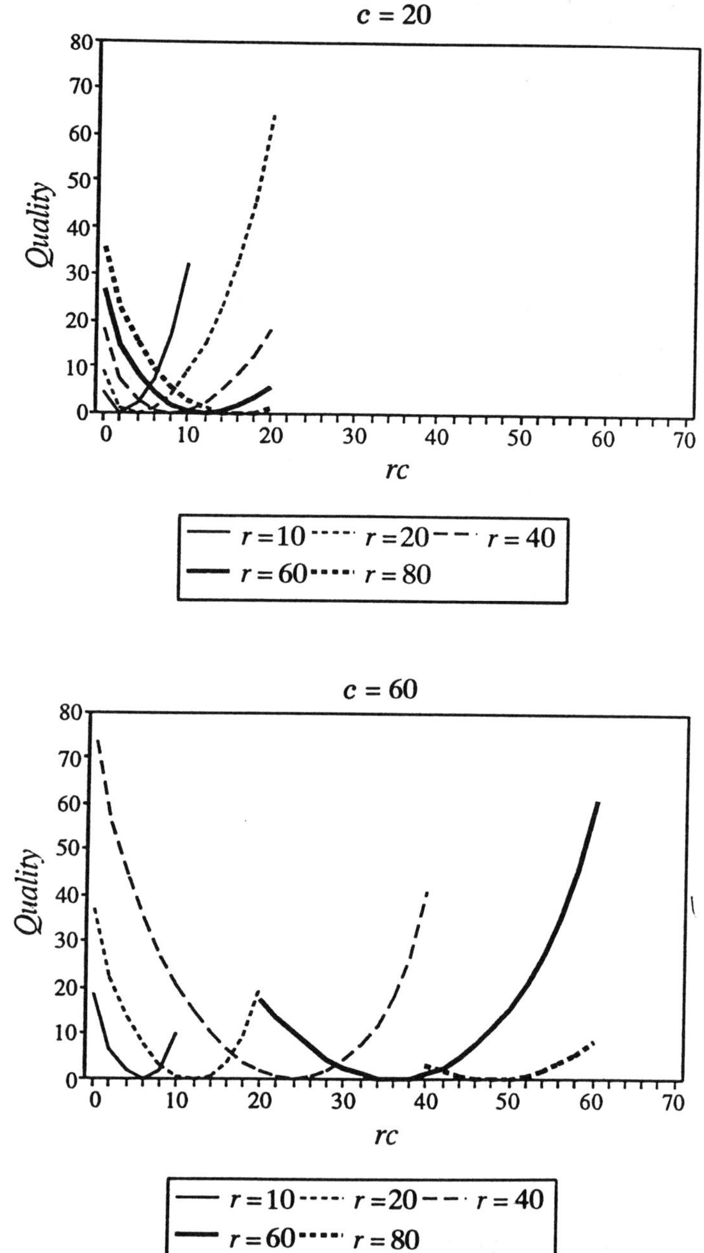

Figure 5.1. *Quality*$_{G2}$ *for c = 20 and 60 as a function of rc for various values of r (K = 100).*

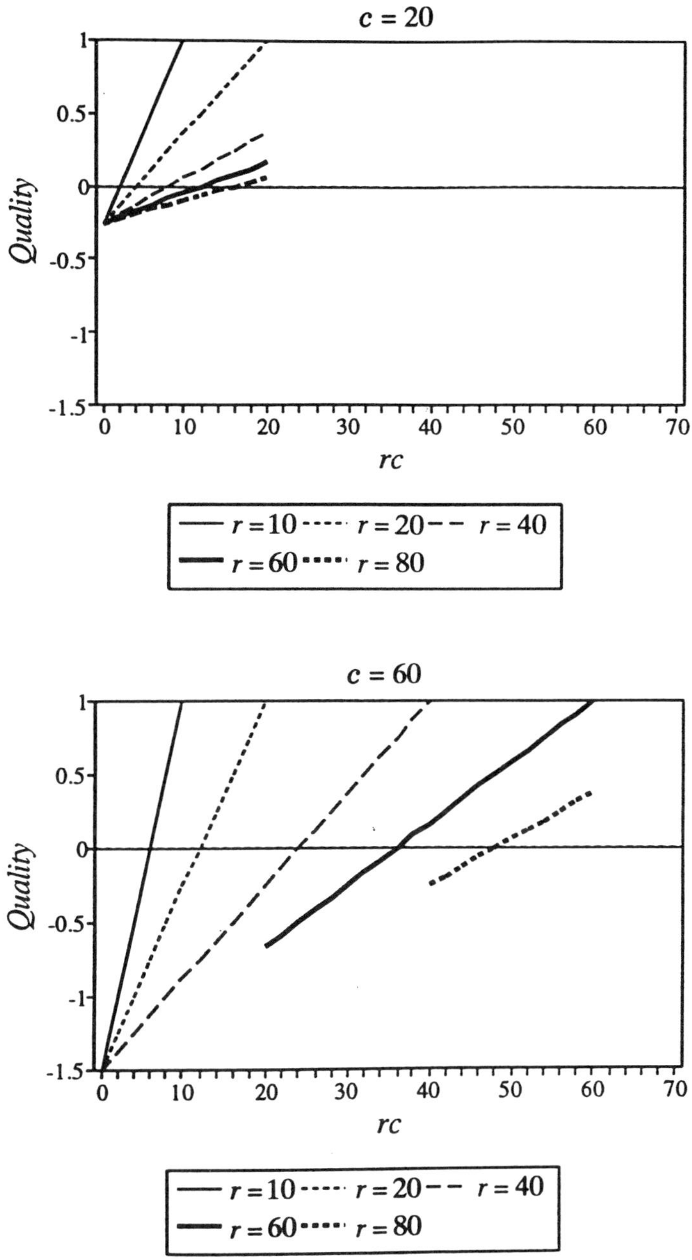

Figure 5.2. *Quality*$_{\text{Cohen}}$ *for c = 20 and 60 as a function of rc for various values of r (K = 100).*

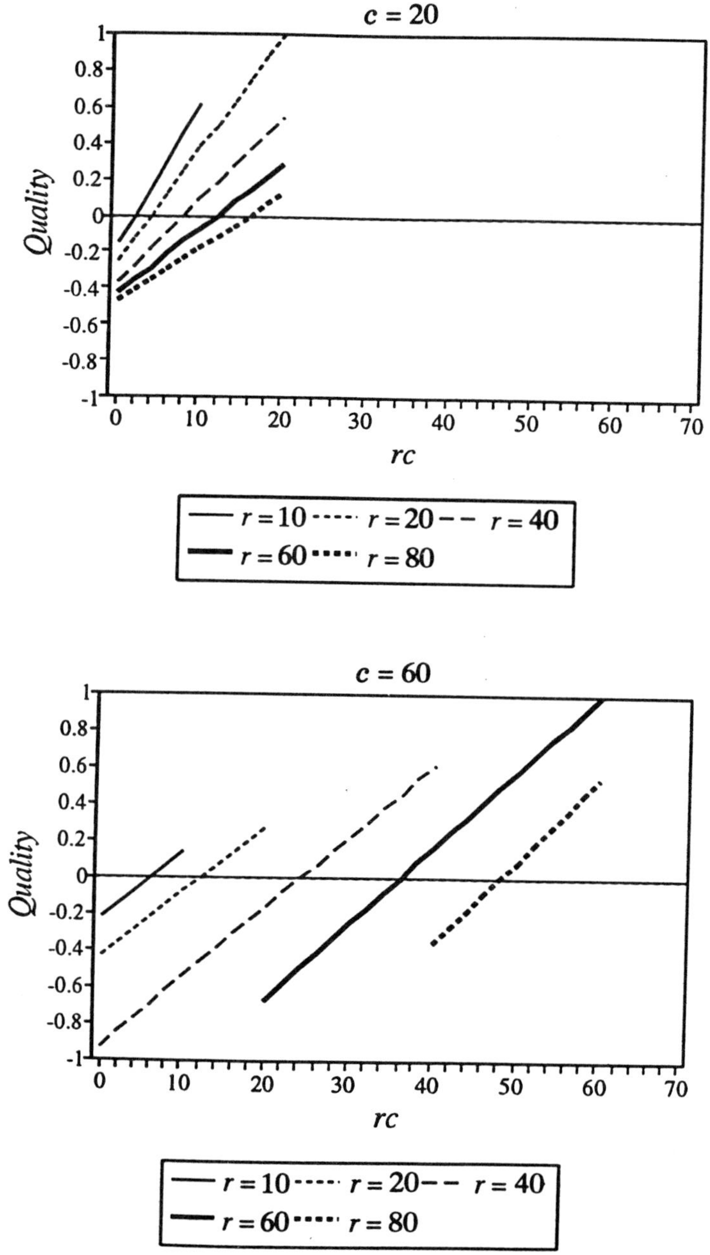

Figure 5.3. *Quality*$_{Coleman}$ *for c = 20 and 60 as a function of rc for various values of r (K = 100).*

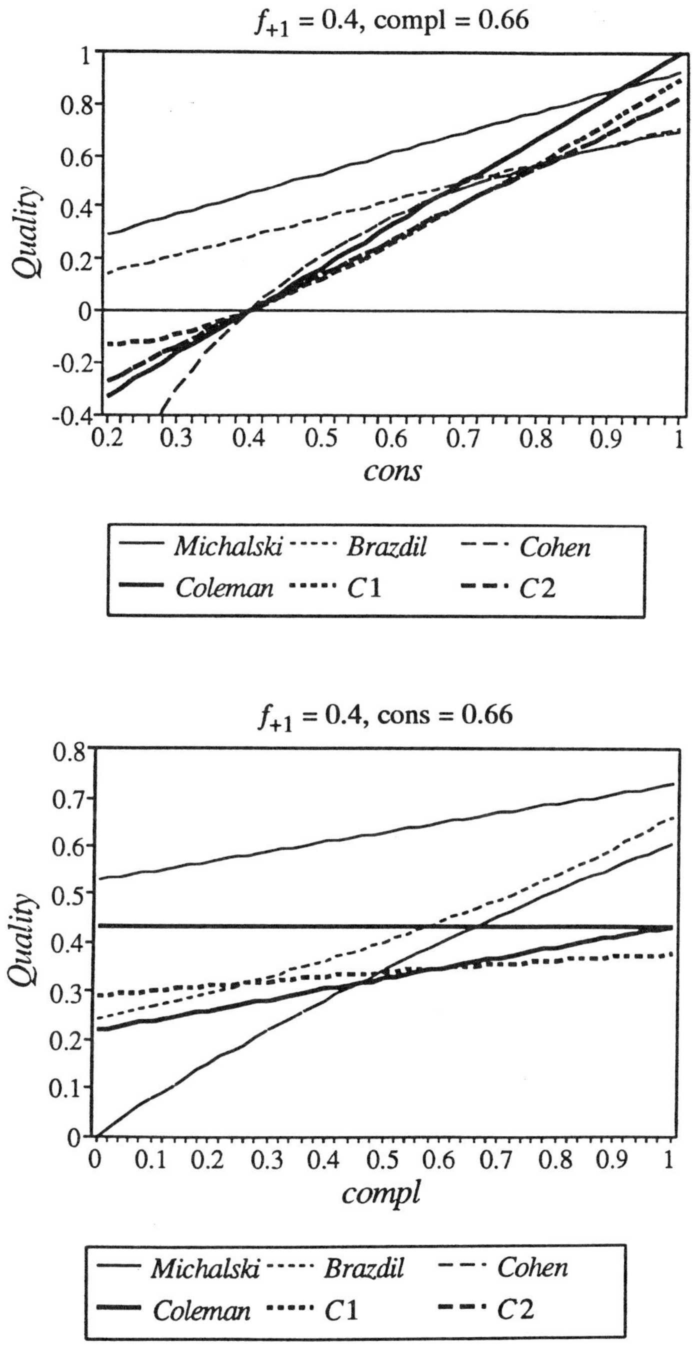

Figure 5.4. *Rule qualities as functions of (a) cons for* $f_{+1} = 0.4$, *compl* $= 0.66$; *(b) compl for* $f_{+1} = 0.4$, *cons* $= 0.66$. *Quality*$_{\text{Michalski}}$ *has* $w_1 = 0.8$, $w_2 = 0.2$.

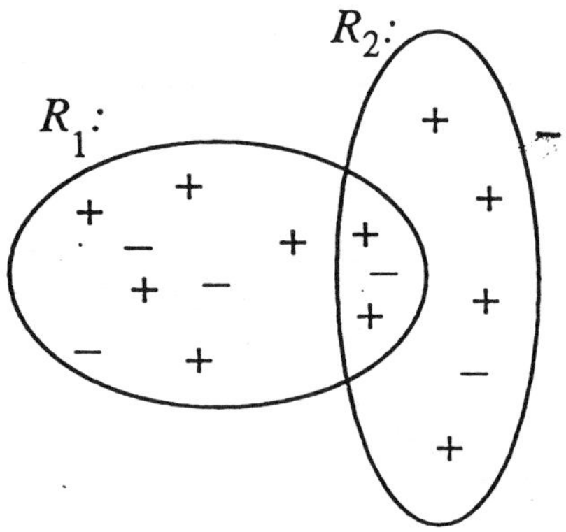

Figure 5.5. *Two overlapping rules: cons(R_1) = 0.64, cons(R_2) = 0.75, cons($R_1 \vee R_2$) = 0.69.*

D. Figure 5.4 indicates that the behavior of the agreement-based qualities as a function of consistency is almost identical and also, those of empirical qualities are comparable. We may find that all formulas for rule qualities are applicable and comply with the property of monotony toward consistency and completeness.

Moreover, Figure 5.4 shows that certain formulas exhibit similar dependencies. Namely, the following pairs can be detected: formulas for $Quality_{\mathbf{Michalski}}$ and $Quality_{\mathbf{Brazdil}}$, $Quality_{C1}$ and $Quality_{C2}$, $Quality_{\chi2}$ and $Quality_{G2}$.

E. Ranges of values of various formulas are also worth mentioning. The empirical formulas yield values between 0 and 1, as does the χ^2 statistic, whereas the $G2$-likelihood gives nonnegative values. Cohen's statistic yields values between -1 and 1; the lower bound -1 comes with the table

$$\begin{bmatrix} 0 & \frac{K}{2} \\ \frac{K}{2} & 0 \end{bmatrix}$$

Coleman's upper bound is 1 and its lower bound is $-c/\bar{c}$ for $rc = 0$; in particular, the absolute minimum is $-(K-1)$ for $\bar{c} = 1$. Due to normalization the upper bound of the statistics (10) and (11) is 1.

The differences in the ranges of various formulas are not worth mentioning, because a user selects a particular formula for a task and the algorithm compares only the values of the formula. The monotony of these functions is what is important.

6. COMBINING RULE QUALITIES

As soon as a formula for rule quality is selected and an unordered set of decision rules is induced, we may employ several different classification schemes. Some of them are discussed, for example, in Kononenko (1992). The current version of our covering learning algorithm CN4 so far involves the following three strategies:

(a). Ordering an Unordered Set

An unordered set of decision rules may be easily ordered by rule qualities. Classification is then done in the same way as in the ordered mode, that is, the classifier goes through the list of rules starting with the best one (having the greatest quality) and looks for the first rule that matches ('fires') the given object; it is then categorized into the class attached to the rule.

The rules in CN4 are first ordered according to qualities, second (if more rules have identical quality), according to their significance (Clark and Niblett, 1989), third, according to the numbers of covered examples of the class attached to the rule. If no rule 'fires' then the object is classified to the majority class of the entire training set (which is the standard scheme of the default rule in the CNx family).

This scheme is also used, for example, in *Integ.3* (Brazdil and Torgo, 1990). Kononenko (1992) exploits the equivalent strategy with the difference that it checks all rules and selects the best rule that 'fired'.

(b). Weighted Sum of Qualities

A classifier exploiting this scheme goes through the entire set of decision rules and checks whether or not each rule 'fires'. Afterwards, the weighted sum of qualities of fired rules is calculated for each class separately:

$$WeightedQuality(C) = \frac{\Sigma\, compl(R) \,.\, Quality(R)}{\Sigma\, compl(R)} \qquad (14)$$

Here both sums are performed for each rule R of the class C that has fired for a given object. The tested object is then categorized to the class for which *WeightedQuality* is maximum.

If there are more classes with the maximum weighted quality, then a formula similar to (14) is executed for rule significances. The default rule is again applied if no rule has fired.

Clark and Boswell (1991) utilize a simple version of this strategy: frequencies f_{11} for all rules that cover the given object are summed up and the object is categorized into the majority class of the resulting distribution (i.e., the algorithm assumes quality to be equal to 1 for all rules induced).

(c). Combining Qualities

A classifier using this scheme goes through the entire set of rules and looks for rules that cover a given testing object. Qualities of the rules belonging to the same class are then calculated and the object is classified to the class that maximizes the above sum. It can happen that a combination of several rules covers some examples more than once.

The problem of duplicates is solved in CN4 as follows. It keeps a 'track' of examples covered by each rule. Hence, it can easily recalculate the consistency and completeness of any combination of fired rules so that each example is counted only once. Thus, if R_1 and R_2 are the rules (of the same class C) that cover a given tested object, then the consistency of their combination is

$$cons(R_1 \vee R_2) = \frac{\#\ \text{examples of class } C \text{ covered by } R_1 \vee R_2}{\#\ \text{examples covered by } R_1 \vee R_2}$$

that is, the examples covered by both rules are counted only once; see Figure 5.5 for a portrayal. Completeness of the entire collection is calculated similarly and its quality is recalculated according to the selected formula.

The scheme seems intuitively to be the most natural way of processing qualities if several rules fire. Our experiments have revealed that this intuition was correct.

7. EXPERIMENTS

The following seven different formulas for rule quality have been used in our experiments:

A. Michalski's formula (1) for $w_1 = 0.8$, $w_2 = 0.2$;
B. Michalski's formula (1) for $w_1 = 0.5$, $w_2 = 0.5$;
C. Torgo's formula, that is, (1) with (2);
D. Brazdil's formula (3);
E. J-measure;
F. one modification of Coleman's formula (10);
G. another modification of Coleman's formula (11).

Next, all three classification schemes (a, b, c) introduced in the previous section have been analyzed.

All 21 combinations of formulas and schemes have been tested on four AI databases. Each database has been randomly separated into two subsets (70% learning, 30% testing) and this scenario has been executed ten times. The following tables thus involve in each slot an average of classification accuracy acquired from tens runs. The parameters of the CN4 learning algorithms were set up to their default values (i.e. Laplace evaluation function, routine Common Bruha and Kočková (1993) for unknown attribute value processing, *star_size* equal to 5, parameter for χ^2-distribution equal to 0.025) because we wanted to focus on the behavior of the above qualities and schemes rather than on that of the system itself.

The four AI databases are as follows:

Onco:
The oncological data (Pecen, 1993) were used for testing in the Czech Academy of Sciences, Prague, and also in the project CN4 (Bruha and Kočková, 1994). The entire set involves 127 examples. Each example is represented by eight attributes; seven of them are numerical attributes, the only symbolic one involves three values. 4.3% of attribute values are unknown. The task involves three classes; the frequency of the majority clas is 50%.

Soybean:
This well-known dataset has been used in many various experiments in machine learning, namely by Michalski's research group and the AQ family. The set available to the author of this paper involves 290 training examples and 15 classes. Each example is characterized by 24 attributes, all are symbolic ones with average 2.9 values per attribute. The set exhibits 3.7% unknown attribute values. There are four classes exposing the maximum frequency (14%).

ThyroidGland:
This task of diagnosis of thyroid gland disease has been provided by the Institute of Nuclear Medicine of Inselspital, Bern, Switzerland. This database was used at the Department of Advanced Mathematics, University of Bern, and also in the project CN4. The entire set involves 269 patients' data. Each patient is described by ten attributes, three of them are numerical attributes, the rest are symbolic ones; the average number of values per symbolic attribute is 4.9. About 20 percent of attribute values are unknown. The task involves two classes; the frequency of the majority class is 72%.

BreastTumor:
This dataset has been provided by the Jozef Stefan Institute, the research group of Prof. Dr. I. Bratko, Ljubljana. It involves 288 examples and two classes. Each object is represented by ten attributes, four of them are numerical;

TABLE 5.1.

	A	B	C	D	E	F	G
Onco	73.5	50.5	72.8	72.8	72.9	73.0	73.0
Soybean	82.1	77.7	82.1	82.3	82.5	82.5	82.5
ThyroidGland	90.7	71.5	90.7	91.0	91.1	91.1	91.1
BreastTumor	83.5	82.0	84.1	84.1	84.5	84.5	84.5
Average	82.5	70.4	82.4	82.6	82.8	82.8	82.8

TABLE 5.2.

	A	B	C	D	E	F	G
Onco	73.0	72.0	73.0	73.0	73.3	74.3	74.3
Soybean	82.7	81.7	82.7	82.7	82.3	82.9	82.9
ThyroidGland	91.1	91.0	91.1	91.1	91.2	91.5	91.5
BreastTumor	84.5	84.5	84.5	84.5	84.5	85.3	85.3
Average	82.8	82.3	82.8	82.8	82.8	83.5	83.5

TABLE 5.3.

	A	B	C	D	E	F	G
Onco	74.0	72.5	73.5	73.5	73.5	74.9	74.9
Soybean	82.7	81.7	83.0	82.7	83.3	83.6	83.6
ThyroidGland	91.9	91.0	91.5	91.1	91.0	92.2	92.2
BreastTumor	84.1	83.9	84.1	84.0	84.7	85.6	85.6
Average	83.2	82.3	83.0	82.8	83.1	84.1	84.1

TABLE 5.4.

Rule Quality	A	B	C	D	E	F	G
Accuracy	82.8	78.3	82.7	82.7	82.9	83.4	83.4

TABLE 5.5.

Classif. Scenario	(*a*)	(*b*)	(*c*)
Accuracy	80.9	82.9	83.2

symbolic attributes exhibit on average 2.7 values per attribute. 0.7 percent of attribute values are unknown. The majority class has frequency 80%.

The tables comprise the average classification accuracy (in %) for all the above formulas and classification scenarios. Table 5.1 depicts the results for the scenario (*a*), that is, ordering an unordered set by rule qualities. Table 5.2 portrays the classification accuracy for the scenario (*b*). Table 5.3 exhibits the results received by performing the scenario (*c*), that is, combining rule qualities.

8. CONCLUSION

The chapter introduces various formulas for rule quality as well as three different scenarios for classification. The previous section exhibits the results performed for the above four data sets. Tables 5.4 and 5.5 exhibit the overall accuracy (in %) for each formula as well as each classification scenario.

By analyzing all the above results of experiments we came to the following conclusions. One may observe from Tables 5.1–5.5 that all formulas for rule quality introduced in this chapter are applicable. Although the empirical formulas are not backed by any statistical theory, they work quite well. The statistical formulas work better, but the difference in classification accuracy is almost negligible. On the other hand, statistical formulas have been derived from statistical theories and exhibit a so-called independence property, which has the advantage that the threshold of meaningful rules is clearly expressed.

To be more precise, even the statistical formulas used in machine learning are applied mostly in an ad hoc, empirical fashion, since most algorithms do not check the conditions or assumptions of their applicability. Unlike the genuine empirical formulas, however, we are aware of the error we have made when the conditions of applicability are not checked. If needed, the condition checking could in principle be embedded into the algorithm.

The weighted-sum rule quality (1) exhibits, to a certain extent, a quite interesting characteristic. By selecting the weights w_1 and w_2 a user may select between inducing only reliable rules (when supporting consistency of rules) and only robust rules (when supporting completeness). However, our experiments undoubtedly revealed that a larger support of completeness deteriorates the classification accuracy. Hence, in order to achieve a reasonable accuracy one should rather support the rule consistency (compare formula A vs. B). Furthermore, formula C, although it provides dynamic adjustments of weights, has not exhibited striking improvement in this approach.

Testing all the decision rules provides better results than just looking for the first rule that fires. This approach thus gives a "chance" for every rule to "say" something in the decision making. Nevertheless, the quality combining (scenarios (*b*) and (*c*)) overcame in classification accuracy the plain ordering of the set (scenario (*a*)) only by 1–2 percent.

By analyzing this aspect we found out that the rules induced for all four data sets overlap only minimally: about 7 to 12 percent of examples are covered by more than one rule. Consequently, the scenarios that include rule combining should be rather employed in the case of redundant (multiple) rules. Therefore, one of our research plans is to focus on generating redundant rules (and exploiting the above combining scenarios).

The chapter focuses on combining rule qualities from the classification viewpoint only. The learning phase has been considered as more or less independent of the classification scheme. In other words, evaluation heuristics invoked by an inductive algorithm are more or less expected to have no connection with the rule quality invoked by a classification scheme. For instance, the CN2 algorithm exploits the Laplacian estimate formula for evaluating complexes and the $G2$-likelihood statistic as a stopping condition; on the other hand, a classifier using unordered decision rules may exploit any of the above formulas for rule quality.

It would be natural that the heuristics of a learning algorithm and the rule quality of a classification scheme be selected together according to the demands of a 'client', the user of a classifier. The most appropriate strategy exploits any rule quality formula both within classification and within inductive learning as evaluation heuristics. The above scenario is embedded in POSEIDON (Bergadano et al., 1992). We plan to adopt this scenario in the CN4 learning algorithm in the future.

REFERENCES

Bergadano, F. et al. (1988). *Measuring Quality of Concept Descriptions*, EWSL-88. Glasgow.

Bergadano, F., Matwin, S., Michalski, R. S. and Zhang, J. (1992). Learning two-tiered descriptions of flexible concepts: The POSEIDON system. *Mach. Learn.* **8**, 5–43.

Bishop, Y. M. M., Fienberg, S. E., and Holland; P. W. (1991). *Discrete Multivariate Analysis: Theory and Practice.* MIT Press, Cambridge, MA.

Brazdil, P. B., and Bruha, I. (1992). A note on processing missing attribute values: A modified technique. In *Workshop on Machine Learning, Canadian Conference on Artificial Intelligence*, Vancouver.

Brazdil, P. and Torgo, L. (1990). Knowledge acquisition via knowledge integration. In *Current Trends in Knowledge Acquisition.* IOS Press.

Bruha, I., and Kočková, S. (1993). A support for decision making: Cost-sensitive learning system. *Artif. Intelli. Med.*

Cestnik, B., Kononenko, I., and Bratko, I. (1988). Assistant 86: A knowledge-elicitation tool for sophisticated users. In *Progress in Machine Learning* (I. Bratko and N. Lavrac, eds.), Proc. EWSL 88, pp. 31–46. Sigma Press.

Clark, P., and Boswell, R. (1991). *Rule Induction with CN2: Some Recent Improvements*, EWSL-91. Porto.

Clark, P., and Niblett, T. (1989). The CN2 induction algorithm. *Mach. Learn.* **3**, 261–283.

Cohen, J. (1960). A coefficient of agreement for nominal scales. *Educ. Psychol. Meas.* **20**, 37–46.

Kočková, S., and Bruha, I. (1993). *CN4: An Extension of CN2 Covering Algorithm*, Tech. Rep. 549. Institute of Computer Science of ASCR, Prague.

Kononenko, I. (1992). Combining decisions of multiple rules. In *Artificial Intelligence V: Methodology, Systems, Applications* (B. du Boulay and V. Sgurev, eds.). Elsevier, Amsterdam.

Kononenko, I., and Bratko, I. (1991) Information-based evaluation criterion for classifier's performance. *Mach. Learn.* **6**, 67–80.

Michalski, R. S. (1980). Pattern recognition as rule-guided inductive inference. *IEEE Trans. Pattern Anal. Mach. Intell.* **PAMI-2**(4) 349–361.

Michalski, R. S., Mozetic, I., Hong, J. and Lavrac, N. (1986). The multi-purpose incremental learning system AQ15 and its testing application to three medical domains. *Proc. 5th AAAI*, pp. 1041–1045. Philadelphia, PA.

Mingers, J. (1989). An empirical comparison of selection measures for decision-tree induction. *Mach. Learn.* **3**, 319–342.

Nunez, M. (1988). *Economic Induction: A Case Study*, EWSL 88, pp. 139–145. Glasgow.

Pearl, J. (1978). On the connection between the complexity and the credibility of inferred models. *Int. J. Gen. Syst.* **4**, 255–264.

Pecen L. (1993). Private communication. Prague.

Quinlan, J. R. (1986). Induction of decision trees. *Mach. Learn.* **1**, 81–106.

Quinlan, J. R. (1987). *Generating Production Rules from Decision Trees*, Tech. Rep. 87.7. Sydney University, Sydney, Australia.

Quinlan, J. R. (1989). Unknown attribute values in ID3. In *Proceedings of the Sixth International Conference on Machine Learning*, pp. 164–168.

Quinlan, J. R. (1992). *C4.5: Programs for Machine Learning.* Morgan Kaufmann, San Mateo, CA.

Smyth, P., and Goodman, R. M. (1990). Rule induction using information theory. In *Knowledge Discovery in Databases* (G. Piatetsky and W. Frawley, eds.). MIT Press, Cambridge, MA.

Tan, M., and Schlimmer, J. C. (1990). Two case studies in cost-sensitive concept acquisition. In *Proceedings of the Eighth National Conference on Artificial Intelligence.*

Torgo, L. (1993). *Controlled Redundancy in Incremental Rule Learning*, ECML-93. pp. 185–195. Vienna.

Weiss, S. M., Galen, R. S., and Tadepalli, P. V. (1987). Optimizing the predictive value of diagnostic decision rules. In *Proceedings of the Sixth National Conference on Artificial Intelligence*, Seattle, AAAI-87, pp. 521–526.

CHAPTER 6

DIPOL—A Hybrid Piecewise Linear Classifier

B. SCHULMEISTER
F. WYSOTZKI

1. INTRODUCTION

In most real-world problems, the assumption that the forms for the underlying probability distributions of the classes are known does not hold. Because of that, the aim of the development was to derive a distribution-free (nonparametric) classification procedure that is able to handle a wide variety of problems and can be used in practical settings. The introduced hybrid algorithm can be regarded as:

- a nonparametric statistical approach, with the special option of clustering classes and carrying out the subtask of classification on a symbolic level; or
- a neural network approach, with the special choice of the initial conditions of hidden units (number and weights) and fixed Boolean weights in the second layer.

These two ways of interpreting the algorithm emphasize that there are many relations between statistical and neural network algorithms. In neural net algorithms and some statistical algorithms many parameters have to be determined in order to obtain an optimal solution of the classification problem. The disadvantage of neural network design is the shortage of rigorous principles on which to base a choice of architecture and initial weights, but many heuristic principles have been invented and explored. Here another path

Machine Learning and Statistics, The Interface, Edited by G. Nakhaeizadeh and C. C. Taylor.
ISBN 0-471-14890-3

is followed. Section 2 describes a linear regression approach, which is used to generate the initial positions of the hyperplanes for the pairwise discrimination of the classes. In section 3 a learning procedure for decreasing the classification error is defined. The procedure represents an accumulated correction of the decision hyperplane positions. If the training set is updated, that is, if some new patterns are available for training, the procedure can also be used for incremental learning. Also, to handle problems with non-convex, especially non-singly connected class regions, a clustering procedure is proposed for finding subclasses of the classes in an initial step in section 4. Section 5 introduces the classification procedure for a pattern. It is defined as a set of scalar products of two Boolean vectors that contain both the coded information of the convex class or subclass regions and the position of the handled pattern. Section 7 presents three artificial examples demonstrating the strengths and weaknesses of the introduced algorithm.

The algorithm is implemented as the program DIPOL (DIscrimination and POst Learning). The first version developed in 1992 (DIPOL92) was part of the Esprit project *StatLog* (Michie et al., 1994). It was successfully applied to a great variety of real-world pattern-recognition problems.

There are many other linear and piecewise linear classification algorithms (Duda and Hart, 1973; Nilsson, 1965). Similar principles (determination of the initial hyperplanes by statistical methods, further adaption by learning and introduction of a symbolic level as in DIPOL (although different in detail) are used in Unger and Wysotzki (1981), but the integration of clustering classes with the statistical determination of the initial hyperplane positions by linear regression, accumulated correction, and the classification on a symbolic level into a unified procedure may be considered a new starting point. The results are very encouraging. Note that in comparison to neural networks, the symbolic level of DIPOL replaces additional layers of "hidden units" (for example, used in backpropagation).

2. PAIRWISE LINEAR REGRESSION

Suppose that $X \subset \boldsymbol{R}^m$ (m-dimensional Euclidean space) is a finite set of classified patterns $\boldsymbol{x} = (x_1, \ldots, x_m)$. It is assumed that there is a subdivision of X into n classes $X = k_1 \cup k_2 \cup \cdots \cup k_n$, and for this reason, it is necessary to define $n \times (n-1)/2$ hyperplanes for pairwise discrimination. Linear regression is used for the discrimination of two classes of patterns k_i and k_j, by defining the dependent variable b in the following manner:

$$\text{if } \boldsymbol{x} \in k_i, \quad \text{then } b = +1,$$
$$\text{if } \boldsymbol{x} \in k_j, \quad \text{then } b = -1.$$

Let W be the linear regression function $W: X \to \boldsymbol{R}$ with

$$W(\boldsymbol{x}) = w_0 + w_1 x_1 + \cdots + w_m x_m.$$

Let n_i and n_j be the number of patterns k_i and k_j, respectively, with corresponding mean values

$$\boldsymbol{\mu}_i = \frac{1}{n_i} \sum_{x \in k_i} \boldsymbol{x} \qquad \text{and} \qquad \boldsymbol{\mu}_j = \frac{1}{n_j} \sum_{x \in k_j} \boldsymbol{x}$$

and common mean value

$$\boldsymbol{\mu} = \frac{1}{\hat{n}} \sum_{x \in k_i \cup k_j} \boldsymbol{x}$$

with $\hat{n} = n_i + n_j$.

The system of equations for the determination of the regression coefficients can be written as follows:

$$\mathbf{A}(w_0, \boldsymbol{w})^T = (\mathbf{e}, \mathbf{X})(w_0, \boldsymbol{w})^T = \mathbf{b}.$$

where $\mathbf{e}$ consists of $\hat{n}$ ones.

Without loss of generality, the patterns can be ordered in such a way that the first n_i rows of the matrix $\mathbf{X}$ consist of the patterns from class k_i and the following rows consists of the n_j patterns of k_j. Then the first n_i components of the vector $\mathbf{b}$ are 1 and the following n_j components are -1.

In most cases, the number of equations is much larger than the dimension of the pattern space ($\hat{n} \gg m$), that is, the system of equations is overdetermined. Because of that, the minimum-squared-error of the residual is determined:

$$\|\mathbf{r}\|^2 = \|\mathbf{b} - \mathbf{A}(w_0, \boldsymbol{w})^T\|^2 \to \mathit{Minimum}$$

This leads to the following system of equations with symmetric system matrix:

$$\mathbf{A}^T\mathbf{A}(w_0, \boldsymbol{w})^T = \mathbf{A}^T\mathbf{b}$$

In detail, (see Duda and Hart, 1973, for example)

$$\begin{pmatrix} 1 & 1 & \cdots & 1 & 1 & \cdots & 1 \\ x_1^1 & x_1^2 & \cdots & x_1^{n_i} & x_1^{n_i+1} & \cdots & x_1^{\hat{n}} \\ x_2^1 & x_2^2 & \cdots & x_2^{n_i} & x_2^{n_i+1} & \cdots & x_2^{\hat{n}} \\ \vdots & \vdots & \ddots & \vdots & \vdots & \ddots & \vdots \\ x_m^1 & x_m^2 & \cdots & x_m^{n_i} & x_m^{n_i+1} & \cdots & x_m^{\hat{n}} \end{pmatrix} \begin{pmatrix} 1 & x_1^1 & \cdots & x_m^1 \\ 1 & x_1^2 & \cdots & x_m^2 \\ \vdots & \vdots & \ddots & \vdots \\ 1 & x_1^{n_i} & \cdots & x_m^{n_i} \\ 1 & x_1^{n_i+1} & \cdots & x_m^{n_i+1} \\ \vdots & \vdots & \ddots & \vdots \\ 1 & x_1^{\hat{n}} & \cdots & x_m^{\hat{n}} \end{pmatrix} \begin{pmatrix} w_0 \\ w_1 \\ w_2 \\ \vdots \\ w_m \end{pmatrix}$$

$$= \begin{pmatrix} 1 & 1 & \cdots & 1 & 1 & \cdots & 1 \\ x_1^1 & x_1^2 & \cdots & x_1^{n_i} & x_1^{n_i+1} & \cdots & x_1^{\hat{n}} \\ x_2^1 & x_2^2 & \cdots & x_2^{n_i} & x_2^{n_i+1} & \cdots & x_2^{\hat{n}} \\ \vdots & \vdots & \ddots & \vdots & \vdots & \ddots & \vdots \\ x_m^1 & x_m^2 & \cdots & x_m^{n_i} & x_m^{n_i+1} & \cdots & z_m^{\hat{n}} \end{pmatrix} \begin{pmatrix} 1 \\ \vdots \\ 1 \\ -1 \\ \vdots \\ -1 \end{pmatrix}$$

and summarized

$$\begin{pmatrix} \hat{n} & & (\hat{n}\boldsymbol{\mu})^T & \\ & \sum_{r=1}^{\hat{n}} (x_1^r)^2 & \cdots & \sum_{r=1}^{\hat{n}} x_1^r x_m^r \\ \hat{n}\boldsymbol{\mu} & \vdots & \ddots & \vdots \\ & \sum_{r=1}^{\hat{n}} x_m^r x_1^r & \cdots & \sum_{r=1}^{\hat{n}} (x_m^r)^2 \end{pmatrix} \begin{pmatrix} w_0 \\ \boldsymbol{w} \end{pmatrix} = \begin{pmatrix} n_i - n_j \\ n_i\boldsymbol{\mu}_i - n_j\boldsymbol{\mu}_j \end{pmatrix}$$

The first equation can be solved for w_0 in terms of $\boldsymbol{w}$

$$w_0 = \frac{n_i - n_j}{\hat{n}} - \boldsymbol{\mu}^T \boldsymbol{w}$$

and substituted in the remaining equations. After a few manipulations, the following system of equations is obtained:

$$\mathbf{B}\boldsymbol{w} = \frac{2n_i n_j}{\hat{n}} (\boldsymbol{\mu}_i - \boldsymbol{\mu}_j)$$

where

$$(\mathbf{B})_{st} = (\mathbf{S}_w)_{st} + \frac{n_i n_j}{\hat{n}} (\mathbf{S}_b)_{st},$$

$\mathbf{S}_w$ is the within-class scatter matrix

$$(\mathbf{S}_w)_{st} = \sum_{r=1}^{n_i} (x_s^r - \mu_{is})(x_t^r - \mu_{it}) + \sum_{r=n_i+1}^{\hat{n}} (x_s^r - \mu_{js})(x_t^r - \mu_{jt})$$

and $\mathbf{S}_b$ is the between-class scatter matrix

$$(\mathbf{S}_b)_{st} = (\mu_{is} - \mu_{js})(\mu_{it} - \mu_{jt}).$$

The set $\{\boldsymbol{y} \in \boldsymbol{R}^m \mid W(\boldsymbol{y}) = w_0 + \boldsymbol{w}^T\boldsymbol{y} = 0\}$ is a hyperplane in $\boldsymbol{R}^m$ and divides the pattern space into two halfspaces. The vector $\boldsymbol{w}$ corresponds to the normal vector of the hyperplane and points into the region of class k_i (the region of class k_i is on the positive side of the hyperplane and the region of class k_j is on the negative side). The distance from the origin to the hyperplane is given by $w_0/\|\boldsymbol{w}\|$, and $W(\boldsymbol{x})/\|\boldsymbol{w}\|$ corresponds to the distance from the pattern $\boldsymbol{x}$ to the hyperplane.

The position of the discriminating hyperplane depends on:

- the positions of the mean values $\boldsymbol{\mu}_i$ and $\boldsymbol{\mu}_j$;
- the scattering of the classes; and
- the numbers of patterns of the classes and their ratio.

The position of the discriminating hyperplane is invariant with respect to translations of the training set in the pattern space (if there is a fixed vector added to each pattern of the training set, the hyperplane is moved in the same way). Because of that, in the following it is assumed that the training set does not contain the zero vector.

The system of equations is solved numerically by Cholesky decomposition of $\mathbf{A}^T\mathbf{A} = \mathbf{L}\mathbf{L}^T$ ($\mathbf{L}$ a lower triangle matrix) and backsubstitution. Pivoting is used to indicate whether the matrix $\mathbf{A}^T\mathbf{A}$ is positive definite or positive semidefinite only. In the positive semidefinite case a special solution from the linear solution manifold is chosen (where the remaining components of the solution vector are 0). The solution of the minimum-squared-error problem is rather susceptible to rounding error. If the condition of the matrix is not too bad (the problem is correct with respect to the error level v if $v\sqrt{m+1}\, cond(\mathbf{A}) < 1$), the numerical stability of the solution is ensured by an iterative improvement.

A pattern $\boldsymbol{x}$ is correctly classified if

$$W(\boldsymbol{x}) > 0 \quad \text{for } \boldsymbol{x} \in k_1.$$
$$W(\boldsymbol{x}) < 0 \quad \text{for } \boldsymbol{x} \in k_2.$$

If the mean values of both classes coincide, the solution vector $\boldsymbol{w} = 0$ is obtained (e.g., this case can occur if one class consists of two subclasses which are located symmetrically with respect to the other class in the pattern space or in XOR-problems). Because of $w_0 = (n_i - n_j)/\hat{n}$, all patterns are assigned to class k_i if $n_i \geqslant n_j$, and all patterns are assigned to class k_j if $n_i < n_j$ (i.e., if $n_i = n_j$ without loss of generality patterns are assigned to class k_i, and in all other cases the patterns are assigned to the class with the highest a priori probability).

It is important to mention that this minimum-squared-error solution also minimizes the mean-squared approximation error to the Bayes discriminant function

$$g(\boldsymbol{x}) = P(k_i \mid \boldsymbol{x}) - P(k_j \mid \boldsymbol{x}),$$

in the limit as the number of patterns approaches infinity.

Another property of the minimum-squared-error solution is that it places emphasis on points where the density of patterns is large, rather than on points near the decision hyperplane. Thus, the regression function does not necessarily minimize the classification error. Especially, there is no guarantee that this hyperplane separates the classes in the separable case, but a useful solution is obtained in both the separable and the nonseparable cases. By all means, the hyperplane determined this way represents a good initial position for the following learning procedure.

3. LEARNING PROCEDURE

A criterion function F is defined which is used to decrease the classification error. For all misclassified patterns, the weighted squared distance from the corresponding decision hyperplane are summed up:

Suppose, that $\{\boldsymbol{y} \in \boldsymbol{R}^m \mid W(\boldsymbol{y}) = 0\}$ defines the decision hyperplane between the classes k_i and k_j. Let m_i be the set of all misclassified patterns of class k_i, that is $\boldsymbol{x} \in k_i$ and $W(\boldsymbol{x}) < 0$, let m_j be the set of all misclassified patterns of class k_j, that is, $\boldsymbol{x} \in k_j$ and $W(\boldsymbol{x}) > 0$ ($M = m_i \cup m_j$):

$$F(W) = \begin{cases} 0, & \text{if } M = \emptyset \\ \dfrac{1}{2} \sum\limits_{x \in M} \dfrac{W(\boldsymbol{x})^2}{\|\boldsymbol{x}\|^2}, & \text{if } M \neq \emptyset \end{cases}$$

The learning procedure consists of minimizing the criterion function by a gradient descent algorithm separately for each decision surface. Since the gradient

$$\nabla_W F(W) = \sum_{x \in M} \frac{W(\boldsymbol{x})}{\|\boldsymbol{x}\|^2} (1, \boldsymbol{x})^T$$

defines the direction of maximum increase in F, it is used to form the iterative minimization procedure. The procedure requires a step size parameter, different values are appropriate for different problems, and for different stages of learning. In general, the sequence $\{\rho_l\}_{l=0}^{\infty}$ with $\rho_l = 1/l$ and $\lim_{l \to \infty} \rho_l = 0$ is a useful compromise:

$$\boldsymbol{w}^{(l+1)} = \boldsymbol{w}^{(l)} - \rho_l \sum_{x \in M} \frac{W^l(\boldsymbol{x})}{\|\boldsymbol{x}\|^2} \boldsymbol{x},$$

$$w_0^{(l+1)} = w_0^{(l)} - \rho_l \sum_{x \in M} \frac{W^l(\boldsymbol{x})}{\|\boldsymbol{x}\|^2}.$$

The coefficients of the discriminating hyperplane are adjusted, only if the correction is moving the hyperplane in a good direction, that is:

- The value of the criterion function decreases; and
- The classification error does not increase.

$F(W)$ is the "weighted" sum of the squared distances of the misclassified patterns from the hyperplane: each summand is multiplied by $\|\boldsymbol{w}\|^2/\|\boldsymbol{x}\|^2$.

There is another interpretation of $F(W)$:

M is the set of all misclassified patterns with respect to the hyperplane

$$\{\boldsymbol{y} \in \boldsymbol{R}^m \mid W(\boldsymbol{y}) = w_0 + \boldsymbol{w}^T \boldsymbol{y} = 0\}.$$

For each $\boldsymbol{x} \in M$, the sets

$$H_x = \{v \in \boldsymbol{R}^m \mid V(\boldsymbol{v}) = w_0 + \boldsymbol{v}^T \boldsymbol{x} = 0\}$$

also define hyperplanes in $\boldsymbol{R}^m$, in which $\boldsymbol{x}$ is the normal vector of the hyperplane, the absolute value of $w_0/\|\boldsymbol{x}\|$ corresponds to the distance of the hyperplane H_x from the origin and the absolute value of $V(\boldsymbol{w})/\|\boldsymbol{x}\| = W(\boldsymbol{x})/\|\boldsymbol{x}\|$ corresponds to the distance of $\boldsymbol{w}$ from the hyperplane H_x. This means that $F(W)$ is the sum of the squared distances of the vector $\boldsymbol{w}$ from all hyperplanes, defined by $(w_0, \boldsymbol{x})$, $\boldsymbol{x} \in M$. Diminishing the mean distance of $\boldsymbol{w}$ from these

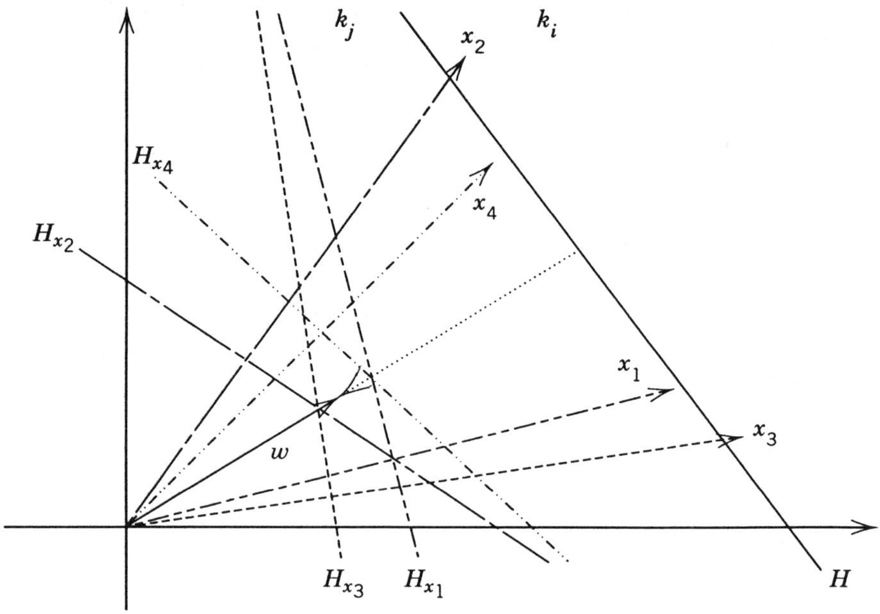

Figure 6.1. *See text for description.*

hyperplanes causes a decrease of the criterion function. Figure 6.1 tries to illustrate the criterion function with a 2-D example. Let $m_i = \{\boldsymbol{x}_1, \boldsymbol{x}_2\}$ and $m_j = \{\boldsymbol{x}_3, \boldsymbol{x}_4\}$ be the sets of misclassified patterns of the classes k_i and k_j, respectively, with corresponding hyperplanes H_{x_r}, $r = 1, 2, 3, 4$. The distances of these hyperplanes from the normal vector $\boldsymbol{w}$ of the hyperplane H discriminating the classes k_i and k_j are represented by bold lines.

A heuristic understanding of the learning procedure can be obtained by the following (see Fig. 6.2):

Let $\boldsymbol{x}$ be the only misclassified pattern with respect to the hyperplane

$$H = \{\boldsymbol{y} \in \boldsymbol{R}^m \mid W(\boldsymbol{y}) = w_0 + \boldsymbol{w}^T \boldsymbol{y} = 0\},$$

without loss of generality, $M = m_j = \{\boldsymbol{x}\}$, $W(\boldsymbol{x}) > 0$. This gives $F(W) = 1/2(W(\boldsymbol{x})^2/\|\boldsymbol{x}\|^2)$ and $\nabla_W F(W) = W(\boldsymbol{x})/\|\boldsymbol{x}\|^2(1, \boldsymbol{x})^T$.

If $\rho = 1$ is chosen, then the correction of $\boldsymbol{w}$ results in

$$\boldsymbol{w}_1 = \boldsymbol{w} - \frac{W(\boldsymbol{x})}{\|\boldsymbol{x}\|^2}\boldsymbol{x} = \boldsymbol{w} - \frac{W(\boldsymbol{x})}{\|\boldsymbol{x}\|}\frac{\boldsymbol{x}}{\|\boldsymbol{x}\|}$$

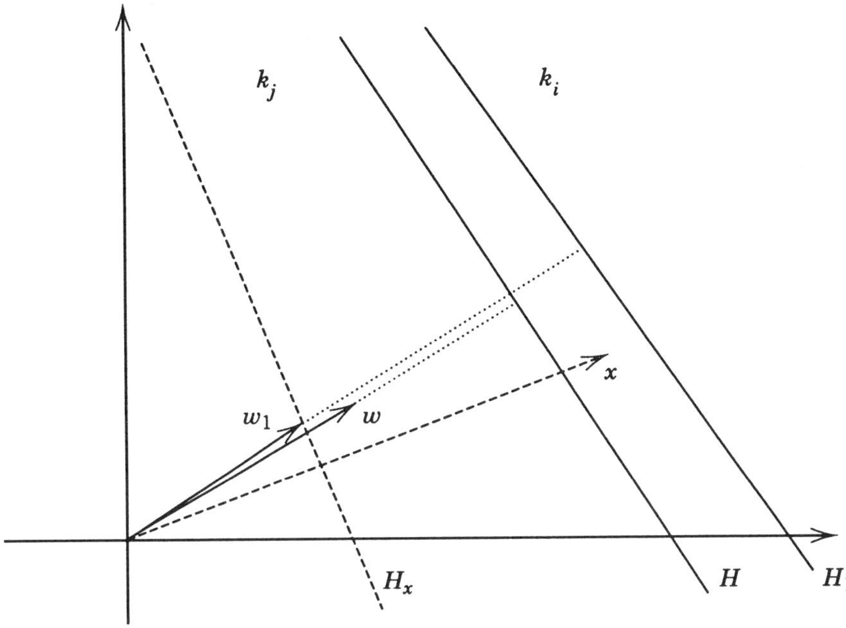

Figure 6.2. *See text for description.*

and

$$w_0 + \boldsymbol{w}_1^T \boldsymbol{x} = 0.$$

This means that the vector $\boldsymbol{w}$ is moved by its distance from the hyperplane H_x in the direction of the unit normal vector $\boldsymbol{x}/\|\boldsymbol{x}\|$ of that H_x, that is, $\boldsymbol{w}_1$ is located on the hyperplane H_x. Furthermore,

$$w_{01} = w_0 - \frac{W(\boldsymbol{x})}{\|\boldsymbol{x}\|^2},$$

so the adjusted hyperplane becomes $H_1 = \{\boldsymbol{y} \in \boldsymbol{R}^m \mid W_1(\boldsymbol{y}) = w_{01} + \boldsymbol{w}_1^T \boldsymbol{y} = 0\}$ and the vector $\boldsymbol{x}$ is correctly classified with respect to this hyperplane H_1:

$$\begin{aligned} W_1(\boldsymbol{x}) &= w_{01} + \boldsymbol{w}_1^T \boldsymbol{x} \\ &= w_0 - \frac{W(\boldsymbol{x})}{\|\boldsymbol{x}\|^2} + \boldsymbol{w}_1^T \boldsymbol{x} \\ &= -\frac{W(\boldsymbol{x})}{\|\boldsymbol{x}\|^2} < 0. \end{aligned}$$

This behavior of the gradient $\nabla_W F(W)$ of the criterion function $F(W)$ forms the basis of the learning procedure. The gradient of the criterion function adjusts the position of the hyperplane, if it is possible, on average.

The procedure represents an accumulated correction of the position of the decision hyperplane, or *learning by epoch* as an alternative to stochastic approximation, or *learning by sample*. If the training set is updated, that is, if new patterns become available for training which are misclassified, the procedure can also be used for incremental learning. There are again two possibilities:

- The accumulated correction on the basis of all misclassified patterns; and
- Learning with single new misclassified patterns to adapt the discriminating hyperplanes to the new situation sequentially.

Remark. At the beginning of the iteration process it does not matter with respect to this criterion function which scaling of the hyperplane coefficients is used. But during the iteration process, the scaling of the hyperplane coefficients must not be changed.

Suppose $a > 0$, $v_0 = aw_0$, $\boldsymbol{v} = a\boldsymbol{w}$. Then it follows $V(\boldsymbol{x}) = v_0 + \boldsymbol{v}^T\boldsymbol{x} = a(w_0 + \boldsymbol{w}^T\boldsymbol{x}) = aW(\boldsymbol{x})$, that is, $W(\boldsymbol{x}) = 0$ and $V(\boldsymbol{x}) = 0$ describe the same hyperplane, and $F(V) = a^2F(W)$. Furthermore,

$$\nabla_V F(V) = a\nabla_W F(W)$$

and

$$\begin{aligned} v_0^{(l+1)} &= v_0^{(l)} - \rho_l \sum_{\boldsymbol{x}\in M} \frac{V^l(\boldsymbol{x})}{\|\boldsymbol{x}\|^2} \\ &= a\left(w_0^{(l)} - \rho_l \sum_{\boldsymbol{x}\in M} \frac{W^l(\boldsymbol{x})}{\|\boldsymbol{x}\|^2}\right) = aw_0^{(l+1)} \\ \boldsymbol{v}^{(l+1)} &= \boldsymbol{v}^{(l)} - \rho_l \sum_{\boldsymbol{x}\in M} \frac{V^l(\boldsymbol{x})}{\|\boldsymbol{x}\|^2}\boldsymbol{x} \\ &= a\left(\boldsymbol{w}^{(l)} - \rho_l \sum_{\boldsymbol{x}\in M} \frac{W^l(\boldsymbol{x})}{\|\boldsymbol{x}\|^2}\boldsymbol{x}\right) = a\boldsymbol{w}^{(l+1)} \end{aligned}$$

If there are costs of misclassification, these costs can be included explicitly in the learning procedure. Let $cost(k_i, k_j)$ be the costs of misclassification of class k_i into class k_j, then the criterion function is modified in the following way:

$$F(W) = \frac{1}{2}\left(cost(k_i, k_j) \sum_{\boldsymbol{x}\in m_i} \frac{W(\boldsymbol{x})^2}{\|\boldsymbol{x}\|^2} + cost(k_j, k_i) \sum_{\boldsymbol{x}\in m_j} \frac{W(\boldsymbol{x})^2}{\|\boldsymbol{x}\|^2}\right)$$

4. CLUSTERING CLASSES

It is easy to see that the decision regions determined by pairwise regression and the learning procedure for all classes are convex. This restriction definitely limits the flexibility of the classifier. In particular, every decision region must be singly connected. Because of that, the procedure is suitable only for problems for which the distributions are unimodal.

To also handle problems with non-convex, especially non-singly connected class regions (for examples to solve an XOR-like problem), a clustering procedure can be carried out for the classes before the regression where the classes are handled independently of each other.

To solve the clustering problem a standard minimum-squared-error algorithm is used (Duda and Hart, 1973). From any initial partition of a class k_i of n_i patterns into I clusters k_i^r $(r = 1, \ldots, I)$ with l_r patterns and with mean vectors $\boldsymbol{s}_r$

$$\boldsymbol{s}_r = \frac{1}{l_r} \sum_{\boldsymbol{x} \in k_i^r} \boldsymbol{x}$$

the criterion function

$$J = \sum_{r=1}^{I} \sum_{\boldsymbol{x} \in k_i^r} \|\boldsymbol{x} - \boldsymbol{s}_r\|^2$$

is calculated. If no reasonable initial partition is known a general approach to an initial partition is the following:

$$\boldsymbol{x}_j \in k_i^r, \; (j = 1, \ldots, n_i, r = 1, \ldots, I), \qquad \text{if } r = j \; (mod\ I) + 1$$

Patterns are moved from one cluster to another if such a move will improve the value of the criterion function J. The mean vectors are updated after each pattern move. Like hill-climbing algorithms in general, this approach guarantees local but not global optimization. Different initial partitions and different sequences of presenting the patterns can result in different solutions. In general, it performs sufficiently well.

Another problem in finding an adequate clustering of a class (that means, a clustering which allows a linear discrimination of all classes or subclasses) is the data-based choice of an appropriate distance measure. In some data sets there is no problem, but if the attributes are measured on different scales this problem can be serious. Note also that an appropriate distance measure can be class dependent. In other words, it is well known that the Euclidean distance used for inappropriately scaled attributes does not result in a natural clustering seen by a human in 2-D data sets without effort. In higher dimensions of the attribute space, it is often quite impossible to find a scaling that is appropriate for all classes. The consequences are:

- If too few subclasses are used, the algorithm may not learn a piecewise linear classifier; and
- If too many subclasses are used, this leads to overfitting of the training data.

Nevertheless, often the introduction of the clustering results in two main effects for the classification:

- It is possible to get an insight in the structure of the classes; and
- It is possible to handle non-convex class regions.

The clustering requires fixing the number of clusters of the classes a priori. This is the main shortcoming DIPOL. Until now, there is no conception to fully automate clustering. But the underlying problem (the data-based choice of an appropriate distance measure) is inherent in all cluster algorithms and other classification algorithms based on distance or similarity assessments like prototype or nearest-neighbor methods. It is related to the attribute extraction problem. Since what constitutes a valuable attribute depends strongly on the application being considered, no general guide to the solution of the problem is available.

In the case of clustering, the number of two-class problems increases correspondingly.

5. CLASSIFICATION PROCEDURE

If discriminating hyperplanes have been computed then any pattern $\boldsymbol{x}$ (member of the training set or not) can be classified. The case without clustering is described.

For the pairwise discrimination of n classes $N = n(n-1)/2$ hyperplanes W^l are calculated. The following N-dimensional vector V_k is defined for each class k: if the function W^l discriminates the classes k_i and k_j, then the $l-th$ component $V_{k,l}$ is equal to 1, if $k = k_i$, $V_{k,l}$ is equal to -1, if $k = k_j$, and $V_{k,l}$ is equal to 0 in all other cases:

$$V_{k,l} = \begin{cases} 1, & \text{if } k = k_i \\ -1, & \text{if } k = k_j \\ 0, & \text{otherwise} \end{cases}$$

These vectors contain the coded information about the convex class regions depending on the N functions W^l. On the basis of the set of discriminant functions, the vector function sw is defined for each pattern $\boldsymbol{y} \in \boldsymbol{R}^m$

$$sw: \boldsymbol{R}^m \rightarrow \{1, 0, -1\}^N$$

with the components

$$(sw(\boldsymbol{y}))_l = sign(W^l(\boldsymbol{y})), \ \boldsymbol{y} \in \boldsymbol{R}^m$$

The vector $sw(\boldsymbol{y})$ contains the coded information about the position of the pattern $\boldsymbol{y}$ depending on the N functions W^l. For each class k the function (Z is the set of integers)

$$S_k: \boldsymbol{R}^m \rightarrow Z$$

is defined as the scalar product S_k of the two coded vectors $\boldsymbol{V}_k$ and $sw(\boldsymbol{y})$

$$S_k(\boldsymbol{y}) = \sum_{l=1}^{N} V_{k,l}(sw(\boldsymbol{y}))_l.$$

A pattern $\boldsymbol{y}$ is uniquely classified by the discriminating hyperplanes into the class k if

$$S_k(\boldsymbol{y}) = n - 1,$$

that is, with respect to the $n - 1$ hyperplanes, which discriminate the class k from the other $n - 1$ classes, the pattern $\boldsymbol{y}$ is placed in the halfspace belonging to class k (the coded vectors $\boldsymbol{V}_k$ and $sw(\boldsymbol{y})$ have the same value ($+1$ or -1) for all components 1 with $V_{k,l} \neq 0$). For all other classes $j, j \neq k, S_j(\boldsymbol{y}) < n - 1$ holds, because at least with respect to the hyperplane, which discriminates class j from class k the pattern $\boldsymbol{y}$ is placed in the halfspace of class k (the coded vectors $\boldsymbol{V}_k$ and $sw(\boldsymbol{y})$ do not have the same value ($+1$ or -1) for all components 1 with $V_{k,l} \neq 0$). A pattern $\boldsymbol{y}$ is not uniquely classified if

$$\max_j S_j(\boldsymbol{y}) < n - 1.$$

In this case, the pattern is classified on the basis of the minimum of the distance to the class regions. If clustering is used, the number of two-class problems increases and the classification procedure is modified because no discriminating hyperplanes are determined for two subclasses of the same class.

6. NEURAL NET INTERPRETATION

The set of concept descriptions that can be modeled by neural nets is universal in the sense that any concept can be implemented to any degree of accuracy by a sufficiently large network (Hornik et al., 1989). However, there is no guarantee that an optimal architecture and optimal parameters are found. The training of a net can take an enormous amount of time because the algorithms work in a highly experimental fashion.

If DIPOL is regarded as a neural net, the combination of the clustering with the regression and the learning procedure solves the problems of choosing an appropriate number of nodes in the first hidden layer and of determining appropriate initial weights (that is, the number and the initial positions of the discriminating hyperplanes) in a reasonable manner. Because of that the training time is very short in comparison to other neural net algorithms. Another advantage of DIPOL, it preserves the interpretability of the constructed concept description—a subdivison of the attribute space by hyperplanes. The structure of DIPOL is shown in Figure 6.3. Note that the architecture has some similarity with the well-known MADALINE-system (Widrow, 1962). But in MADALINE there is only one neuron in the second layer performing a majority decision by simply adding up the decisions (1 or −1) of the first layer neurons. Instead of this DIPOL is able to compute any Boolean class description, that is, subclasses by conjunctions and classes by disjunctive

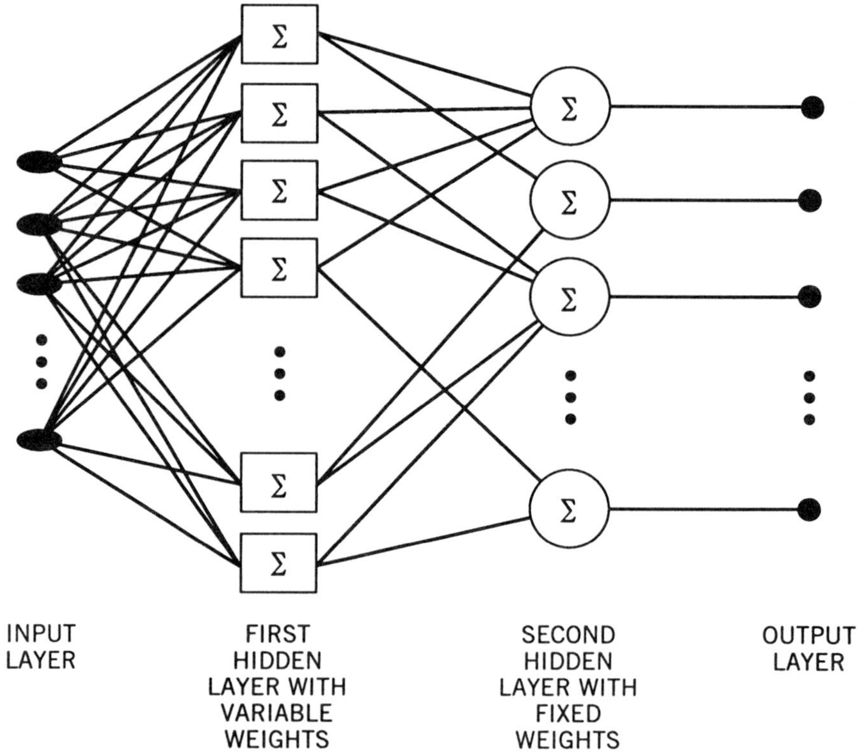

Figure 6.3. *Schematic representation of DIPOL.*

normal forms. The encircled nodes compute the scalar products as described in Section 5. (The final maximum decision is not depicted in Figure 6.3).

7. EXAMPLES

The first version of the classification learning procedure DIPOL developed in 1992 was part of the Esprit project *StatLog*. The aim of this project was to provide a review of different approaches to classification learning, to compare their performance on a wide range of data sets and to draw conclusions on their applicability to real-world problems, for example, technical and medical diagnosis, image recognition and credit assignment. DIPOL92 came off very well in comparison to the other 21 algorithms. Taking the top five places across all 23 data sets it occurs more frequently than all other algorithms. The results are reported in Michie et al. (1994).

Here, three artificial 2-D data sets are investigated to demonstrate the advantages and disadvantages of the algorithm.

Example 1

The first data set shown in Figure 6.4 presents a typical case where the regression does not find the separating hyperplane in the separable case. The

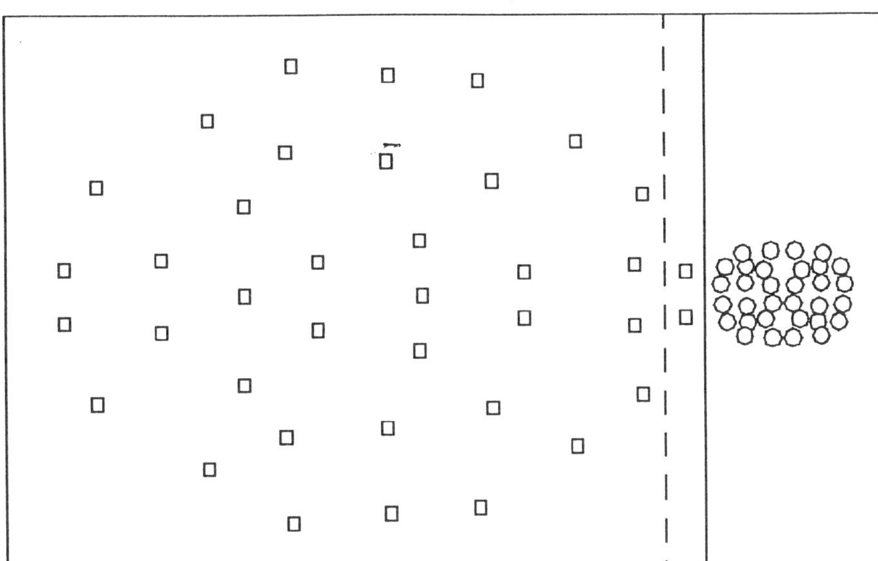

Figure 6.4. *2-D example for demonstrating the performance of DIPOL's learning procedure in the separable case—the dotted line represents the discriminating hyperplane after regression and the solid line represents the discriminating hyperplane after the adjustment.*

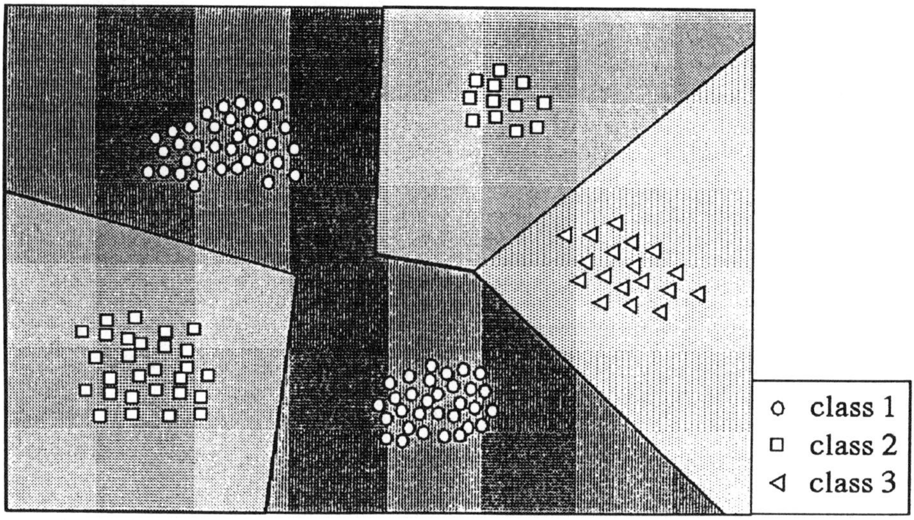

Figure 6.5. *Two classes of this three-class problem consist of two subclasses. The figure shows the class regions constructed by DIPOL, using eight hyperplanes for the pairwise discrimination of the subclasses.*

learning procedure adjusts the coefficients of the hyperplane in one iteration step.

Example 2

The distributions of class 1 and class 2 are not unimodal. Figure 6.5 shows the class regions constructed by DIPOL. The algorithm uses eight hyperplanes for the pairwise discrimination of classes and subclasses that do not belong to the same class. Only the pieces of the six hyperplanes that are relevant for the boundaries of the class regions are drawn.

Example 3

The data set ART has 3,000 patterns and was divided into a training set with 2,000 patterns and into a test set with 1,000 patterns. There are two variants: ART-1 and ART-2. Attribute 1 of ART-2 corresponds to attribute 1 of ART-1 multiplied by 5. The patterns were all generated randomly and labeled according to the predefined subdivision, shown in Figure 6.6 (6.6a shows ART-1 in the original form; in Figure 6.6b, both classes of ART-1 are clustered into three subclasses (clusters A, B, and C belong to class 0, clusters D, E, and F belong to class 1), and in Figure 6.6c, the same configuration is drawn for ART-2).

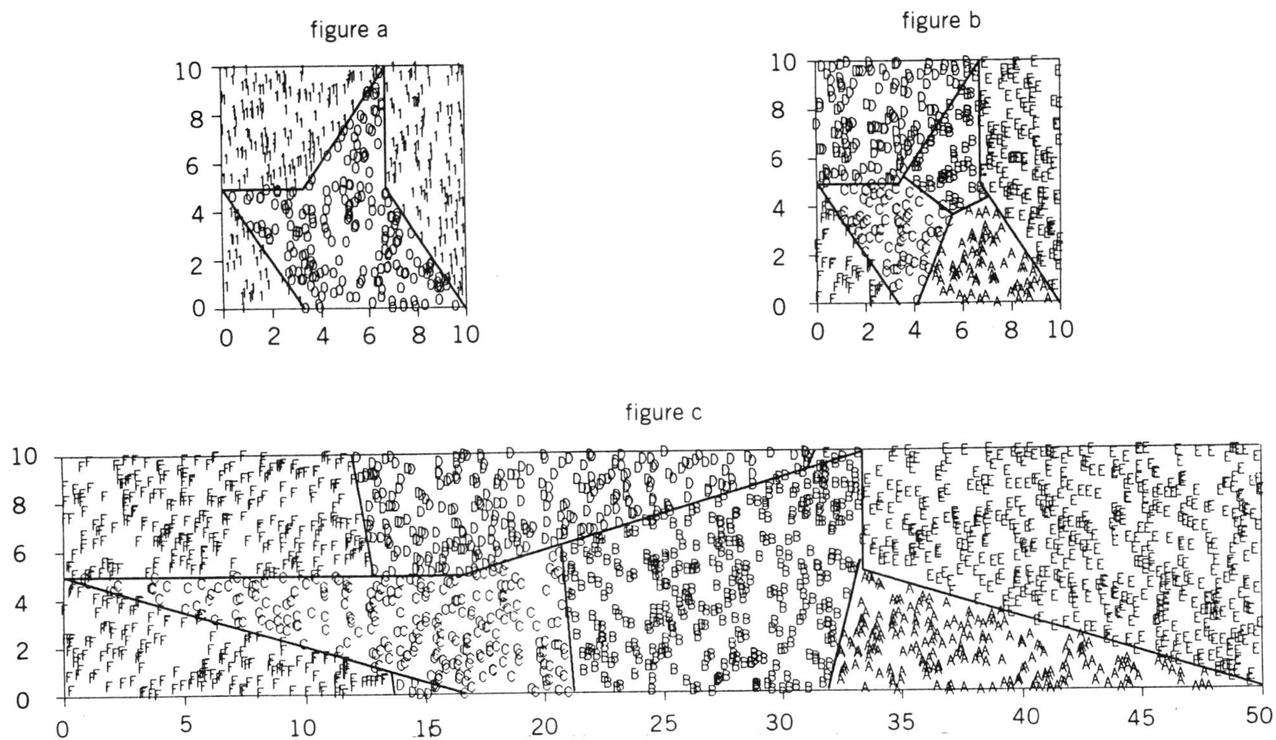

Figure 6.6. *Training set of ART.*

TABLE 6.1. Comparison of Error Rates of Reclassification of the Training Data and the Classification of the Test Data Using DIPOL with Several Clusterings

ART-1	Train	Test	ART2	Train	Test
1–1	29.1 %	31.0 %	*1–1*	29.0 %	31.9 %
2–2	13.4 %	13.4 %	*2–2*	12.8 %	12.9 %
3–3	1.3 %	2.5 %	*3–3*	9.1 %	10.9 %
4–4	4.3 %	5.0 %	*4–4*	9.2 %	11.6 %
5–5	1.4 %	2.7 %	*5–5*	9.8 %	11.5 %
6–6	2.4 %	4.8 %	*6–6*	9.0 %	10.2 %
7–7	2.8 %	4.7 %	*7–7*	7.4 %	9.1 %
8–8	2.5 %	4.1 %	*8–8*	3.8 %	6.2 %
9–9	3.0 %	4.5 %	*9–9*	3.4 %	4.8 %
10–10	3.8 %	5.0 %	*10–10*	3.4 %	5.0 %

This example brings out the main weakness of DIPOL (and of all other classification algorithms based on distance or similarity assessments)—scaling the attributes can have a great effect on the results. Table 6.1 shows the results of the classification by using several clusterings for both the data sets ART-1 and ART-2. Clustering both classes of ART-1 into three subclasses gives low error rates for both the training data and the test data, whereas the same clustering of ART-2 leads to higher error rates. It is clear that this behavior is caused by the scaling of attribute 1—the clustering (3–3) leads to an adequate subdivision of ART-1, whereas the subclasses of ART-2 do not allow a linear discrimination.

Nevertheless, DIPOL finds in both cases solutions with an acceptable accuracy whereas decision tree algorithms and statistical algorithms cannot (although for other reasons—decision trees can split the attribute space in axis-parallel hyperrectangles only, and statistical algorithms are suitable for unimodal distributions of the classes only).

8. CONCLUSIONS

The chapter describes the construction of a piecewise linear classifier which uses:

- the well-known statistical regression technique to obtain initial positions of hyperplanes for pairwise discrimination of the classes;
- A learning procedure for decreasing the classification error (if possible);
- A classification procedure on the basis of Boolean vectors; and
- An optional clustering of classes for the treatment of non-convex class regions.

The classification results are very good, for a wide variety of problems.

ACKNOWLEDGMENTS

We thank our colleague M. Bleich for constructing the data sets and preparing the figures. The paper is based upon work within the project WISCON supported by the German Ministry of Research and Technology.

REFERENCES

Duda, R. O., and Hart, P. E. (1973). *Pattern Classification and Scene Analysis.* Wiley, New York.

Hornik, K., Stinchcombe, M., and White, H. (1989). Multi-layer feedforward networks are universal approximators. *Neural Networks* **2**, 359–366.

Meyer-Brötz, G., and Schürmann, J. (1970). *Methoden der automatischen Zeichenerkennung.* Akademie-Verlag, Berlin.

Michie, D., Spiegelhalter, D. J., and Taylor, C. C., eds. (1994). *Machine Learning, Neural and Statistical Classification.* Ellis Horwood, Chichester, UK.

Nilsson, N. J. (1965). *Learning Machines. Foundations of Trainable Pattern-classifying Systems.* McGraw-Hill, New York.

Unger, S., and Wysotzki, F. (1981). *Lernfähige Klassifizierungssysteme.* Akademie-Verlag, Berlin.

Widrow, B. (1962). Generalization and information storage in networks of ADALINE "neurons." In *Self-Organizing Systems* (M. C. Yovits, G. T. Jacoby, and G. D. Goldstein, eds.), pp. 435–461. Spartan Books, Washington, D.C.

CHAPTER 7

Combining Classification Procedures

R. J. HENERY

1. GENERALIZATION BY COMBINATIONS

The main aim of this chapter is to describe methods for generalization that involve *combinations of classification algorithms*. The three examples that will be given can be crudely summarized as dealing with:

1. *Rescaling*. Rescaling is a simple case of supervised algorithms being applied in series. The output from one classification algorithm is used as input to a second classification algorithm.
2. *Stacked Generalization*. Several classification algorithms are applied in parallel to the same data set, and their outputs are combined to form the inputs for a second level of generalization.
3. *Correspondence Analysis*. Unsupervised learning methods are applied to the outputs of several classification procedures to look for clusters or similarities within algorithms and within data samples.

1.1. Rescaling

A frequent problem in classification is that some classes are over-predicted, leading to bias. Such bias may be reduced in those cases where the output of the classifier is a class-conditional probability rather than a class allocation. Statistical classifiers, and here we include multi-layer perceptrons and some decision trees such as CART (Breiman et al., 1984), produce probability

Machine Learning and Statistics, The Interface, Edited by G. Nakhaeizadeh and C. C. Taylor.
ISBN 0-471-14890-3

estimates, and these may be post-processed to (i) inspect for bias, and (ii) rescale to remove bias. The general rescaling procedure can be described as applying logistic regression methods to the output of the classifier in an attempt to reduce bias. In like spirit, Ripley (1993) used linear discriminants as post-processor on the MARS (Friedman, 1991) output for the glass data set.

Reduction of bias through rescaling is demonstrated on two discriminants (linear and quadratic discrimination) when applied to the DNA dataset (Murphy and Aha, 1992).

1.2. Stacked Generalization

In comparative studies, several classification procedures are compared by evaluating their performance on a given data set, and the winner is declared to be that with the lowest error rate on the test set. Wolpert (1992) maintains that it is often more desirable to *combine* procedures rather than *compare* them, and he proposes a general scheme where the outputs of the various procedures form the inputs of the next level of generalization. In statistics, this idea has been taken up by LeBlanc and Tibshirani (1993) in the context of classification and regression, and by Breiman (1992) for regression. Stacked generalization involves taking predictions from several procedures and using these predictions as the basis for the next stage of classification. In this chapter, a prediction is taken to be a predicted class conditional probability, rather than the predicted class. The problems involved will be illustrated by applying nine different classifiers to a practical data set taken from industry.

1.3. Correspondence Analysis

A potential problem with many discrimination procedures is that the classes may not be homogeneous. For example, in the recognition of handwritten digits, the digit zero may be written either as 0 or as Ø. To improve discriminant performance, it is then often necessary to identify potential subclasses and, if necessary, to treat the subclasses as distinct for the purposes of recognition. This raises the issue of doing some kind of unsupervised learning on the data set before the supervised stage. This is the procedure used in DIPOL92 (Schulmeister and Wysotzki, 1993), for example. Here we reverse the situation. We consider unsupervised learning applied, *a posteriori*, to the output from several supervised algorithms applied to the same data set. The purpose here is to check the output to see whether class probability estimates are clustered within classes, so pointing to the existence of subgroups.

2. CLASS-CONDITIONAL PROBABILITIES

A second unifying theme to our discussion will be that, at the level of making generalizations, the raw material will be the estimated class probabilities rather than the predicted classifications or even the given features.

Fukunaga and Ando (1977) observe that the *true* class-conditional probabilities are the best features for classification in the Bayes sense, but they do not present a practical algorithm based on this observation. This chapter is concerned with using the *estimated* class-conditional probabilities as near-optimal features for combining, comparing, or improving the performance of classifiers. For example, the estimated class-conditional probabilities are used to build log-linear models for the *true* class-conditional probabilities, that is, logistic regression techniques are used to reduce the bias of statistical procedures.

Some of the newer classification procedures produce class-conditional probabilities directly, and often have the merit of being nonparametric or distribution-free. As examples we may cite the decision tree CART (Breiman et al., 1984) and the non-linear regression procedure ACE (Breiman and Ihaka, 1984). Also, the outputs from some neural networks such as the multi-layer perceptron can be directly interpreted as class-conditional probabilities, and indeed the multi-layer perceptron may be regarded as an extension to logistic regression (Ripley, 1994). Kernel density methods, such as the widely used ALLOC80 method of Hermans and colleagues (1982) also produce direct estimates of class probabilities. As a general reference to these and many other statistical procedures, the reader should consult McLachlan (1992).

We are less interested in classification methods that do not produce class probabilities. For example neural nets with linear output functions (Hertz et al., 1991) produce "probabilities" that may lie outside the interval $(0, 1)$. Also, the methods proposed here do not work so well, and the rescaling procedure may even fail altogether, if the output probabilities are either 0 or 1, or more generally a small number of discrete values. In particular, the nearest neighbor method (Fix and Hodges, 1951) and some of the older decision trees (Weiss and Kulikowski, 1991) produce classifications as end-product, and these would translate into posterior probabilities that are either 0 or 1. Even the statistically based decision tree CART (Breiman et al., 1984) tends to concentrate probability estimates on a small set of discrete probabilities, these being the class estimates at the nodes. In brief, we assume that the estimated class probabilities are continuous over the open interval $(0, 1)$.

2.1. Role of Cross-Validation

In general we need unbiased estimates of probability to make reliable generalizations, and the simplest way of achieving this is to train the classification rule on one data set (the training set) and test the rule on a second independent data set (the test set). Classification performance, and this includes the sampling properties of class-conditional probabilities, is only reliable for the test set. If we are prepared to spend much more time in the training stages, we get more efficient (and more symmetric) use of the training data by using the leave-one-out method. Less demanding computationally, but possibly even better than leave-one-out according to Breiman (1992), is to use ten-fold cross-validation.

In order to estimate the class probabilities for a given observation, the important thing is that this particular observation is not part of the training set.

2.2. Classification—Notation

Classification is the allocation of an individual to one of q known classes, $C_1, \ldots, C_q$ on the basis of measurements on a d-dimensional vector of attributes $\mathbf{X} = (X_1, \ldots, X_d)'$. For the purposes of this chapter, we will speak of statistical discrimination as providing estimates $\hat{P}(C_i|\mathbf{X})$ of the conditional probability that the individual belongs to a class C_i given observation $\mathbf{X}$, for each class $i = 1, \ldots, q$. We will say that the discrimination procedure is biased if $\hat{P}(C_i|\mathbf{X})$ is a biased estimate of the true probability $P(C_i|\mathbf{X})$, and we are particularly concerned with cases in which the amount of bias is a function of the true probability $P(C_i|\mathbf{X})$ only. Most classical classification procedures are capable of producing conditional probabilities, although they generally operate by calculating discriminant scores and explicit calculation of probabilities is avoided. For example, Fisher's linear discriminants (Fisher, 1938) leads to a discriminant score for each class that is a linear combination of the attributes $X_1, \ldots, X_d$: if conditional probabilities are required, they are calculated from the logistic function. Quadratic discrimination (Smith, 1947) uses a quadratic functional of the attributes, namely, the Mahalanobis distance measure from each population: and if conditional probabilities are required, they too may be calculated via the logistic function. Generalized discriminants (Randles et al., 1978) and regularized discriminant analysis (Friedman, 1989) provide a spectrum of models somewhere between linear and quadratic discrimination, but both estimate probabilities as with quadratic discriminants, the only difference being in how the class covariance matrices are calculated.

Logistic regression (Cox, 1966) is able to model the class probabilities directly. It has the advantage that the necessary parameters (and consequently the class probabilities) are estimated by maximum likelihood. Apart from the method of estimating the model parameters, logistic regression and linear discrimination are structurally identical: in any given case, the two differ only in the choice of parameters.

2.3. Class-Conditional Probabilities

Given a set of N observations, of known origin, with n_1 examples from class C_1, n_2 examples from class $C_2, \ldots,$ and n_q examples from class C_q, a statistical procedure allows us to estimate the class conditional probabilities for any observation on the d-dimensional vector $\mathbf{X}$. We suppose that the original d-dimensional variable $\mathbf{X}$ has been manipulated, in some complex and highly non-linear fashion, into estimated class probabilities by the statistical discrimination procedure. Let

$$W_i = \hat{P}(C_i|\mathbf{X})$$

denote the estimated probability that the true class is C_i given the value of $\mathbf{X}$. We can regard the estimated probabilities $\mathbf{W} = (W_1, \ldots, W_q)$ as estimates of the Bayes-optimal features for this problem (Fukunaga and Ando, 1977).

3. RESCALING

Usually, however, the discrimination procedure is biased, that is, $W_i = \hat{P}(C_i|\mathbf{X})$ is a biased estimate of the true probability $P(C_i|\mathbf{X})$. We consider a particular model for this bias, namely, that it is, in fact, a function of the true probability $P(C_i|\mathbf{X})$ only. In this simplest of cases, the bias is proportional to the true probability, and we produce a new estimate

$$\hat{P}(C_i|\mathbf{W}) = \gamma_i W_i \tag{1}$$

where γ_i accounts for the systematic bias. The new notation $\hat{P}(C_i|\mathbf{W})$ for the estimated class probabilities denotes the fact that we are now using the previous estimated probabilities W_i as (what we hope are the Bayes-optimal) features.

More generally, the aim would be to use the estimated probabilities W_i derived from the statistical discriminant as input to a second statistical discriminant procedure the purpose of which is to produce an unbiased estimate $Q(C_i|\mathbf{W}) = \hat{P}(C_i|\mathbf{W})$ of the conditional probability for class C_i, given the observed estimated probabilities $\mathbf{W}$, where the notation $Q(C_i|\mathbf{W})$ suggests that the new probabilities are functions of the old estimates W only.

In order that we may investigate the bias and error rates of the statistical procedure, we will also be looking at the set of estimated class probabilities for a test data set whose true class membership is known. Consider first the confusion matrix M, which is a square matrix of size q, where q is the number of classes. The rows of M refer to the actual class; the columns give the predicted class. The entry M_{rs} counts the total number of true class C_r examples that are allocated as belonging to class C_s by the discriminant procedure, $1 \leqslant r \leqslant q, 1 \leqslant s \leqslant q$. The row marginal totals of the confusion matrix are just the sample sizes of the classes:

$$\sum_{s=1}^{q} M_{rs} = n_r.$$

Let $\mathbf{X}_i$ denote the vector of observations for the ith example in the test data, and let

$$W_{ij} = \hat{P}(C_j|\mathbf{X} = \mathbf{X}_i)$$

denote the estimated probability that the ith example belongs to class C_j given the value of $\mathbf{X} = \mathbf{X}_i$. The elements of W_{ij} form a matrix whose rows sum to unity. It would seem reasonable also that the columns should sum to the class frequencies:

$$\sum_{i=1}^{n} W_{ij} = n_j; \qquad j = 1, \ldots, q. \tag{2}$$

as equation (2) is automatically satisfied for maximum likelihood methods (*cf.* equation (7) below). Note that for procedures like nearest neighbor, where W_{ij} is either 0 or 1,

$$\sum_{i=1}^{n} W_{ij} = \sum_{r=1}^{q} M_{rj}; \qquad j = 1, \ldots, q, \tag{3}$$

as both sums count the total number of observations allocated to class j. So equation (2) is the entirely reasonable requirement that the predicted number in class C_j should equal the observed number. In general then, we expect that an unbiased procedure will predict the correct proportions for the various classes. Problems of bias may be picked up at this stage, by checking the marginal totals of the confusion matrix for test data.

Not all methods guarantee that condition (2) holds, and the estimated probabilities W_{ij} are biased in this respect. As a consequence, the allocation of classes, which is done on the basis of highest posterior probability or lowest expected cost, will also be biased. It is possible to reduce bias *a posteriori* by rescaling the probabilities $\mathbf{W}$ so that they satisfy equation (2). In the simplest case, the new estimated probabilities $Q(C_i|\mathbf{W})$ are proportional to the old estimated probabilities $W_i = \hat{P}(C_i|\mathbf{X})$. In the more general model, the new estimated probabilities $Q(C_i|\mathbf{W})$ will be logistic functions of the estimated probabilities $\mathbf{W}$, and by estimating the parameters of the model via maximum likelihood these new functions will be unbiased, in the sense that certain consistency conditions like equation (2) will hold.

3.1. Estimating the Scaling Factors

Thus we might think in terms of undoing the bias of the estimated class-conditional probabilities by scaling all probabilities for a given class by a constant factor so that the left side of equation (2) is correctly given. As a first try, we could define the scaling factors γ_j by

$$\gamma_j = \frac{n_j}{\Sigma_i W_{ij}} \tag{4}$$

and rescale the estimated probabilities W_{ij} to give the pseudo-probabilities Q_{ij}:

$$Q_{ij} = \gamma_j W_{ij} = \gamma_j \hat{P}(C_j|\mathbf{X} = \mathbf{X}_i) \tag{5}$$

because then the Q_{ij} would satisfy the relation

$$\sum_i Q_{ij} = n_j: \qquad j = 1, \ldots, q.$$

Such a simple solution would not work however, as the Q_{ij} quantities would not, in general, be probabilities summing to unity over the rows. It is necessary to re-normalize the Q_{ij}, so that the unbiased class conditional probabilities are to be modeled by

$$p_{ij} = \frac{Q_{ij}}{\Sigma_j Q_{ij}}. \tag{6}$$

With the original W_{ij} replaced by p_{ij}, we find that the new W_{ij} no longer satisfy equation (2), though it may be shown that the discrepancy is reduced (Agresti, 1990). Rescaling through equations (4) and (5) and then renormalized by equation (6) is the basis of an iterative method for finding the parameters γ_j, the method being known as iterated proportional scaling or proportional fitting (Bishop et al., 1975).

3.2. Rescaling as Logistic Regression

Consider the use of logistic regression to model the logarithms of the probability density functions for the classes as linear functions of the features $\mathbf{W} = (W_1, \ldots, W_q)'$, which are the outputs of the statistical discrimination procedure. Generally, the W_i variable will be the estimated probability for class C_i, or equivalently the log-odds for class C_i. If we choose $W_i = \hat{P}(C_i|\mathbf{X})$, the W_i will sum to unity, so the effective dimensionality of $\mathbf{W}$ is $q - 1$. Thus, taking class C_q as reference class,

$$\log \frac{P(C_i|\mathbf{W})}{P(C_q|\mathbf{W})} = \alpha_i + \beta_i' \mathbf{W},$$

where α_i and the q-dimensional vector β_i are the parameters of the model for class C_i, $i = 1, \ldots, q - 1$. To simplify later formulas, we define $\alpha_q = 0$ and $\beta_q = 0$. We obtain a form of logistic regression if we take the logarithm of equation 1 and make the substitutions $\alpha_i \to \log \gamma_i$ and $W_i \to \log W_i$, and further note that the vector of coefficients β_i is zero except for the i'th component, which is unity.

In practice, the parameters are estimated by maximum *conditional* likelihood. The model implies that, given attribute values $\mathbf{W}$, the conditional class probability for class C_i takes the form:

$$P(C_i|\mathbf{W}) = \frac{\exp(\alpha_i + \beta_i' \mathbf{W})}{\Sigma_{j=1,\ldots,q} \exp(\alpha_j + \beta_j' \mathbf{W})}.$$

Given independent samples from the q classes, the conditional likelihood for the parameters α and β is defined to be

$$L(\alpha, \beta) = \prod_{\{C_1 \text{ sample}\}} P(C_1|\mathbf{W}) \ldots \prod_{\{C_q \text{ sample}\}} P(C_q|\mathbf{W}).$$

The maximum likelihood estimates of α and β are the values that maximize this conditional likelihood. The equation that results from differentiating with respect to α_i is

$$n_i = \sum_{\{\text{all samples}\}} P(C_i|\mathbf{W}) \tag{7}$$

and the equation that results from differentiating with respect to the j'th component β_{ij} of β_i is

$$\sum_{\{C_i \text{ sample}\}} W_j = \sum_{\{\text{all samples}\}} W_j P(C_i|\mathbf{W}) \tag{8}$$

for $i < q$. These equations are interpretable as equating certain sample and model expected values. Equation (7), for example, equates the number of observations in class C_i to the number expected according to the model probabilities $P(C_i|\mathbf{W})$: equations (8) provide additional constraints by which the rescaled probabilities are matched up to empirical data. Usually the equations must be solved by iterative methods, as proposed by Cox (1966) and Day and Kerridge (1967). These methods belong to the general class of iterative reweighted least squares (IRLS) methods (McCullagh and Nelder, 1989). Logistic regression is asymptotically unbiased and efficient when used in appropriate circumstances, that is, when the true log-odds ratio $\log P(C_i|\mathbf{X})/P(C_j|\mathbf{X})$ is linear in X for every pair i,j. For finite data sets, however, logistic regression may be biased, especially if the number of parameters to be estimated is large.

3.3. Estimation of Parameters α

The most common sources of bias are (i) model bias, for example using parametric procedures when the underlying assumptions are false, and (ii) overfitting, for example using too many parameters on small data sets. The two proposed methods for estimating α and β are designed to cater for these two cases.

3.3.1. One-stage procedure for model bias. If the distributed assumptions of the statistical discriminant are false, using quadratic discriminant analysis on non-normal data for example, the incorrectness of the model should be apparent from the training data itself. In this case, which we refer to as *model bias*, we assume that the estimated class-conditional probabilities are systematically biased due to the systematic departure of the assumed distributions from the true distributions. It is possible then to use the training data for selecting the parameters α and β. In section 4.5, we will give an example involving linear discrimination on the DNA dataset. The model bias there may be due to the false assumption of equal variances over the classes.

3.3.2. Two-stage procedure for overfitting. If overfitting is the problem, the bias is not evident on the training set, and the rescaling must be done on an independent set of data, as is done in the example in section 4.3. When the fitted model has a very large number of parameters compared with the number of data, therefore, the recommended procedure for reducing bias in statistical discrimination procedures is to train the algorithm using a two-stage procedure, involving two independent datasets, which we will call the training and tuning datasets.

1. Use the training data set to estimate all the usual parameters for that procedure.
2. Estimate the class-conditional probabilities $\hat{P}(C_i|\mathbf{X})$ for all the examples in the tuning data set, and select the coefficients α and β so that the rescaled probabilities $Q(C_i|\mathbf{W})$ are unbiased in the tuning set.

It need hardly be said that the size of these data sets, that is, the relative proportions in the training and tuning data sets, is of the greatest importance in determining the effectiveness of the overall procedure.

No general advice can be given, as the optimal proportions will depend on the algorithm and on the data set. We would suggest something like a 90%/10% split for the rescaling model of section 4. The situation is analogous to the problem of determining the optimal proportion of test data to use in selecting parameters for the classification procedure itself, and Devroye (1987) gives some general guidance on asymptotic sizes of training and test data sets. Roughly speaking, the more parameters that must be chosen by reference to results from the tuning data set, the higher proportion required in the tuning data.

4. RESCALING EXAMPLE

4.1. DNA Data set

The DNA data set is available by ftp from ftp.stams.strath.ac.uk in directory pub/Statlog. In its original form, available in the University of California, Irvine repository (Murphy and Aha, 1992) in directory

pub/machine-learning-databases/undocumented/sigillito

the actual sequence of DNA is given. In the *StatLog* project (Michie et al., 1994), the data were transformed into a classification problem. The sequence was divided into segments of length 60: each element in a sequence is a categorical variable with three levels, which was coded into 3 binary variables, giving 180 binary variables in total. The problem is to predict the class of the DNA sequence (one of three possible classes: *ei*, *ie*, or *neither*) using these 180 binary valued attributes, $X_1, \ldots, X_{180}$. The so-called class value is in fact an event that occurs at the middle of the DNA sequence, so that it is very likely that attributes near the middle of the sequence (say attributes $X_{61}, \ldots, X_{120}$) are more relevant than the attributes at the ends of the sequence.

A training set of 2,000 examples was used to construct the discrimination rule; and the resulting rules were then tested on an independent test set of 1,186 examples. When a separate tuning data set was required for the rescaling parameters, a sample of size 200 was taken from the training set, leaving 1,800 examples for the initial training of the discrimination rule. Of the greatest importance for our purposes is the fact that the training, tuning and test sets were chosen by random sampling from one large data set. In particular, the class proportions in the tuning set are unbiased estimates of the prior class proportions.

To provide a baseline for the various algorithms quoted below, note that logistic regression, when trained on the 2,000 example data set, achieved an error rate of 0.061 on the test set. Linear discriminants had a lower error rate of 0.059. This slightly surprising result is probably due to the very large number of irrelevant attributes, which causes the logistic procedure to overtrain (evidenced by the very low error-rate of 0.008 for the training data).

4.2. Quadratic Discriminants

As initial discriminant procedure, we take quadratic discriminants (QDA). This procedure uses a very large number of parameters—many thousands in the DNA example that follows. It is also very relevant to note that, in the DNA data set, there is the added complication that two of the sample covariance matrices are singular, so some form of smoothing is essential if quadratic discrimination is to work at all. We use a variant of quadratic discriminant, namely, the regularized discriminant of Friedman (1989). In an attempt to reduce the bias inherent in quadratic discriminants, both Randles and colleagues (1978) and Friedman (1989) suggested discriminants that were, in some sense, a linear interpolation between pure linear and pure quadratic discriminants. Generalized discriminant analysis involves a linear interpolation of the inverse covariance matrices (Randles et al., 1978), while regularized discriminant analysis (Friedman, 1989) is based on a linear interpolation of the covariance matrices via two parameters λ and δ (see Friedman, 1989, for an explanation of these parameters). Regularized discrimination covers a wide spectrum of possible models, including pure linear and pure quadratic discrimination. The details are not important for our purposes: all we need is that class-conditional probabilities can be estimated from the model. However, it is very relevant to note that the motivation for using regularization is to reduce bias in quadratic discrimination by using smoothed estimators of the covariance matrix.

We used smoothing parameters of $\lambda = 0.01$ and $\delta = 0.0$, which is to say, about the lowest level of smoothing required to get quadratic discrimination to work. Usually, a choice of smoothing parameters λ and δ is made by searching over a grid of values to minimize the bias, but this is not our main concern. For purposes of illustration we deliberately pick the value of λ so that the resulting procedure is biased, as we wish to demonstrate that the bias may be reduced by subsequent rescaling of the probabilities.

4.3. Quadratic Discriminants on DNA

The main concerns with the DNA data set are that the distributions are non-normal (in fact, multivariate binary) and that the number of parameters required by quadratic discrimination is large in relation to the number of examples (each of the three covariance matrices requires $180 \times 181/2$ parameters). Let us first describe what happens if the quadratic procedure is applied without modification to the full training set of 2,000 examples. Because of the large number of parameters used, it is no surprise that the procedure overfits, with an estimated error rate of 0.0025 on the training set of 2,000 examples. Table 7.1 gives the confusion matrices for both training and test sets. The training set shows little evidence of asymmetry because of the very low error rate. On the other hand, the error rate is 0.2521 on the independent test set of 1,186 examples, and a glance at the test confusion matrix in Table 7.1 shows a very different picture, with some errors much more frequent than others. When the predicted class is *ei*, for example, the prediction is almost certainly correct, whereas when the predicted class is *neither*, the prediction has a relatively low probability ($602/898 = 0.67$) of being correct. This is the phenomenon that we describe as an imbalance of error rates. An alternative way of viewing this phenomenon is in terms of the marginal totals in the

TABLE 7.1. Confusion Matrix Showing Actual and Predicted Classes for the DNA Training Set of 2,000 Examples and for the Test Set of 1,186 examples

	Training			Test		
Predicted	*ei*	*ie*	*neither*	*ei*	*ie*	*neither*
Actual						
ei	462	0	2	135	2	166
ie	0	482	3	0	150	130
neither	0	0	1051	0	1	602

Predictions via quadratic (regularized) discrimination.

confusion matrix, There are 603 examples in class "*neither*," but 989 examples are predicted to be in this class. It is by adjusting the prediction so that the proposed marginal totals are correct that we may hope to redress this imbalance (note that maximum likelihood methods usually end up with the proper marginal totals anyway).

Clearly, QDA is very biased. The question of interest is this however: can we take the estimated class probabilities from this (biased) model, and remove the bias through the scaling factors? Because the training data are overfitted, it will be necessary to use the two-stage procedure, with the complete training data split up into a new training set of 1,800 examples and a tuning set of 200 examples (chosen at random from the original training set).

4.4. Rescaled QDA on DNA

We ran the QDA procedure on a training set of 1,800 examples, and tested it on an independent tuning set of 200 examples. The tuning set confusion matrix is shown in the left side of Table 7.2. Now, to see the benefit of rescaling, we took the QDA probability estimates for the tuning data and rescaled them so that the marginal total probabilities match up with the marginal class totals in the tuning data. When tested on the same tuning data set, the rescaled QDA probabilities result in the confusion matrix on the right side of Table 7.2.

After rescaling, the confusion matrix is more symmetric, but more to the point, the error rate is much reduced. The error rates on the tuning data set are 0.305 (QDA), 0.055 (QDA rescaled). Since we have used the tuning data to adjust our model, the estimated error rate of 0.055 is biased downward, so we checked the error rate on an independent test set of size 1,186, which gave error rates of 0.2791 (QDA) and 0.0666 (QDA rescaled). This shows that rescaling has been able to reduce the bias of QDA very considerably, and for no great additional expense, either in terms of additional data required for tuning or computational load.

TABLE 7.2. Confusion Matrices for QDA on the Tuning Data Set

	Raw QDA			Rescaled QDA		
Class	*ei*	*ie*	*neither*	*ei*	*ie*	*neither*
ei	13	1	34	42	1	5
ie	0	16	26	0	39	3
neither	0	0	110	0	2	108

Left side of figure is for QDA; right side is for the QDA with bias removed by rescaling probabilities.

TABLE 7.3. Error Rates for Linear Discriminants (Raw LDA) and for Linear Discriminants with Rescaled Probabilities (Rescaled LDA)

	Raw LDA		Rescaled LDA	
Data Set	Train	Test	Train	Test
DNA 180	0.0340	0.0590	0.0260	0.0438

4.5. Linear Discriminants on DNA

The number of parameters needed for linear discriminants (LDA) is much less than for the quadratic rule. Using LDA on the full DNA data set, there are only 2 × 181 coefficients to be determined from a fairly large sample (we used a training set of size 2,000). We can check the bias, and make any necessary rescaling adjustments, by reference to the probabilities calculated for the training set. Thus we can estimate all necessary parameters for the linear discriminants, plus the rescaling coefficients, using only the training data. The resulting rule can be checked on an independent test set of size 1,186. The error rates estimated for the training and test sets are given in Table 7.3, where we also give the results from ordinary linear discriminants for comparison.

The differences in error rates are not large, as the number of errors after rescaling (52 out of 1,186) is not significantly lower than the number for raw LDA (70 out of 1,186) ($P = 0.057$ one-tail). At least they tend to confirm that bias (and so error rate) is diminished when the probabilities are rescaled.

What is also of interest is that the error rate of LDA followed by rescaling (52 errors out of 1,186) is lower than for logistic regression applied to the original data (72 errors out of 1,186).

4.6. Comparison with Other Algorithms

In the *StatLog* project (Michie et al., 1994), around 20 classification procedures were applied to the DNA180 data set. For comparison, we quote now the lowest error rates on the test set (size 1,186) among the various types of algorithm. The best decision tree result was for Quinlan's (1992) C4.5 algorithm (error rate 0.076). Among the neural net procedures, the best was radial basis functions or RBF (see Hertz et al., 1991) with an error rate of 0.041. Logistic regression achieved a disappointingly high error rate of 0.061: but this is probably due to the inclusion of many irrelevant variables. The 180-variable DNA data set is available by anonymous ftp from the *pub/Statlog* directory of ftp.stams.strath.ac.uk. The associated document file gives more information, as well as references to previous studies. In the same directory, there are also several other classification data sets that were used in the *StatLog* project mentioned above.

The rescaling procedure was used on linear and quadratic discriminants in several of the *StatLog* data sets. Noticeable improvements are obtained in only three: vehicle, digit, and DNA. In most cases the changes are very minor, with a very slight improvement resulting from the rescaling. In the letters data set, for example, there are 26 classes but the asymmetry seems to affect only a few selected pairs of classes, and it seems that we cannot improve performance on these pairs without corresponding loss in performance in others, so the overall error rate does not change. Perhaps the solution would be to rescale only the affected classes. In any case, the main problem with the letters data set is that the classes are not pure.

5. STACKED GENERALIZATION

Several recent papers discuss ways of combining classification procedures with the aim of improving performance. Stacked Generalization (Wolpert, 1992) involves taking predictions from several procedures and using these predictions as the basis for the next stage of classification. Wolpert proposes a general scheme where the outputs of the various procedures form the inputs of the next level of generalization. In statistics, this idea has been taken up by LeBlanc and Tibshirani (1993) in the context of classification and regression, and by Breiman (1992) for regression. The approach in section 4 may be regarded as a special case of stacked generalization, but with particular reasons advanced for using logistic regression as the method for "combining" the outputs of the classification procedures.

In other contents, Zimmerman and Zysno (1983) combine the predictions of several human classifiers using a fuzzy set model for the human decision process. Granger and Newbold (1971) give a method for combining predictions in the time series context.

We will apply the principle of Stacking Generalizers to the problem of classification, with the important distinction that the outputs from each classifier must be the class-conditional probabilities. Our approach is most easily illustrated for a two-class problem, for then there is only one independent output variable, which, with no loss of generality, we may take as the estimated conditional probability that the test point belongs to class 1 given the observed features X.

5.1. Level 1

We suppose we are given a set of training data, of size n, in which a d-dimensional feature $\mathbf{X}$ is used to predict the classification c, where, for the purposes of this example, we suppose that the true classifications c are either 1 or 2. The format of the training data is shown in Table 7.4.

TABLE 7.4. Level 1 Training Data: Original Features X in *d*-Dimensions, Together with Known Classifications *c*

X_{11}	X_{12}	...	X_{1d}	c_1
X_{21}	X_{22}	...	X_{2d}	c_2
...	...	...	...	...
...	...	...	...	...
X_{n1}	X_{n2}	...	X_{nd}	c_n

Each row represents one training point.

Suppose each classification procedure produces a cross-validated class probability $W = P(C = 1 \mid X)$ for every point in the training set, basing these estimates, in the first instance, on the given attributes **X**. Given g algorithms, the j-th algorithm gives an estimate $W_j, j = 1, \ldots, g$, of the conditional probability of class 1, so that we obtain, for each observation in the training set, a new feature vector **W** of length g. This new feature vector is used, along with the known classes c, as the basis for the next stage in generalization.

5.2. Level 2

The new effective training data, of size n, consists of a g-dimensional features **W**, together with the true classifications c, as shown in Table 7.5.

The procedure can obviously be repeated, with the next level of generalization taking as input the estimated class probabilities of the previous level. At each stage, cross-validation (or leave-one-out) methods are used to find the features for the next stage. In practice, we have never observed, however, any improvement in proceeding beyond the second level.

5.3. Pipe Data Set

We will use the pipe data set as illustration for Stacking (section 5.5) and also later on in the section on Clustering (section 6). This data set concerns the classification of pipes from ultrasound wave measurements. It is a relatively

TABLE 7.5. Level 2 Training Data: Class-Conditional Probability Estimates W from Level 1 in *g*-Dimensions, Together with Known Classifications *c*

W_{11}	W_{12}	...	W_{1g}	c_1
W_{21}	W_{22}	...	W_{2g}	c_2
...	...	...	...	...
...	...	...	...	...
W_{n1}	W_{n2}	...	W_{ng}	c_n

small ($n = 613$ examples) data set, so that cross-validation is necessary. In its original form, 37 measurements were available, but the trials here reported on are for a reduced set with only 13 variables. Some preliminary trials on the original data (as supplied) were carried out to establish suitable parameters for the various algorithms (using 67%–33% train-and-test data sets). The purpose of the initial trials was to choose parameters such as the number of layers and hidden nodes for the neural network. These parameters were then used unchanged in the full-scale trials. All the results quoted in the following refer to tenfold cross-validation runs.

5.4. Classifiers

The algorithms under consideration consisted of:

ACE: Alternating Conditional Expectation (Breiman and Ihaka, 1984);
DISCRIM: Linear Discriminant (Fisher 1936);
KNN: K-Nearest neighbor for K = 1 (Fix and Hodges, 1951);
LOGDISCR: Logistic Discriminant (Cox, 1966);
LVQ: Learning Vector Quantization (Kohonen, 1989);
MARS: Multiple Adaptive Regression Splines (Friedman, 1991);
NNET: Neural Network multilayer perceptron (Ripley, 1994);
QDA: Quadratic Discriminant (Smith, 1947);
TREE: Decision Tree CART (Breiman et al., 1984).

Some details of these, and related algorithms, can be found in the *StatLog Book* (Michie et al., 1994). Public-domain versions of five of the above algorithms have been made available by Ripley (1994): DISCRIM, KNN, LVQ, NNET, and QDA.

5.5. Stacking on Pipe Data Set

In the pipe data set we used the nine different classifiers of section 5.4. Each classifier was applied to the original (level 1) data, using tenfold cross-validation to produce unbiased predictions of the probability that the pipe is class 1. Probabilities are produced by seven algorithms: ACE, DISCRIM, LOGDISCR, MARS, NNET, TREE, and quadratic. Two algorithms (KNN and LVQ) give only a classification: this classification is recoded as 1 if class 1 is predicted and 0 else. In this way, the result of running all nine algorithms is to give nine estimates of class 1 probability for each observation. These nine estimates comprise the level-2 data: they can be used as attributes to predict the class value, which, because we use cross-validation, has not been used to form the class probabilities. The procedures can be iterated, each level of generalization using (only) the predictions of the previous level. The results of two levels of generalization are given in Table 7.6. Level 1 refers to the raw data (the reduced data set of 13 variables). Level 2 uses the nine probability

TABLE 7.6. Error Rates for Three Levels of Generalization

Error Rates	Level 1	Level 2	Level 3
ACE	0.1484	0.1256	0.1305
DISCRIM	0.1550	0.1305	0.1289
KNN	0.1452	0.1631	0.1468
LOGDISCR	0.1566	0.1240	0.1338
LVQ	0.1876	0.1321	0.1403
MARS	0.1305	0.1354	0.1370
NNET	0.1452	0.1305	0.1468
QDA	0.1582	0.1354	0.1419
TREE	0.2219	0.1272	0.1484

Level 1 refers to the raw data; level j uses only class probability estimates from level $j - 1$. The standard error of these estimates is approximately 0.014.

estimates from level 1. Similarly, level 3 uses the nine probability estimates from level 2.

Using the raw data (level 1), MARS is noticeably better than the other procedures, and the decision tree is rather poor. Performance at level 2 is much closer: most procedures have an error rate of around 0.13. Continuing the process of generalization to level 3 does not seem to improve matters, indeed on average the error rates are worse if anything. An error rate of around 0.13 is probably near the limit for this data, and we would conclude, on practical grounds, that MARS on the original data is the best predictor.

Deciding which of these levels of generalization offers the best classifier would not be an easy matter in general. Using high levels of generalization would require enormous amounts of computation, and the resulting classifiers would be very difficult to understand.

6. ANALYSIS OF CLASSIFIER OUTPUT

It is standard practice in statistics to monitor the performance of any regression package, especially by looking for outliers or patterns in the residuals. This would apply to regression-type procedures like discriminant analysis, where it would be possible to give some idea if an observation is likely to have come from any of the given classes (if not it is declared to be an outlier). When we have the output from several classification procedures, we can go further, and make use of all the predictions to form opinions on whether given observations are outliers, or even if some procedures are out of step with the others. Correspondence Analysis (see Hill, 1982, for a description) is such a procedure. It is normally used for analyzing contingency tables, but can be used more generally to analyze tables with positive entries. In particular, if we apply several classification procedures to a given data set, and collect the predicted class conditional probabilities together, we can analyze the probabilities via correspondence analysis.

6.1. Illustrative Example

To set the scene, we begin by constructing a classification example that would be particularly appropriate for correspondence analysis, illustrating how we can establish scores for the classes.

When satellite images are used to identify soil types, a sub-problem is to determine whether plots of land contain dry, damp, or wet soil. Here there are three classes that are naturally ordered by degree of wetness. There is an underlying or latent variable that determines the degree of wetness (the amount of water), but we must assume that any such latent variable in unknown. However, if we can arrive at an objective set of scores for the classes, by whatever means, we hope that these scores might reflect the underlying wetness variable. Usually we must guess at a set of scores, and in this case a natural set of scores would be $(1, 2, 3)$ for the classes (dry, damp, wet), respectively. Let us suppose that we have data for five neighboring plots, and that we have used a statistical classification procedure to estimate the relative class probabilities for the five plots as in Table 7.7.

Now consider the first plot of land. Instead of simply allocating this plot to the class "dry," we calculate the average "wetness" score by weighting the three classes with the estimated probabilities: $0.8 \times 1 + 0.2 \times 2 + 0.0 \times 3 = 1.2$. So the first plot can be given a "wetness" score of 1.2, implying a soil type somewhere between dry and damp (and nearer dry). The second plot would be given a score of 2.0, and so on. In this way, starting from "wetness" scores for the soil types, we can arrive at "wetness" scores for all five plots of land: (1.2, 2.0, 2.1, 2.4, 1.6). These average scores might be correlated with geographical position: the drier areas might be areas of higher ground for example, and it might be possible to correlate plot scores with height above sea level.

TABLE 7.7. Conditional Class Probabilities for Five Plots of Land

	Class Probabilities		
Plot Number	Dry	Damp	Wet
1	0.8	0.2	0.0
2	0.3	0.4	0.3
3	0.2	0.5	0.3
4	0.1	0.4	0.5
5	0.5	0.3	0.2

The plot number can be thought of as distance along a transect.

Now, suppose we turn the problem on its head: suppose we start from the plots of land, allocating them scores by some means or another? Most significantly, suppose we start from the set of scores that we have just calculated for the plots of land, namely, (1.2, 2.0, 2.1, 2.4, 1.6)? We can now calculate average scores for the soil types, using the estimated class probabilities as weights. For example, soil type "dry" has probabilities (0.8, 0.3, 0.2, 0.1, 0.5), for plots 1, 2, 3, 4, 5, whose plot scores are known to be (1.2, 2.0, 2.1, 2.4, 1.7). Bearing in mind that the total probability for "dry" is $0.8 + 0.3 + 0.2 + 0.1 + 0.5 = 1.9$, the mean score for this soil type is,

$$(0.8 \times 1.2 + 0.3 \times 2.0 + 0.2 \times 2.1 + 0.1 \times 2.4 + 0.5 \times 1.6)/1.9 = 1.59.$$

In this way we arrive at scores of (1.59, 1.96, 2.12) for the three classes. Two points are of potential interest: (i) the scores for classes "damp" and "wet" are rather similar (see that the class probabilities for classes "damp" and "wet" in Table 7.7 tend to be rather similar); and (ii) these class scores are not the scores (1, 2, 3) that we started with. If, by starting from the land plot scores, we arrived at mean scores for the soil types identical to the scores that we assumed for the soil types at the outset, then we say that the two sets of scores are mutually consistent. Correspondence analysis gives a means of finding pairs of mutually consistent scores that have maximal correlation. One way of finding such a pair is to start with an arbitrary set of row scores and then iterate the process we have just outlined. One use of such scores is to identify some underlying ordering of the classes or objects.

6.2. Correspondence Analysis

Correspondence analysis gives sets of mutually optimal scores to the rows and columns of the contingency table, and these scores can be used as a basis for comparing rows (*cf.* pipes) and columns (*cf.* classifiers). It tends to be used either as a method of scaling or a basis for collapsing classifications. The starting point of correspondence analysis is a two-way table of frequencies, or, as here, a table of positive entries to be used as weights. Supposing, for example, that the *columns* of the two-table have been allocated scores by some procedure. Using the table entries for each *row* as weights, we calculate the average score for that row. This row score then gives some measure of where that row is positioned in the scale of column scores. In the same way, if we start from a set of *row scores* we can get a corresponding set of column scores by using the column weights. Proceeding iteratively, using the row or column scores arrived at in the previous step to get the column or row scores in the next step, we arrive at a stationary set of scores that are mutually consistent and have maximal correlation. Correspondence analysis works by simultaneously scaling columns (classifiers) and rows (pipes) so that row and column scores have maximal correlation using the frequency table as weights. This leads to an eigenvalue decomposition of distances between columns (classi-

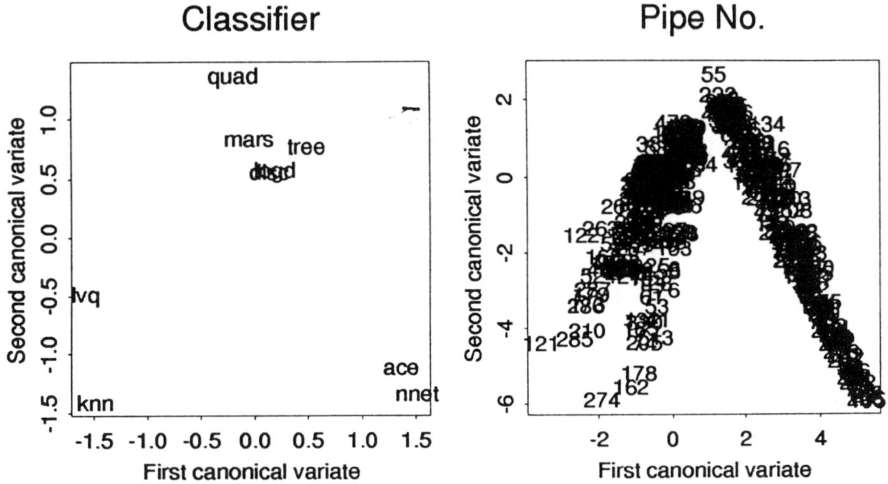

Figure 7.1. *Correspondence analysis of class probabilities from the reduced data set.*

fiers), with distance measure being the chi-square distance between the sets of probabilities allocated by the classifiers. We use the first two column eigenvectors to give a pictorial representation of distances between the classifiers. The corresponding row eigenvectors give a representation of distances between the rows. This is illustrated for the pipe data in Figure 7.1.

Note that the input to the correspondence analysis consists of the various cross-validated probabilities, and does not involve the actual classification, and so is unsupervised. From the left diagram of Figure 7.1 it is apparent that the linear procedures LOGDISCR and DISCRIM are very similar in performance (i.e., produce probabilities that are very similar). Similarly, ACE and NNET are comparable to each other, while KNN is itself most similar to LVQ. In the right-hand plot, where the plotting symbols are the original case numbers, there seem to be three clusters. We proceed to look at this phenomenon in a little more detail.

In case it is the discrete nature of the class probability estimates that is the source of the apparent clustering in Figure 7.1, the three procedures that produce discrete probabilities, namely KNN, LVQ and TREE, were dropped, and correspondence analysis rerun on the remaining six sets of probabilities. Figure 7.2 investigates the resulting distribution of cases. It corresponds to the right diagram of Figure 7.1, but with separate plots for each distinct class. It would appear that there are two main clusters, and these are evident in both left and right diagrams in Figure 7.2. One is a roughly spherical grouping centered near coordinates (0, 0), and we may conjecture that this cluster is more typical of class 1 as it appears particularly strongly in the left diagram. The other grouping is spread linearly along the bottom left, especially in the right

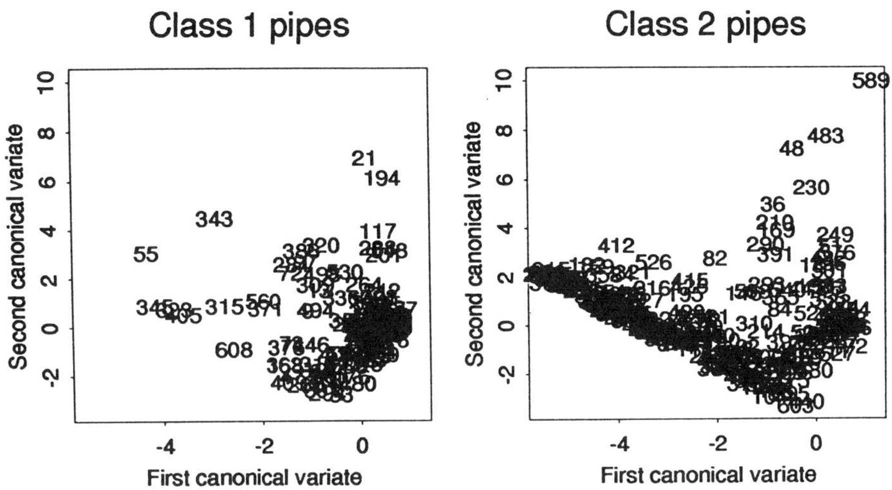

Figure 7.2. *Correspondence analysis by classes, omitting results from* TREE, KNN *and* LVQ.

diagram (i.e., is more appropriate to class 2). However, there are echoes of both these clusters in both diagrams, and there are other, less easily characterized, points scattered over probability space.

6.3. Hierarchical Clustering

We now discuss, very briefly, the question of measuring the similarity of several algorithms when applied to the same data set. The one big advantage of comparing class-conditional probabilities, is that they are necessarily all on the same scale [we assume that all algorithms give strict probabilities on the interval (0, 1)]. The problem of scaling the attributes is then not so important. On the other hand, we could use a wide variety of similarity or distance measures. We might use the chi-square distance measure used in correspondence analysis, for instance. However, we used the simplest of all, namely, the correlations between the probability estimates on the test data. As example, we used the same nine algorithms on the same data set as in section 5. As we are concerned with the correlation between the columns of probability estimates, the base data for the hierarchical clustering procedure consist of the level-2 data discussed in section 5.

Figure 7.3 shows a dendrogram for the algorithm based on the correlation matrix of the probability estimates. It shows similarities between algorithms. Algorithms are joined with each other at a level that corresponds to their similarity; the greater the similarity, the lower the level at which the algorithms are joined. Thus, the two most similar algorithms are DISCRIM and LOGDISCR (i.e., their probability estimates are most highly correlated). ACE

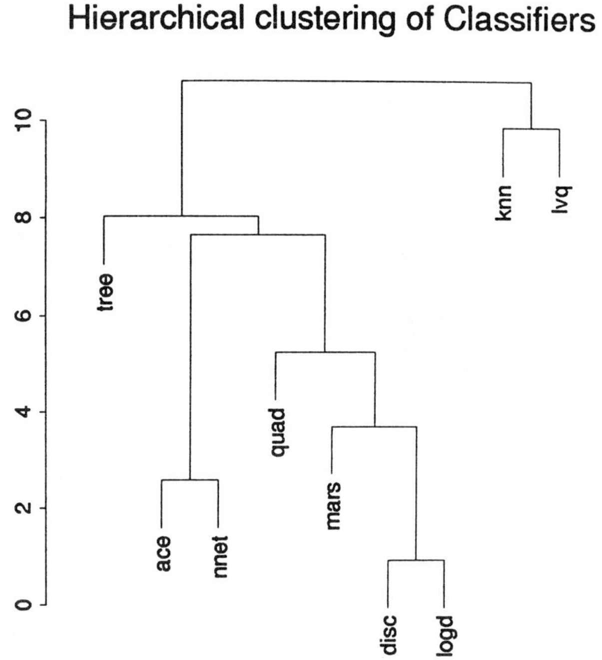

Figure 7.3. *Clustering applied to correlation matrix of class probabilities.*

and NNET are fairly similar to each other, but not very similar to either DISCRIM or LOGDISCR. LVQ and KNN are rather dissimilar, but are nearer each other than to any of the other algorithms.

This method enables algorithms to be compared as to their performance on a single data set. See the *Statlog* book (Michie et al., 1994) for clustering based on performance over several data sets using error rates as a basis for comparison.

7. CONCLUSIONS AND DISCUSSION

Rescaling may be regarded as a special case of stacked generalization, but with particular reasons advanced for using logistic regression as the method for "combining" the outputs of the classification procedures. We have shown, by several practical examples, that post-processing the output probabilities from a statistical discriminant procedure may reduce the error-rate very substantially. One disadvantage of rescaling is that the discriminant rule must be translated to probabilities in the first place. Linear discriminants, for example, must be transformed into conditional probabilities before the rescaling factors can be found. And it is still an open question as to how best to deal with "probabilities" that are not defined in the unit interval (as may occur with some neural nets, for example). Once found, however, the scaling factors translate back to the linear discriminant by adding the logarithm of the scaling factor to each of the constant terms, or by taking a linear transformation of the Mahalanobis distance in the case of quadratic discriminants. A general difficulty in similar situations is that the relative size of training and tuning data sets may well be critical, and very little guidance can be given for this. With very small data sets, it will be necessary to resort to cross-validation techniques to evaluate the error rates, and then the computations may become very tedious. Before using rescaling, it should be borne in mind that the simple expedient of dropping irrelevant attributes is often the most effective way of reducing bias. And we repeat that rescaling cannot be done on probabilities that are 0 or 1. Nor will rescaling make any impression on discriminants based on a single binary attribute.

More extensive studies (not described here) have shown a reduction of bias on several data sets from the *StatLog* project (Michie et al., 1994).

On the question of which generalizers to be used in stacking, it is clear that we need a very wide spread. Suppose, on the contrary, that all the classifiers are in a very narrow family, say, the linear family. Then the generalizations are necessarily confined to the linear family also.

Another approach to stacking, this time in the context of multilayer perceptrons, is the cascade architecture of Fahlman and Lebière (1990). Cascade correlation incorporates predictions from earlier stages into the training of subsequent stages. This may be viewed as an example of stacked generalization, with the same generalizer used to refine previous levels of generalization, with the further difference that in cascade correlation the inputs are used, as well as the predictions, when proceeding to the next level of generalization. Also, cascade correlation does not normally use cross-validated predictions.

REFERENCES

Agresti, A. (1990). *Categorical Data Analysis.* Wiley, New York.

Bishop, Y., Fienberg, S., and Holland, P. (1975). *Discrete Multivariate Analysis: Theory and Practice.* MIT Press, Cambridge, MA.

Breiman, L. (1992). *Stacked Regressions*, Tech. Rep. No. 367. Department of Statistics, University of California, Berkeley.

Breiman, L, and Ihaka, R. (1984). *Nonlinear Discriminant Analysis via Scaling and Ace*, Tech. Rep. No. 40. Department of Statistics, University of California, Berkeley.

Breiman, L., Friedman, J. H., Olshen, R. A., and Stone, C. J. (1984) *Classification and Regression Trees.* Wadsworth & Brooks, Monterey, CA.

Cox, D. R. (1966). Some procedures associated with the logistic qualitative response curve. In *Research Papers on Statistics: Festschrift for J. Neyman* (F. N. David, ed.), pp. 55–77. Wiley, New York.

Day, N. E., and Kerridge, D. F. (1967). A general maximum likelihood discriminant. *Biometrics* **23**, 313–324.

Devroye, L. (1987). Automatic selection of a discriminant rule based upon minimization of the empirical risk. In *Pattern Recognition Theory and Applications* (P. A. Devijver and J. Kittler, eds.). Springer-Verlag, Berlin.

Fahlman, S. E., and Lebière, C. (1990). The cascade correlation learning architecture. In *Advances in Neural Information Processing Systems* (D. S. Touretsky, ed.), Vol. 2, pp. 524–532. Morgan Kaufmann, San Mateo, CA.

Fisher, R. A. (1936). The use of multiple measurements in taxonomic problems. *Ann. Eugen. (London)* **7**, 179–188.

Fisher, R. A. (1938). The statistical utilisation of multiple measurements. *Ann. Eugen. (London)* **8**, 376–386.

Fix, E., and Hodges, J. L. (1951). *Discriminatory Analysis, Nonparametric Estimation: Consistency Properties*, Rep. No. 4, Proj. No. 21-49-004. USAF School of Aviation Medicine, Randolph Field, TX.

Friedman, J. H. (1989). Regularized discriminant analysis. *J. Amer. Stat. Assoc.* **84**, 165–175.

Friedman, J. H. (1991). Multivariate adaptive regression splines (with discussion). *Ann. Stat.* **19**, 1–141.

Fukunaga, K., and Ando, S. (1977). The optimum nonlinear features for a scatter criterion in discriminant analysis. *IEEE Trans. Inf. Theory* **IT-23**, 453–459.

Granger, C., and Newbold, P. (1977). *Forecasting Economic Time Series.* Academic Press, New York.

Hermans, J., Habbema, J., and Schaefer, J. (1982). The alloc80 package for discriminant analysis. *Stat. Software Newsl.* **8**, 15–20.

Hertz, J., Krogh, A., and Palmer, R. (1991). *Introduction to the Theory of Neural Computation.* Addison-Wesley, Reading, MA.

Hill, M. (1982). Correspondence analysis. In *Encyclopaedia of Statistical Sciences* (S. Kotz, N. L. Johnson, and C. B. Read, eds.), Vol. 2, pp. 204–210. Wiley, New York.

Kohonen, T. (1989). *Self-Organization and Associative Memory*, 3rd ed. Springer-Verlag, Berlin.

LeBlanc, M., and Tibshirani, R. (1993). *Combining Estimates in Regression and Classification.* Preprint. Department of Statistics, University of Toronto.

McCullagh, P., and Nelder, J. A. (1989). *Generalized Linear Models.* 2nd ed. Chapman & Hall, London.

McLachlan, G. J. (1992). *Discriminant Analysis and Statistical Pattern Recognition.* Wiley, New York.

Michie, D., Spiegelhalter, D. J., and Taylor, C. C. (1994). *Machine Learning, Neural and Statistical Classification.* Ellis Horwood, Chichester, UK.

Murphy, P. M., and Aha, D. W. (1992). *UCI Repository of Machine Learning Databases. Machine-readable Data Repository*, Department of Information and Computer Science, University of California, Irvine.

Quinlan, J. (1992). *C4.5: Programs for Machine Learning.* Morgan Kaufmann, San Mateo, CA.

Randles, R., Broffitt, J., Ramberg, J., and Hogg, R. (1978). Generalized linear and quadratic discriminant functions using robust estimates. *J. Am. Stat. Assoc.* **73**, 564–568.

Ripley, B. (1993). Statistical aspects of neural networks. In *Chaos and Networks: Statistical and Probabilistic Aspects.* (O. Barndorff-Nielsen, D. Cox, J. Jensen, and W. Kendall, eds.). Chapman & Hall, London.

Ripley, B. (1994). Neural networks and related methods for classification. (with discussion). *J. R. Stat. Soc. Ser. B* **56**, 410–456.

Schulmeister, B. and Wysotzki, F. (1993). Learning of a piecewise linear classifier by combination of statistical methods with principles of neural nets. In *Information Systems and Data Analysis. Studies in Classification, Data Analysis and Knowledge Organization, Vol. 4* (H. Bock, W. Lenski and M. Richter, eds.). Springer-Verlag, Berlin.

Smith, C. (1947). Some examples of discrimination. *Ann. Eugen. (London)* **18**, 272–283.

Weiss, S. M., and Kulikowski, C. A. (1991). *Computer Systems that Learn: Classification and Prediction Methods from Statistics, Neural Networks, Machine Learning and Expert Systems.* Morgan Kaufmann, San Mateo, CA.

Wolpert, D. (1992). Stacked generalization. *Neural Networks* **5**, 241–259.

Zimmerman, H.-J., and Zysno, P. (1983). Decisions and evaluations by hierarchical aggregation of information. *Fuzzy Sets Syst.* **10**, 243–260.

CHAPTER 8

Distance-based Decision Trees

C. C. TAYLOR

1. INTRODUCTION

A recent ESPRIT project (*StatLog*) aimed to evaluate and compare various classification techniques on large-scale, real data sets. Trials were carried out on about 20 data sets with 20 algorithms taken from machine learning, statistical, and neural network origins. The results and analysis are presented in Michie et al. (1994). During the course of the project a number of points emerged:

1. There had been inadequate collaboration and communication between the ML and statistical communities, resulting in different terminology and some duplication of work.
2. There were clearly different priorities. The statistical methods were more concerned with performance as measured by error rate, whereas the results from the ML methods were often easier to evaluate and interpret. Feng and Michie (1994, p. 51) state, "ML requires that classifiers should not only classify but should also constitute explicit concepts, that is, expressions in symbolic form meaningful to humans and evaluable in the head."
3. The concept of non-equal costs of incorrect classifications did not feature in the ML algorithms, though, of course, this could be incorporated in future developments.

Machine Learning and Statistics, The Interface, Edited by G. Nakhaeizadeh and C. C. Taylor.
ISBN 0-471-14890-3

4. Preprocessing, that is, feature selection and feature extraction was often much more important than correct choice of algorithm for a given data set. In this part of data analysis, the expertise of the user is critical. However, preprocessing and subsequent improvement in performance of an algorithm may be at the expense of interpretability of the classification rule. In this case the distinction between the interpretability of the ML algorithms and the rest is reduced, since all of them may be dealing with variables that cannot be simply evaluated.

In the following two sections we focus on point 2 above and show that, in some cases at least, algorithms can be adapted so that the rule is described in a manner that is meaningful and evaluable.

2. REFORMULATION OF CLASSIFICATION RULES FOR EVALUATION

This section is organized as follows: We describe each algorithm in its traditional manner, and then we describe how the rule might be modified so it is meaningful to humans and evaluable in the mind. In essence, these examples all use the concept of "distance" from an observation in the test set to a given benchmark (possibly an observation in the training set). The methods continue in a recursive partitioning manner and thereby can result in a decision tree rule for classification.

2.1. Decision Tree

We begin with a familiar method and a simple example, which we will be re-used in subsequent subsections. The data are a random subset of data set 36 in Andrews and Herzberg (1985), which examines the relationship between chemical subclinical and overt nonketotic diabetes in 145 patients (see above for more details). For ease of presentation, we have used only 50 patients and two of the six variables; Relative weight and Glucose area, and the data are shown in Figure 8.1.

The classifications of 50 patients are one of overt diabetic (1), chemical diabetic (2), and normal (3) are labeled on the graph. In this example, it can be seen that Glucose Area (y-axis) is more useful in separating the three classes, and that class 3 is easier to distinguish than classes 1 and 2. The decision tree shown in Figure 8.2 used the default pruning parameters given in *Splus* (StatSci Europe, 1991) and is also equivalent to the partitioning shown in Figure 8.1.

Note that the tree could also be written as a "distance" decision tree, so that the first split is replaced by $D_1^2(x_2) < D_2^2(x_2)$ where $D_1^2(x) = (x - 351)^2$ and $D_2^2(x) = (x - 522)^2$, are the squared distances to two *representative* Glucose Area values, equally spaced around the split. Similarly, the second split can also be stated in terms of distance, and so on. Note that these representative points are not uniquely defined by the original splits, since any two points that

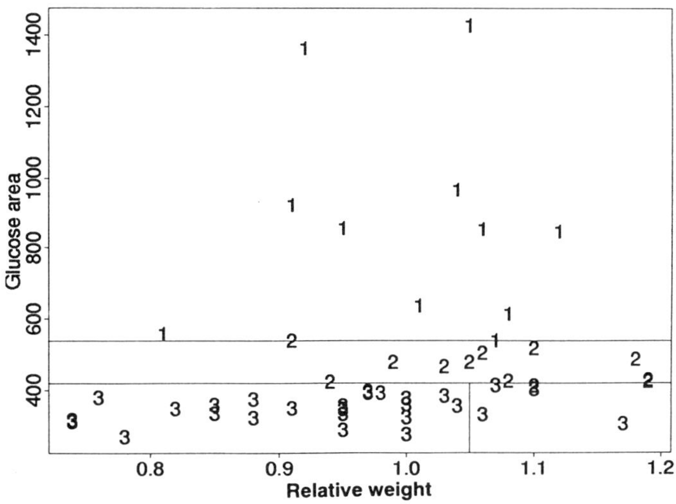

Figure 8.1. *Recursive partitioning of diabetic data.*

Figure 8.2. *Decision tree for diabetic data.*

are equidistant from the splitting value will suffice. However, the term "representative" implies a desire for a value close to the mean/median or mode for the data (where possible) on that side of the split.

2.2. Linear Discriminant Function

This is the most widely used classification rule by the statisticians. The simplest interpretation is that we choose the class A_i to maximize the class-dependent

Normal distribution density which, in p dimensions, is given by

$$P(x \mid A_i) = f_i(x) = \frac{1}{\sqrt{(2\pi)^p \, |\Sigma|}} \exp\left\{ -\frac{1}{2}(x - \mu_i)^T \Sigma^{-1} (x - \mu_i) \right\}$$

In practice we have to estimate the class mean μ_i by $\bar{x}_i$ and the pooled covariance matrix Σ by

$$S = \frac{1}{n - q} \sum_{i=1}^{q} (n_i - 1) S_i$$

where there are q classes with n_i, $i = 1, \ldots, q$ observations in class A_i, and S_i, the covariance matrix for class A_i, is given by

$$S_i = \frac{1}{n_i - 1} X^T X - \bar{x}_i^T \bar{x}_i$$

This is equivalent to choosing class A_i to minimize the Mahalanobis distance

$$D^2(i) = (x - \bar{x}_i)^T S^{-1} (x - \bar{x}_i)$$

The classification rule is extremely simple, although the metric space needs to be understood in terms of the covariance structure of the data. The derived rule *could* be described by a binary "distance" decision tree with $2(q - 1)$ leaves and an additional $(q - 1)(q - 2)/2$ nodes, with a structure that looks somewhat

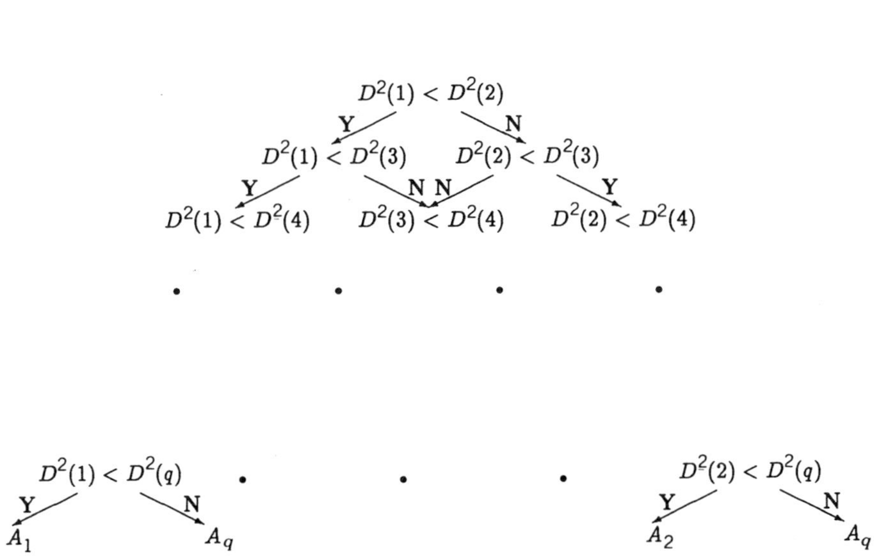

Figure 8.3. *"Distance" decision tree for the linear discrimination rule.*

like that in Figure 8.3. Unfortunately the tree structure will contain branches that cross from the fourth row onwards. For the linear discriminant rule, this is hardly worthwhile but having described such a tree, it can clearly be made more flexible and extended to allow for different distance measures $D^2(i)$ — this would then be adaptable for bimodal distributions. For example, in the above data set, linear discriminant will not do well, since the assumption of equal covariance matrices is clearly violated, so the distance above could be replaced by the more general

$$D^2(i) = (x - \bar{x}_i)^T S_i^{-1}(x - \bar{x}_i)$$

which is similar to, but not quite the same as (since the determinant of the covariance matrix is not used), that used for the quadratic discriminant function.

2.3. Nearest Neighbor and Learning Vector Quantiser (LVQ)

2.3.1. Nearest Neighbor. The nearest-neighbor classifier assigns a new observation to the class to which the nearest training observation belongs. A difficulty with the method is what is meant by "nearest," that is, how to measure distance. In many cases the attributes are measured on different scales, and will need to be standardized by perhaps the standard deviation or range, possibly class dependent. However, once a suitable metric has been obtained, this method is also a partitioning method, in that the entire sample space can be allocated to one or other of the classes. To see this let $V_i = \{x | d(x, x_i) = \min_j \ d(x, x_j)\}$, $i = 1, \ldots, n$ denote the Dirichlet cell (or Voronoi polygon) of points that lie closer to x_i than to any other point in the training set. Then we allocate a new observation to class A_k if

$$x \in \bigcup_{\{i: x_i \in A_k\}} V_i$$

When we are in two dimensions or less, then this partition of space can be readily visualized, and an example is given, using the above data, in Figure 8.4.

This method is much better suited to bimodal distributions than Linear Discriminants. By comparing Figure 8.4 with the decision tree regions in Figure 8.1, it appears that the method may be more succinctly able to model boundaries that are not perpendicular to the axes. However, in higher dimensions it is hard to visualize, and perhaps even harder to evaluate or understand the classification regions. In principle, we could describe the decision regions using a "distance" decision tree, such as that used above for the linear discriminant function. In a simple-minded approach, this will give $2(n - 1)$ leaves and an additional $(n - 1)(n - 2)/2$ nodes, so it will be a massive tree! The method of condensing attempts to reduce the size of the training data sample by judiciously selecting examples which are most useful in the classifi-

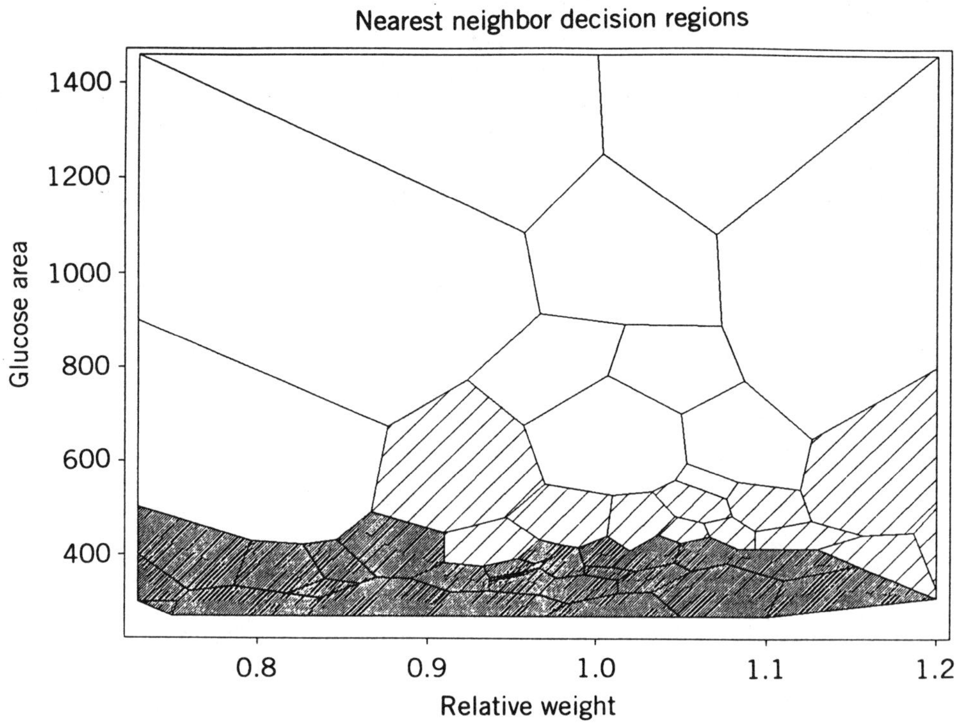

Figure 8.4. *Decision regions for nearest-neighbor classifier.*

cation process. For example, we could select observations i such that V_i is contiguous to V_j (for some $j \neq i$) only if x_i and x_j do not belong to the same class. In the Glucose area example considered above, this would reduce n from 50 to 31 observations. However, a more substantial reduction in tree size could be obtained by creating a cascading sequence of representative points. An example would be to consider *cliques* of neighboring V_i, that is, let $V_j^{(1)}$, $j = 1, \ldots, m_1$ be defined by

$$V_j^{(1)} = \bigcup_{i \in N(j)} V_i \qquad j = 1, \ldots, m_1$$

where $N(j)$ denotes the neighbors of V_j, and the cliques are chosen so as to minimize the number of pairs (i, j) for which $V_i^{(1)} \cap V_j^{(1)} \neq \phi$ while maintaining $\cup V_j^{(1)} = \cup V_i$. Similarly, in a recursive manner, define

$$V_j^{(r+1)} = \bigcup_{i \in N(j)} V_i^{(r)} \qquad j = 1, \ldots, m_{r+1}$$

where $N(j)$ denotes the neighbors of $V_j^{(r)}$. Then we have a recursive partitioning method based on the nearest-neighbor method, which will typically result in a smallish tree. Optimally choosing the cliques is not trivial, and having done so, there may be no fast way of determining j such that $x \in V_j^{(r)}$, but, in principle, the method is feasible.

An alternative approach is to choose representative points using a Kohonen net in a supervised manner, which is the approach used in LVQ, described in the following section.

2.3.2. Learning Vector Quantiser (LVQ). Learning Vector Quantisation (LVQ) is a public domain algorithm of a method due to Kohonen (1989), although it has a longer history (Hertz et al., 1991). Learning Vector Quantization is an example of competitive learning often used for data compression.

Kohonen's (1989) procedure is just the use of competitive learning in a supervised learning problem. In both Kohonen and LVQ, classification is based on a nearest-neighbor rule in the output space, the difference being in the dimensionality of the output space. Assume that a number of "codebook vectors" m_i (free parameter vectors) are placed into the input space to approximate various domains of the input vector x by their quantized values. Usually several codebook vectors are assigned to each class of x values, and x is then decided to belong to the same class to which the nearest m_i belongs. Let

$$c = arg \min_i \{ \|x - m_i\| \}$$

define the nearest m_i to x, denoted by m_c.

Values for the m_i that approximately minimize the misclassification errors in the above nearest-neighbor classification can be found as asymptotic values in the following learning process. Let $x(t)$ be a sample of input and let the $m_i(t)$ represent sequences of the m_i in the discrete-time domain. Starting with properly defined initial values, the following equations define the basic LVQ1 process:

$$m_c(t+1) = m_c(t) + \alpha(t)[x(t) - m_c(t)] \qquad \text{if } x \text{ and } m_c \text{ belong to the same class,}$$

$$m_c(t+1) = m_c(t) - \alpha(t)[x(t) - m_c(t)] \qquad \text{if } x \text{ and } m_c \text{ belongs to different classes,}$$

$$m_i(t+1) = m_i(t) \qquad \text{for } i \neq c$$

Here $0 < \alpha(t) < 1$, and $\alpha(t)$ may be constant or decrease monotonically with time. In the above basic LVQ1 it is recommended that α should initially be smaller than 0.1; linear decrease in time is used in this package.

The number of codebook vectors to use is not a simple matter, but they can be thought of as representative points, and a distance-based decision tree

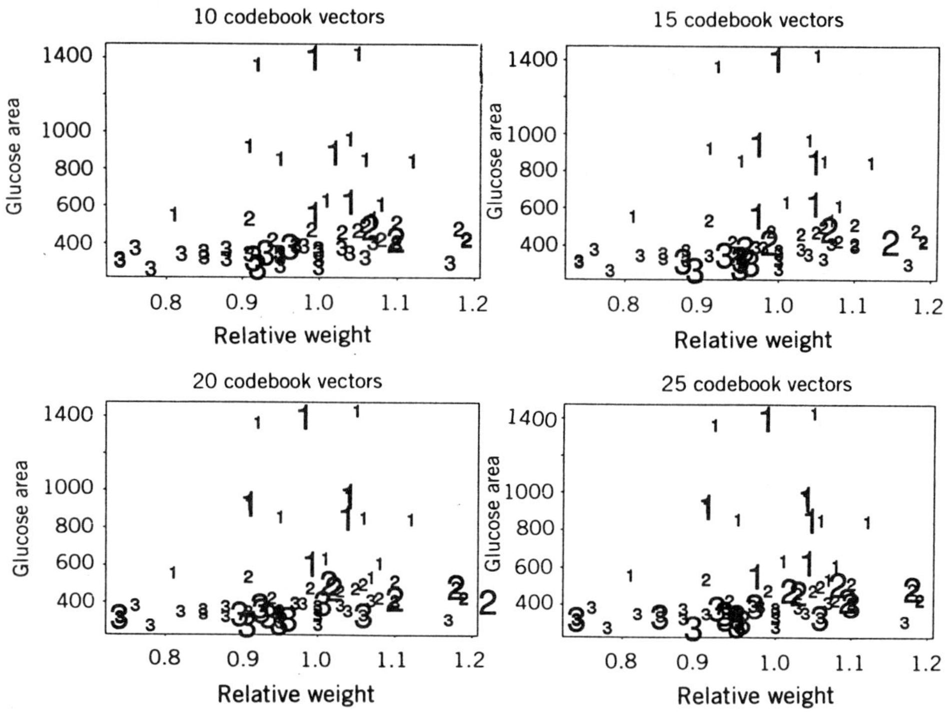

Figure 8.5. *Codebook vector positions for 10–25 codebook vectors. The large numbers are the codebook vectors, and the small numbers are the observations.*

subsequently determined. A similar cascading approach to that suggested above can be used. For illustrative purposes, we show the positions of the codebook vectors for the above data set in Figure 8.5.

3. DISTANCE DECISION TREES

Since we have now restated all of the above algorithms as "distance" decision trees, we can now formulate this approach in its own right. The basic method requires a set of "representative" points and a corresponding set of distance measures. There are many possibilities whereby the representative points may be just carefully chosen data values, and the distance metric can be scaled by the standard deviations, or just considering one attribute at a time (resulting in partitions perpendicular to the axes). Of course the distance metric does not have to be the same at every node of the tree. A simple example, which can be interpreted as just a hybrid method, is to use the linear discriminant function followed by standard decision tree methods applied to the LDF partitions. This would allow for bimodality and still result in a ML rule.

TABLE 8.1. Confusion Matrix for Linear Discriminants

	1	2	3
1	7	0	0
2	3	10	6
3	1	2	21

The column represents the true class, and the row the predicted class.

Another example is a recursive linear discriminant rule. This is illustrated for the above data set. A linear discriminant rule results in choosing A_i to minimize

$$D_1(x) = -64.4 + 92.5x_1 + 0.042x_2$$
$$D_2(x) = -58.0 + 99.2x_1 + 0.021x_2$$
$$D_3(x) = -43.9 + 87.5x_1 + 0.016x_2$$

This results in the following confusion matrix shown in Table 8.1 with 12 incorrectly classified.

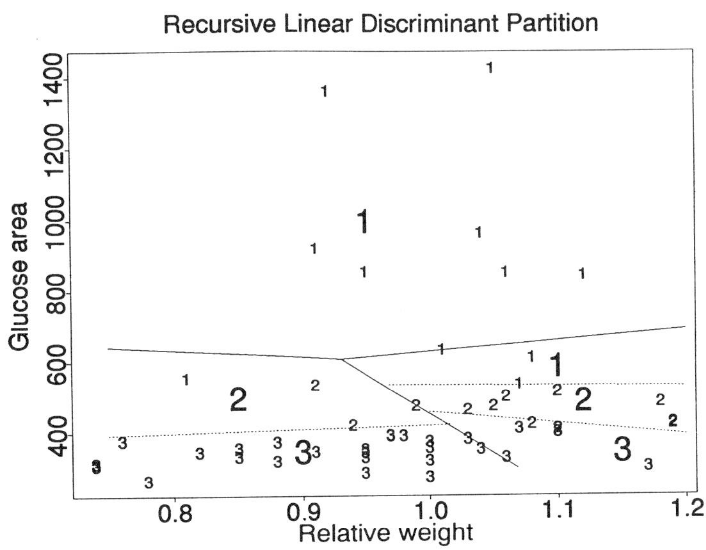

Figure 8.6. *Recursive partitioning using Linear Discriminant Function. The continuous lines are the initial partition, and the dotted lines subsequent partitions. The large numbers are the classes allocated to each region, and the small numbers are the observations.*

We then consider those allocated to class 2 by this rule (there are 3 from class 1, 10 from class 2, and 6 from class 3). If we now classify these by a linear discriminant rule, none are incorrectly classified. Those allocated to class 3 (1 from class 1, 2 from class 2, and 21 from class 3) can be similarly treated, though the linear discriminant rule requires at least two in each class, so we cannot deal with solely misclassified observations. In a sense this is an enforced pruning mechanism; this is discussed further in the next section. The final partition is shown in Figure 8.6. If necessary, this could be further refined by applying a nearest-neighbor rule within each partition. Note that the regions shown in Figure 8.6 have one misclassification, whereas the linear discriminant and quadratic discriminant rules have 12 and 4 misclassifications, respectively (out of 50 observations).

3.1. Pruning

Just as for the standard ML decision trees, pruning will be all important for classification accuracy and interpretability. As mentioned above, the linear discriminant cannot be applied unless there is more than one observation in each class. In applying the linear discriminant function, formal tests [Wilks' lambda and the greatest roots test (Mardia, et al., 1979)] exist to determine whether the split is desirable. However, note that there are circumstances when it would appear useless to split at a node, when subsequent splits are completely transparent. An example is a "random" XOR problem, as shown in Figure 8.7.

The formal tests may, however, still be useful at the pruning stage, together with the usual methods of cross-validation to obtain trees of the right size. In some situations, it may be desirable to "over-prune" so that the resulting rule is easier to interpret.

Further tests which may prove useful are those based on stepwise discrimination. These could be applied at each stage of the recursive partition and might help to simplify the distance metric. The whole issue of distance metrics is a difficult one, but appeal to the more robust nature of the linear discriminant may help.

3.2. Costs and Priors

Note that we have not considered costs or priors in any of the previous sections. These can enter if our final partition is not purely of one class, since we can then estimate a conditional probability of each class at that node. However, the accuracy of these conditional probability estimates would need to be assessed.

They can also enter if we define some trade-off between distance and cost—such as is necessary in the nearest-neighbor rule if costs are to be considered. This latter approach is rather arbitrary, and further work is required.

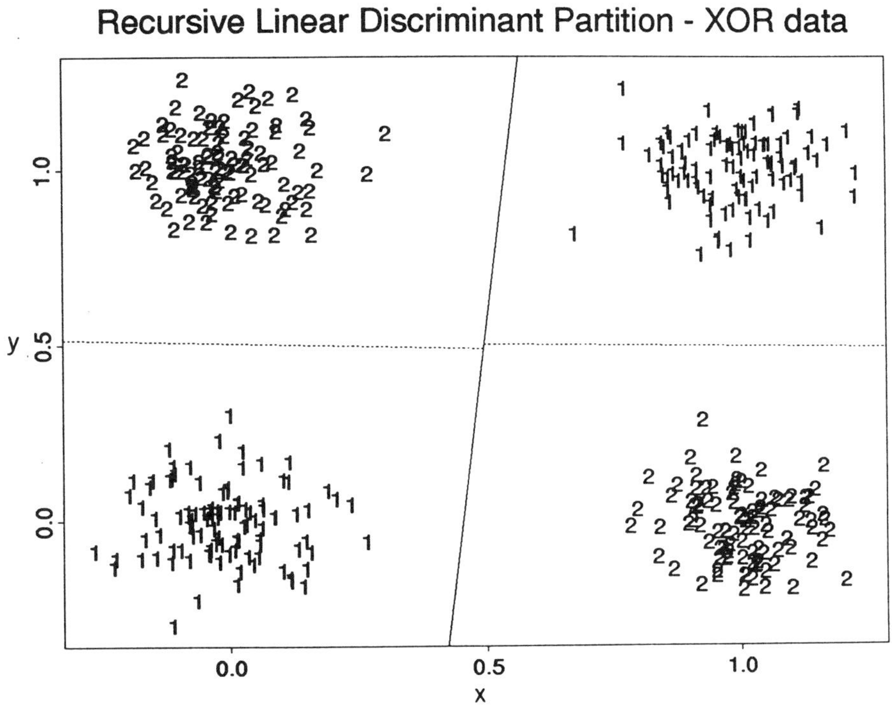

Figure 8.7. *Recursive partitioning using Linear Discriminant Function for some noisy XOR data with 100 observations at each node. The continuous lines are the initial partition which still results in a 50% error rate, and the dotted lines are subsequent partitions which have consequent error rates of 0%.*

4. EXPERIMENTS USING NORMAL DISTRIBUTION ASSUMPTIONS

In this section we pursue some of the above ideas on some real data sets. The performance results (in terms of error rate) are not startlingly different. However, the experiments do illustrate that, apart from presenting some statistical approaches in an ML framework, there is great potential for hybrid methods that can use the most appropriate "splitting" method at each node of the current tree. All the experiments are based on variants of discriminant analysis, assuming equal covariance matrices (LDA) or unequal covariance matrices (QDA), with feature selection (in the case of LDA), and (in one case) testing whether to use QDA or LDA. Unix shell scripts were used as a front-end to SAS (SAS Institute, Inc., 1990) programs; we limited our experiments to what was readily available.

In all the following examples, the data are divided into a *training set* and a *test set*. The basic approach is the same as in the previous section, whereby observations belonging to a node have "no memory." That is, they do not know how they got there, or if they are close to a split and, hence, constitute some sort of boundary decision.

4.1. Satellite Data

This is the same data set as that used in Michie et al. (1994) (see section 9.3.6). Briefly, the data constitute intensities from four spectral bands of a LANDSAT satellite image from 3 × 3 pixel neighborhoods, together with the class for the central pixel. There are 6 classes, and (9 × 4=) 36 attributes for 4,435 (train) and 2,000 (test) observations. In the *StatLog* experiments none of the classical statistical methods performed well, with QDA ranking 14th and LDA ranking 19th out of 22 algorithms. Here, we tried two approaches:

A recursive LDA, with two splits; and
A QDA split, followed by LDA.

Note that both of these could be simply represented by a (non-binary) tree. The results show a marked improvement in both cases over the simple raw algorithms (see Table 8.2).

4.2. Probability Trees

Since we are using normal-based assumptions, we can associate a probability of class membership for each observation at each mode. This can be used in the next step of the splitting procedure as follows.

At the current step, given a "node" of the tree, every observation has a probability of belonging to that node. So, we do not partition the observation space in the usual sense. However, the given set of probabilities could be used to assign each observation to a given node by simply choosing the maximum. Since each node is associated with a class, this gives a classification for each observation.

TABLE 8.2. Error Rates for Recursive Splits on Satellite Data, and Rank Performance Compared with *StatLog* Results

Split 1				Split 2			
	Error Rate				Error Rate		
Splitting Type	Train	Test	Rank (test)	Splitting Type	Train	Test	Rank (test)
LDA	0.149	0.171	19	LDA	0.101	0.139	9
QDA	0.109	0.152	14	LDA	0.089	0.135	6

From Michie et al., 1994.

This initial step is to run QDA or LDA which gives, for observations x_i, $i = 1, \ldots, n$ (where there are n observations),

$$p_i^{(1)}(j) = p^{(1)}(x_i \in A_j)$$

as the probability (at step 1) of belonging to class $A_j, j = 1, \ldots, q$. Alternatively, we can say that the probability that the observation x_i belongs to node k is $n_i^{(1)}(k)$, and at step 1 we have $n_i^{(1)}(k) = p_i^{(1)}(k)$, $k = 1, \ldots, q$. At step r we obtain

$$p_i^{(r)}(j) = \sum_{k=1}^{q^{r-1}} n_i^{(r-1)}(k) p(x_i \in A_j | \text{data at node } k) \tag{1}$$

where the data at node k are the (training) data used to estimate the conditional probabilities in equation (1), and these data are given by a subset of data given by the set $\{x_l : n_l^{(r-1)}(k) = \max_j n_l^{(r-1)}(j)\}$. At step r there will be q^r nodes, and we thus grow the tree by evaluating the probabilities at node l by

$$n_i^{(r)}(l) = n_i^{(r-1)}(k)\, p_i^{(r)}(j) \qquad \text{where } l = j + q(k-1)$$

The motivation of this method is that, even after the data have been split, there is useful information of two kinds which can be further used, namely:

1. The probability of class membership for each observation. That is, we treat observations at the same node according to their proximity to the splitting boundary.
2. We try to use the information at one node to help choose the splitting boundary at another node. The success of this aspect will depend on the configuration of the modes of each distribution. Note that *pure* nodes (where the observations all belong to one class) do not add any useful information to where the other nodes should be split. The algorithm makes this explicit by using the relative proportions at the node as *prior* probabilities. So if the data at node k all belong to class j^*, then

$$p(x_i \in A_j | \text{data at node } k) = \begin{cases} 1 & \text{if } j = j^* \\ 0 & \text{otherwise} \end{cases}$$

The *test set* is "classified" (in the sense that probabilities are initially obtained) using the above tree. To do this we merely use the x_i in the test set, but always estimate class probabilities using the training data (or a subset of it) as described above. So, for example, at step 1 we obtain

$$p^{(1)}(\text{test data } x_i \in A_j | \text{data in training set})$$

TABLE 8.3. Error Rates for Recursive Splits on Credit Management Data, and Rank Performance Compared with *StatLog* Results

Type of Split	Split 1			Split 2			Final Split		
	Error Rate		Rank	Error Rate		Rank	Error Rate		Rank
	Train	Test	(test)	Train	Test	(test)	Train	Test	(test)
A	0.031	0.033	13	0.041	0.039	15	0.031	0.037	15
B	0.031	0.033	13	0.040	0.040	15	0.024	0.028	7
C	0.051	0.050	21	0.906	0.912	23	0.023	0.027	7

From Michie et al., 1994.

4.2.1. Credit Management Data. This data set was also analyzed in the *StatLog* project and is discussed in Michie et al. (1994) (see section 9.2.1). The data are from the general area of credit management, but little is known due to confidentiality. There are seven attributes and two classes with 16,000 observations in the training set and 4,000 observations in the test set (there is a misprint on Table 9.2 of Michie et al. 1994). We give results for three approaches:

A. A recursive LDA, with 3 splits (as in section 4.1);
B. A probability tree using LDA, with 5 splits; and
C. A probability tree using either QDA or LDA, with 4 splits, the decision being taken on the basis of a test for the equality of covariance matrices.[1]

The results are presented in Table 8.3. Note that the splitting method that updates probabilities (method B) performs better than the method that has "no memory" (method A). Note also that, in the second split using method C, the quadratic discriminant (which was actually chosen at every step by the algorithm) dramatically *increases* the error rate to a ridiculous value. Both the train and test error rates mimic this behavior, which occurs in the even (2nd, 4th,...) splits. The reason is probably because of the unequal covariance matrices and the fact that the "training data" used to estimate the parameters are rather different from the rest of the data set. This quirky behavior probably contraindicates the use of recursive QDA without further investigation.

4.2.2. Belgian Power Data. This data set is related to classification of instability in large-scale power systems. It was used in Michie et al. (1994) (see section 9.5.5), and was donated to that project by T. van Cutsem and L. Wehenkel, University of Liège, Institut Montefiore, Sart-Tilman, B-4000 Liège,

[1]Using the POOL-TEST option with SAS.

TABLE 8.4. Error Rates for Recursive Splits on Belgian Power Data, and Rank Performance Compared with *StatLog* Results

Type of Split	Split 1			Split 2			Final Split		
	Error Rate		Rank	Error Rate		Rank	Error Rate		Rank
	Train	Test	(test)	Train	Test	(test)	Train	Test	(test)
A	0.022	0.029	8	0.030	0.039	15	0.015	0.039	15
B	0.022	0.029	8	0.014	0.022	6	0.009	0.031	10
C	0.022	0.026	7	0.023	0.033	10	0.014	0.021	6

From Michie et al., 1994.

Belgium. There are two classes, and 1,250 observations, with 28 attributes in both the training and test sets. We give results for three approaches, each with three steps:

A. A recursive LDA (as in section 4.1);
B. A probability tree using LDA; and
C. A probability tree using LDA using a stepwise selection procedure to choose the important variables at each stage.[2]

The results are presented in Table 8.4. Note that there is a small discrepancy with the *StatLog* results, which used a crude variable selection procedure to eliminate highly correlated variables.

In this case, the results got substantially worse when method A was used; method B is clearly preferred. This raises the question of whether the more commonly used decision tree algorithms [such as CART (Breiman et al., 1984) and *C4.5* (Quinlan, 1993)] could be simply improved by passing probabilities down the branches of the tree. Note that, although the error rate for the *training* data is very low by the third step, the error rate for the *test* data has started to worsen. This indicates that, by the third step, the algorithm has overtrained and some pruning is required. Within the general framework of multivariate analysis and normal distribution assumptions one possibility is to use Wilks' lambda statistic as a pruning method. Applying this to the training data as a pruning strategy gives a slightly better error rate of 0.019 for the training data and 0.029 for the test data. However, it should be stressed that overtraining was generally not found to be a problem in the rest of our experiments, in that it was usually obvious when to stop growing the tree on the basis of the training data alone.

[2]SAS command STEPDISC.

The stepwise procedure (method C) gave quite good results here. Note that, if we had used only the most important variables at each split, this would give the usual splits parallel to the axes. However, this refinement was not found to give good results—it would seem that the normal distribution assumptions cease to hold, and splitting criterion other than that which maximizes a t-test statistic are preferred. Our results are based on a standard approach or forward and backward selection, which makes use of an F statistic to decide which subset of variables to use.

4.3. Discussion

It would be premature to draw any conclusions based on the limited results reported here. However, it seems that trees that pass probabilities down the branches, along with tentative classifications, seem to work better than trees that have no memory.

Many of the standard decision-tree algorithms have slightly different splitting criteria; there seems to be little to choose between them. However, in the above experiments, the final result clearly does critically depend on the "distance" measure used. What is required is a formal procedure to select automatically among the different choices.

The "tree" adaptation of the standard discriminant methods (based on normal distribution assumptions) is obviously most suited to multimodal data, or other violations to these assumptions. Unlike the approach used by Hastie and Tibshirani (1996), there is no explicit modeling of the distribution as a mixture of (an unknown number of) normal distributions, but both approaches are based on a modification of LDA. Another related approach is that of Smyth et al. (1990), who also learn a set of probabilistic rules using information-theoretic techniques, giving a hybrid classifier.

5. CONCLUSIONS

The claim that "Distance-Based" Decision Trees can satisfy the ML criteria does not seem too far-fetched. The essential idea is that the new observation is "most like" one of the existing observations and should be assigned the same class. The issue of what is implied by "likeness" needs to be measured by a suitable distance metric. However, although it is easy to understand, this rule does not provide much understanding as to *why* each classification takes place, but then the same comment might be said for many large decision trees.

There are a number of issues however, that need further attention. Apart from the performance relative to other classifiers, and a characterization of which type of data set it will do well on, we need efficient methods to obtain appropriate distance measures. This is further being investigated.

REFERENCES

Andrews, D. F., and Herzberg, A. M., eds. (1985). *Data: A Collection of Problems from Many Fields for the Student and Research Worker.* Springer-Verlag, New York.

Breiman, L., Friedman, J. H., Olshen, R. A., and Stone, C. J. (1984). *Classification and Regression Trees.* Wadsworth & Brooks, Monterey, CA.

Feng, C., and Michie, D. (1994). Machine learning of rules and trees. In *Machine Learning, Neural and Statistical Classification* (D. Michie, D. J. Spiegelhalter, and C. C. Taylor, eds.), pp. 50–83. Ellis Horwood, Chichester, UK.

Hastie, T., and Tibshirani, R. (1996). Discriminant analysis by Gaussian mixtures. *J. Royal Stat. Soc., Ser. B* **58**, 155–176.

Hertz, J. A., Krugh, A., and Palmer, R. G. (1991). *Introduction to the Theory of Neural Computation.* Addison-Wesley, Redwood City, CA.

Kohonen, T. (1989). *Self-Organization and Associative Memory*, 3rd ed. Springer-Verlag, Berlin.

Mardia, K. V., Kent, J. T., and Bibby, J. M. (1979). *Multivariate Analysis.* Academic Press, London.

Michie, D., Spiegelhalter, D. J., and Taylor, C. C., eds. (1994). *Machine Learning, Neural and Statistical Classification.* Ellis Horwood, Chichester, UK.

Quinlan, J. R. (1993). *C4.5: Programs for Machine Learning.* Morgan Kaufmann, San Mateo, CA.

SAS Institute, Inc. (1990). *SAS User's Guide: Statistics* (6.04 ed.). SAS Institute, Inc., Cary, NC.

Smyth, P., Goodman, R. M., and Higgins, C. (1990). A hybrid rule-based/Bayesian classifier. In *Proceedings of the European Conference on Artificial Intelligence, 1990,* pp. 610–615.

StatSci Europe (1991). *S-plus User's Manual,* Tech. Rep. StatSci Europe, Oxford.

CHAPTER 9

Learning Fuzzy Controllers from Examples

C. BAROGLIO
A. GIORDANA
G. LO BELLO
R. PIOLA

1. INTRODUCTION

In this chapter, we present a methodology that deals with the development of control programs for industrial applications, with a particular interest for robotics. This work has been funded by the European Community BRA 7274 *B-Learn II* project, which is focused on the application of machine-learning techniques to the control of industrial robots; however, the methodology that will be presented can successfully be used for regression tasks in general.

Robotics is a challenging task for artificial intelligence. From a purely scientific point of view, the reason is that it offers a chance for producing machines that exhibit a human-like behavior, adapting it to the circumstances instead of running a fixed program. More practically, the expected benefit is the possibility of increasing the flexibility of control, allowing better performance, and decreasing, at the same time, the costs of control design.

From a machine-learning perspective, a first challenge is to check whether the traditional learning techniques scale successfully to learning a behavior, that is, approximating a control function. The advantage would be an immediate reduction of the controller design phase to the recording of a set of examples of execution, in which a human operator plays the part of the controller to show what has to be done. The harder task of generalizing the examples into a knowledge base would be left to a learning program. However, since the functions to be approximated have continuous, numerical domains

Machine Learning and Statistics, The Interface, Edited by G. Nakhaeizadeh and C. C. Taylor.
ISBN 0-471-14890-3

and co-domains, some means of coping with non-symbolic information needs to be found.

In the literature, function approximation has been approached by different communities, that have applied various tools such as neural networks (Rumelhart and McClelland, 1986; Rumelhart et al., 1985), regression trees (Breiman et al., 1984), or fuzzy logic (Berenji, 1992b) to the problem, and has alternatively been seen as a supervised or as a reinforcement learning task.

In the *B-Learn II* project, we have developed an integrated methodology, which combines many approaches. More specifically, we use supervised learning in order to generate an initial approximator, that is then fine-tuned by means of gradient descent (the best-known neural network learning method). Eventually, the obtained controller can be further tuned by means of reinforcement learning.

Fuzzy controllers have been chosen as the function approximators to be learned, because, on one hand, they can be represented by a set of if–then rules (just like the rule bases symbolic learning algorithms usually deal with), but, on the other hand, they can cope with continuous domains. Therefore, they are a means of relating numerical and symbolic information. Another reason for utilizing fuzzy controllers is that this technology is now widely used in industrial applications (Bonissonc and Chiang, 1993), thc samc kind of applications in which we are interested.

Nonetheless, in this chapter only part of the work is presented. First, the reinforcement learning (RL) step is neglected [those readers who are interested in RL can see Baroglio et al. (1994) for a description of the complete methodology]. Second, even though various symbolic learning techniques have been compared on the synthesis task, only the work done using the system CART, proposed by Breiman et al. (1984), is reported here.

Our method has been tested both on a robotics application, using a robot KUKA-IR 361 engaged in the "peg-into-hole" task, and on a prediction task, using the Mackey-Glass temporal series. In the former application, a set of examples of insertion was generated using a PID-controller (instead of a human controller) to select the actions to be taken by the robot. Then, part of the examples were used to synthesize an approximation of the behavior of the PID controller. We have checked the accuracy of the obtained controller, comparing it with one of the other approximators, such as Time Delay Neural Networks, showing that it has a better performance.

In the following we will start by introducing the problem of function approximation, to focus on approximation by means of a fuzzy knowledge base, describing the structure and the operation of a fuzzy controller. We will, then, describe how a fuzzy controller can be learned by means of CART, and an optimization step, named *tangling*, will be proposed. After that, the gradient descent tuning step will be presented and, eventually, experiments and experimental results both on the peg-into-hole and on the Mackey-Glass time series will be reported.

2. APPROXIMATING CONTROL FUNCTIONS

The task of learning to approximate continuous functions has been investigated in many fields, using alternative approaches such as statistics (Breiman et al., 1984; Specht, 1988, 1990), connectionism (Barto et al., 1990; Rumelhart and McClelland, 1986), fuzzy logic (Berenji, 1992b; Zadeh, 1992) and symbolic machine learning (Quinlan, 1986; Sammut et al., 1992).

In applicative domains such as robotics and automated controls, two approximators attracted great attention: Multi-Layer Perceptrons (MLPs) and Fuzzy Controllers (FCs). The reason for their popularity is mostly due to their robustness and inherent simplicity, which allows applications to be quickly developed without requiring specific skills from the user. Furthermore, low-cost hardware supporting real-time computation is now available for both architecture types. Note, however, that MLPs and FCs are only two examples of two of the alternative approaches to function approximation we find in the connectionist paradigm, which involves a larger zoo of neural networks.

Apart from regression and prediction trees, all the existing approximators (independently from the approach they originate from) can be represented by some multi-layered network, in which nodes (also called neurons) computing a simple function (or activation function) are connected by means of weighted links. The differences between the various kinds of approximators reside in the types of activation functions used and in the learning algorithms, that is, the methods to construct the network and to update the weights.

The activation functions can be grouped into two major families: *Open Field Functions* (OFFs) and *Locally Receptive Field Functions* (LRFFs). The OFFs exhibit an anti-symmetry property with respect to a hyperplane splitting the function domain into two semi-spaces, whereas LRFFs have an axis of radial symmetry. A good example of the first family is the well-known sigmoid function, widely used in MLP-like networks. Examples of the second family are the multidimensional Gaussian functions used in Radial Basis Function Networks (RBFNs) (Poggio and Girosi, 1990) or in Statistical Neural Networks (SNN) (Specht, 1988, 1990), The choice of an activation function belonging to one or the other family entails extremely different properties.

As has been proven in Hornik et al. (1989), a universal function approximator can be constructed using only three layers of nodes (input, hidden, and output), provided the activation function in the hidden layer be nonlinear. In Figure 9.1 we compare the structure of two networks of this kind: one based on sigmoidal and the other on bell-shaped functions. From a topological point of view, they look identical. Nevertheless, the way they encode the target function is very different. Considering the activation function $O_i = \sigma(\Sigma_{j=1}^{N} w_{ij} I_j)$ of the MLPs hidden units, we see that each neuron, apart from a narrow region where the sigmoid transient occurs, splits the input space into two semi-spaces where its output is close to either 1 or 0. Only the semi-space where the output is close to 1 contributes to the value of the target function.

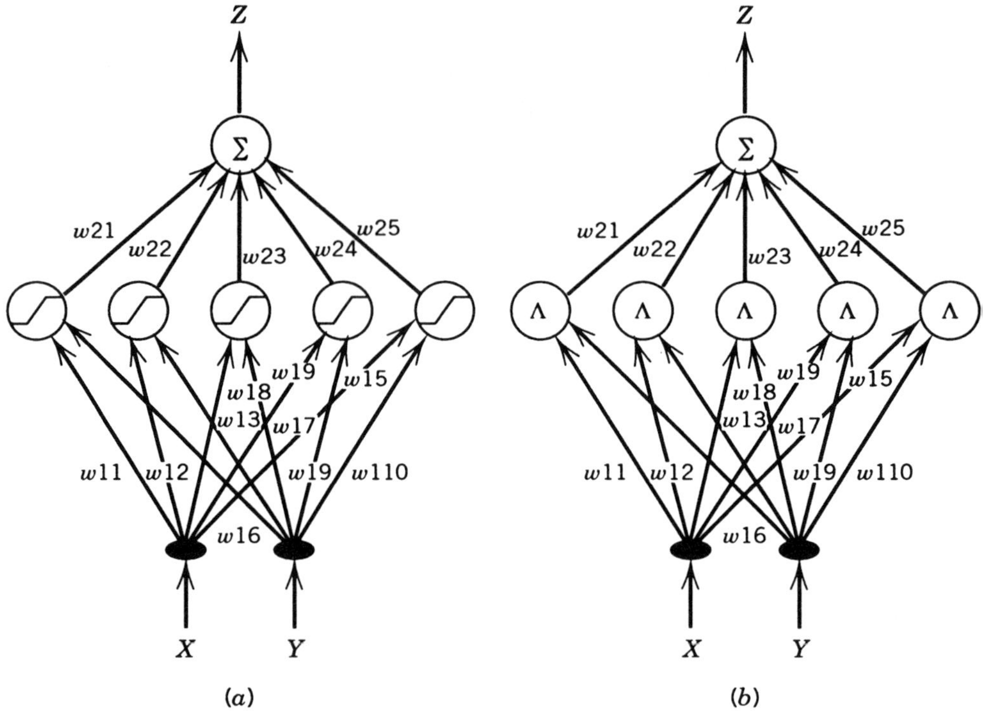

Figure 9.1. *Multilayer perceptron (a) and Radial Basis Function Network (b) architecture. The fundamental difference is in the activation function in the hidden layer neurons. In both cases the output neuron performs the sum* $\Sigma_{i=1}^{N} w_i o_i$.

Consider now the LRFN, in which each neuron is associated to a closed region in the input space (hyper-ellipse), where its response is significantly greater than zero, and dominates over every other node. The contribution to the approximation of the target function comes essentially from this region. Figure 9.2 shows a graphical representation of these concepts.

In order to understand the different behaviors of the two network types, suppose we modify a weight between two nodes in the MLP, as is done by the backpropagation updating rule during the training phase. The effect involves an *infinite* region of the input space and can affect a large part of the co-domain of the target function. On the contrary, changing the amplitude of the region of the input space, in which the activation function of a neuron in an LRFN fires, or shifting its position will have an effect *local* to the region dominated by that neuron. More generally, this *locality property* of the LRFN allows the network layout to be incrementally constructed (see, for instance, del R. Millán, 1994) adjusting the existing neurons and/or adding new ones. As every change has a local effect, the knowledge encoded in the other parts of the network is not lost, so it will not be necessary to go through a global revision process.

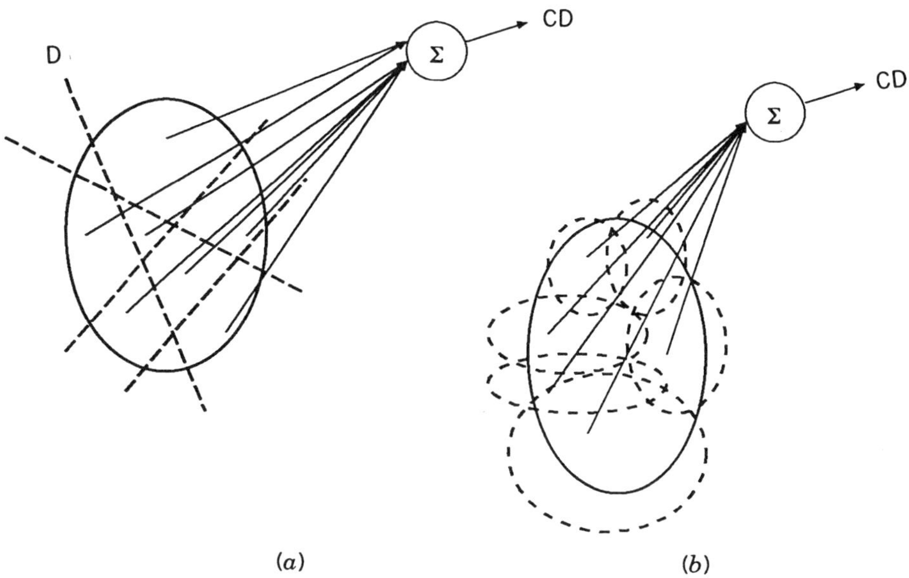

Figure 9.2. *Comparison between the methods used by MLPs and RBFNs to approximate the target function.*

Another property related to locality is the tendency to *undergeneralize* exhibited by LRFNs. Depending on the way the hidden layer (the layout) is constructed, it is possible to have large regions of the input space that are not covered by any neuron. This means that, when current input situation falls in such regions, the output of the network will constantly remain close to the null value (usually zero), because no neuron will fire. On the contrary, every neuron of a MLP covers an infinite region of the input space, so the network is likely to return a non-null value in any situation. As a consequence, a MLP, when trained, has the tendency to overgeneralize, whereas a LRFN always tends to undergeneralize, in the sense that its output will be null for every situation that is far from the examples processed during the training phase. As we will discuss later on, under(over)generalization can be either an advantage or a disadvantage, depending on specific cases.

Finally, an important property of the LRFNs [as opposed to MLPs, which, with few exceptions (Towell et al., 1990), are black-boxes with respect to their interpretation] is the possibility of giving an immediate, *symbolic*, interpretation of the hidden neuron semantics. In fact, a closed region in the input space can be approximated using a hyper-rectangle which, in turn, can be described by a propositional formula (see Figure 9.3).

This property has already been exploited for *manually* encoding fuzzy controller's layouts using qualitative domain knowledge (Bonissone and Chi-

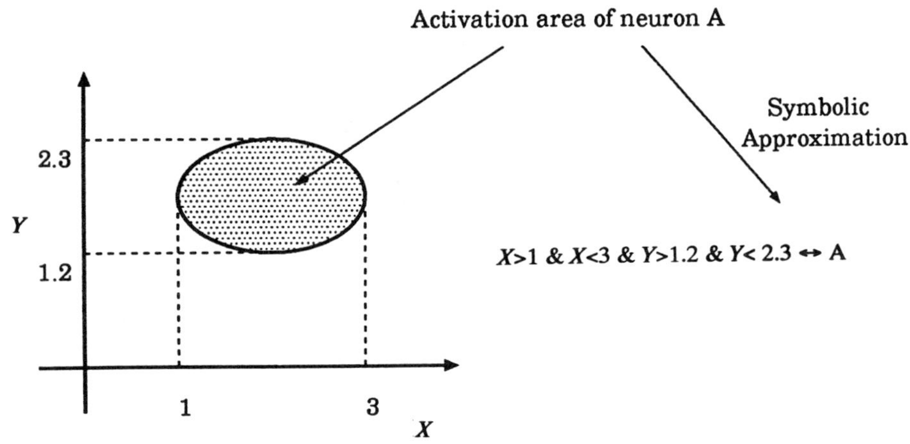

Figure 9.3. *The closed region where LRF neuron is dominant can be roughly approximated using a conjunctive logical assertion.*

ang, 1993); later on, we will show how symbolic learning algorithms can be used *automatically* to build excellent layouts for LRFNs.

In the following we will focus on a specific kind of function approximator that has been thoroughly investigated by the authors and in the project: the Fuzzy Controller, that belongs to the LRFN family. However, for a better understanding of the concepts that will be recalled, first a brief introduction about some Fuzzy Logic features relevant to control is reported.

2.1. Fuzzy Logic and Controllers

Fuzzy logic originates from a proposal by Zadeh (1965) aimed at setting a logical framework for modeling human qualitative reasoning. For most of the concepts we use in everyday life, such as "tall" or "fast", it is not possible to give a sharp semantics. Let's consider, for instance, the word *hot*. There is not a specific temperature over which an object is hot, and below which it is cold. Rather, an object can be hot to various degrees: it can be cold (not hot), warm, pretty hot, quite hot, hot, extremely hot, depending on its temperature, on the circumstances, and on the actor. So, for instance, the way an oven is hot is different than that in which the day is hot, or someone's forehead is. For any specific situation one could, however, consider representing the truth value of "hot", using a function that returns values proportional to the temperature only, that is, a function defined on some continuous, numerical feature. Intuitively, this is the idea that has been developed within the framework of *Fuzzy Logic*.

The main difference between Fuzzy Logic and better-known binary logic stands in the way the truth values of predicates are computed. While tradition-

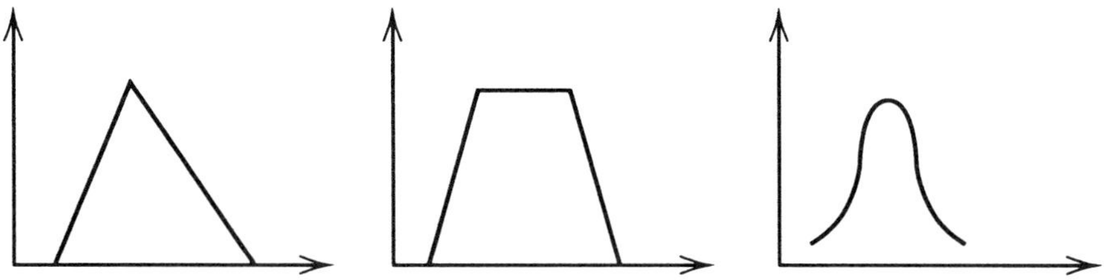

Figure 9.4. *Three common shapes for membership functions.*

ally a predicate value can only be 1 (true) or 0 (false), depending on whether the values of its variables satisfy predicate semantics, in the case of Fuzzy Logic a degree of truth is introduced. This degree is a real value computed by a *membership function*, having as a co-domain the interval $[0, 1]$. Therefore, it is possible for a particular term to satisfy a predicate only partially. A predicate having such a membership function is said to be *fuzzy* and each interval of the domain on which the membership function returns values greater than zero is known as a *fuzzy set*. Each fuzzy set is identified by means of a name or *linguistic variable*.

Membership functions can have any shape, although the most common are the ones reported in Figure 9.4.

Fuzzy sets can be combined by applying operators to their membership functions; the operators are: fuzzy *and* ($\wedge$), *or* ($\vee$), and *not* ($\neg$). Briefly, if A and B are two fuzzy sets, and μ_A and μ_B are their membership functions, then $A \wedge B$ will have membership $min(\mu_A, \mu_B)$, $A \vee B$ will have membership $max(\mu_A, \mu_B)$, and $\neg A$ will have membership $1 - \mu_A$.

However, despite its initial appeal, Fuzzy Logic did not find a rigorous formalization and encountered many difficulties in being accepted by the Artificial Intelligence community, but, on the other hand, it found a very successful application in *fuzzy controllers* (FC), originated by the seminal work of Mamdani and Assilian (1975).

In a control problem there is: a physical system to be maneuvered so that a task be performed, a set of sensors that continuously perceive the values of features meaningful to the task and, last, a controller that generates a set of control values using the perceived inputs. By referring to the typical FC architecture described by Berenji (1992a) (see Figure 9.5), the controller's kernel is a fuzzy knowledge base, which is made of a set of if–then rules of the kind:

$$R: if\ (X_1 \in A_1) \wedge (X_2 \in A_2) \wedge \cdots \wedge (X_n \in A_n)\ then\ (Z \in C)$$

where $X_1, X_2, \ldots, X_n$ are variables that range on input features, Z is the output

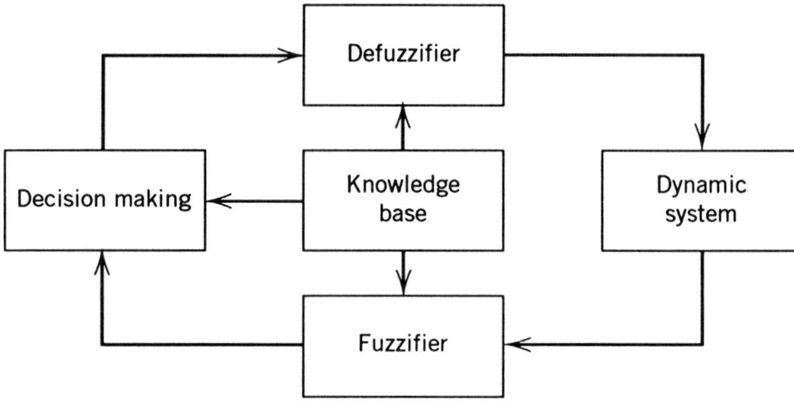

Figure 9.5. *General architecture of a fuzzy controller.*

variable, and ranges on the co-domain of the function to be approximated, while $A_1, A_2, \ldots, A_n$ and C are fuzzy sets. The operation is as follows:

1. The numerical, input values are matched against all the fuzzy sets in the antecedents of the rules in the knowledge base, computing the membership values. This step is known as *fuzzification.*
2. The membership values returned are combined rule by rule, by using the *and* operator (minimum, product, t-norm, etc.). If the result is greater than zero, the rule is said to fire, and the computed value is said to be the strength for the rule.
3. A control output for each rule is computed by applying an inverse mapping to the strength. This phase is known as *defuzzification.*
4. The contributions of all rules are combined by means of a normalized weighted sum.

Usually, FCs are designed and encoded manually, in a way similar to what is done for building expert systems. According to many authors (Bonissone and Chiang, 1993), even in this way it is possible to generate high-performance controllers.

2.2. Fuzzy Controllers and Neural Networks

Like all function approximators, FCs, as explained in section 2, can easily be realized by means of a neural network. More specifically, the model that is used, though originating from a very different approach, bears strong similarities to RBFNs, sharing the same learning algorithms.

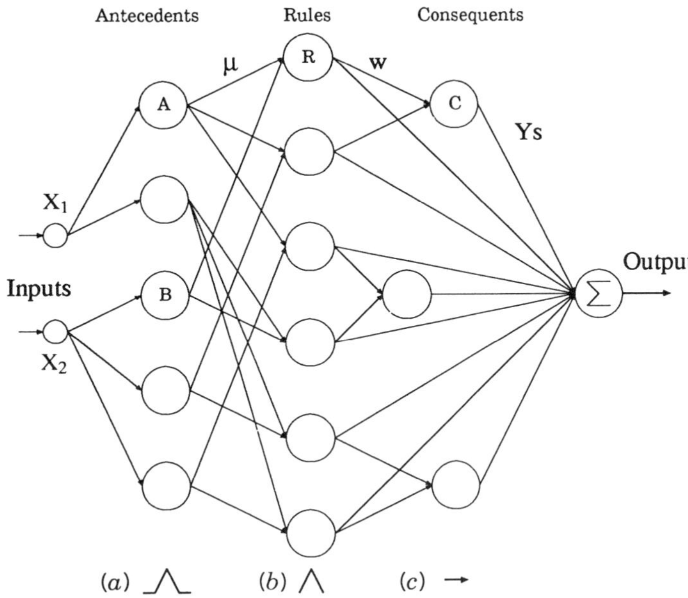

Figure 9.6. *Network corresponding to the fuzzy controller used in current experiments. Gaussian units have a unidimensional Gaussian activation function. Product units compose the input values using arithmetic product or* softmin. *An average sum unit performs the weighted sum of the activation values received from the product units.*

Actually, FCs do not correspond to a single network model but to a wide family that contains the topology of Figure 9.1b as a special case. Nevertheless, in our research, we focused on a specific topology, derived from that experienced by Berenji (1992b) in control problems similar to the ones tackled in this chapter. Figure 9.6 shows it. This choice is particularly appealing because it is always possible to *re-convert* a trained network into a fuzzy knowledge base. The reasons that render it possible will be evident when the model and the learning rule will have been introduced.

In this section we will start by describing the model, relating its structure to the general description of a FC we have already given in section 2.1; the learning rules are, instead, reported in section 3.3.

Starting from the topology, our model can be structured into three layers plus input and output layers. Input signals proceed forward from the input layer through all the others, causing the activation of the correspondents of some of the rules, each of which will return a piece of contribution for the answer to the stimulus.

The neurons in the first hidden layer are feature detectors with a unidimensional Gaussian activation field, that receives only one input signal. Each feature detector actually corresponds to a fuzzy set used in the antecedent of some fuzzy rule, thus, the value computed by the corresponding neuron is the value of the membership function applied to the input. Hence, the feature detectors have the following activation function:

$$\mu_i = e^{-\left(\frac{x^{(j)} - C_i}{\sigma_i}\right)^2} \tag{1}$$

where i is an index used to identify the neuron μ_i is the identifier of the membership for such a neuron (a bell-shaped function of mean C_i and width σ_i), and, last, $x^{(j)}$ is the *j-th* element of the input vector. When the activation value falls below a given threshold (0.01 in the presented experiments) it is set to zero in order to preserve LRFN properties. The task executed by the first hidden layer neurons is the fuzzification step, described in section 2.1.

The neurons in the second hidden layer receive the membership values computed by two (or more) neurons of the previous layer, which are required to process different input signals. To carry on the parallel with the architecture of section 2.1, it should be pointed out that each neuron in this layer corresponds to a rule in the fuzzy knowledge base. Moreover, the first hidden layer neurons connected to a second hidden layer one are those realizing the antecedent fuzzy sets in the rule at issue. If they all fire then the rule fires, and, in this case, the output is obtained by means of the *softmin* composition rule:

$$r_R = \frac{\Sigma_i \, \mu_i e^{-k\mu_i}}{\Sigma_i \, e^{-k\mu_i}}$$

This function is used instead of the regular *min* (which is usually applied in order to compute the fuzzy *and*, the conjunction operator in *Fuzzy Logics*). This choice is due to the fact that the *softmin* is differentiable, allowing the application of gradient descent, but it tends to the regular *min* for $k \to \infty$.

Third hidden layer neurons correspond to the fuzzy sets in the consequents of the rules. Contrary to the antecedent fuzzy sets, they are triangular, and memorized by means of their median (see Figure 9.7). These neurons realize the defuzzification step of the algorithm, computing the equation:

$$\mathscr{F}_R = w_R F_1 + (1 - w_R) F_0$$

Finally, the output neuron performs the linear combination

$$\hat{Y} = \frac{\Sigma_i \, w_i r_i}{\Sigma_i \, r_i} \tag{2}$$

of the weights w_i and the activation values r_i of the corresponding second-layer neurons.

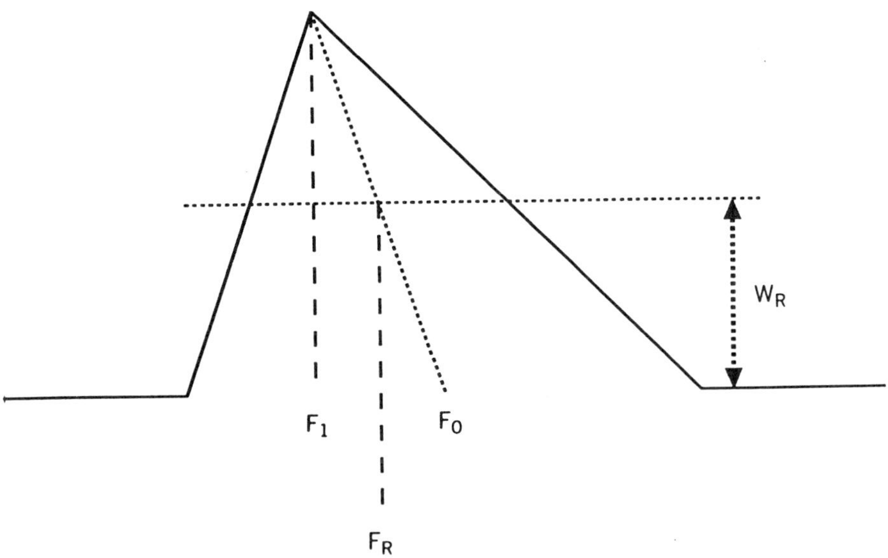

Figure 9.7. *F_1 and F_0 evaluation.*

3. HOW TO LEARN A FUZZY CONTROLLER

Fuzzy controllers produced for real industrial applications are commonly designed and fine-tuned manually, even though several proposals can be found in the literature for learning them on-line [adapting the methods of reinforcement learning (Barto and Sutton, 1981; Watkins and Dayan, 1992)] or off-line, see Rumelhart et al. (1985), with ANNs; other examples are reported in Raju and Zhou (1993) and in Berenji and Khedkar (1992). Nonetheless, in all these cases, the structure of the controller is still supplied by a human expert, as is the case for the topology of a multi-layer perceptron. Our goal is, then, to eliminate the necessity of a human expert in order to generate the controller layout.

Although, in principle, it is possible to build a unique controller with many outputs, we followed the more usual approach of building an independent controller for each of the output signals. To this aim a three-step learning method based on regression trees (Breiman et al., 1984) has been applied: (a) a first set of rules is generated using CART; (b) the set of rules obtained in step (a) is rendered more compact by merging similar input and output fuzzy sets; (c) the fuzzy set parameters are finely tuned by performing a gradient descent with a backpropagation-like algorithm.

In the following we will briefly review the techniques and the algorithms used at the different steps.

3.1. Regression Trees and Fuzzy Controllers

Regression Trees (RTs) were introduced by Breiman et al. (1984) as an extension of decision trees to the task of piecewise function approximation. A decision tree is a structure, which is recursively built starting from a root node, in which each inner node represents a test on a feature in the domain. For instance, if one has to learn a classification theory for apples and oranges, and in the domain there is a feature *color* that ranges on the set ⟨*red, green, orange*⟩, one inner node could be the test *color* = *red.*

A learning procedure able to build a decision tree is an implementation of an algorithm that has to specify a way for selecting features which are useful for distinguishing among the objects belonging to different target classes. At each stage during the tree generation, if one selects any path in the tree and checks the whole learning set to see which examples satisfy all the tests in the path, one will obtain the *extension* of the last node in the selected path. If a node extension contains examples belonging to different classes, then it can be further expanded by selecting some other feature and building a set of children nodes, one for each value in the range of the chosen feature. A node whose extension contains examples of a single class only is named a *leaf* and cannot be expanded anymore. It will be labeled with the name of the class at issue.

After having been built, a decision tree can be used for classifying new examples. One has simply to start from the root and follow the path the example verifies. The leaf that will be reached will give the classification for it.

Decision trees have been applied to classification tasks in discrete, symbolic domains. Regression trees are an extension of this to continuous domains and

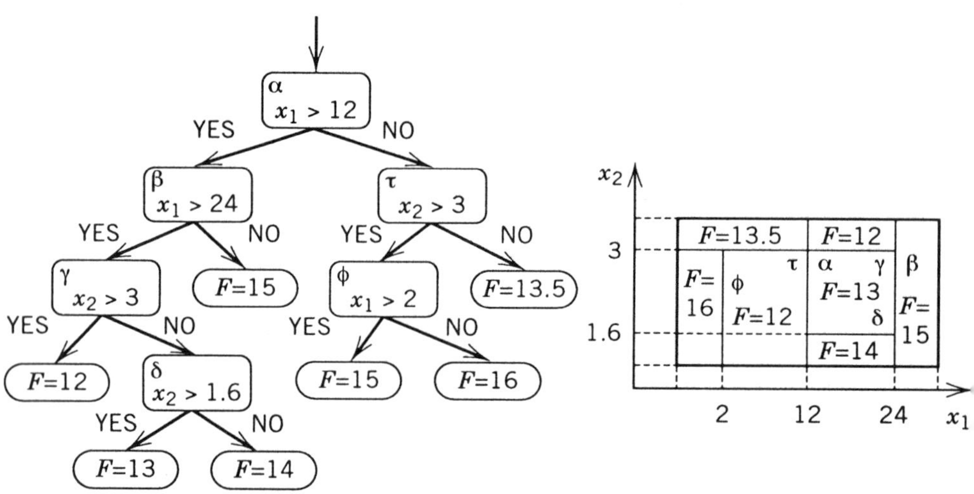

Figure 9.8. *A regression tree and the way it partitions a two dimensional input space into a set of closed regions.*

especially to the task of function approximation. The learning set will be a set of tuples representing the target function, that is, the one to be approximated by points.

As in decision trees, each inner node will be associated with a test, but instead of labeling leaves by means of a class identifier, in RTs leaves are labeled by a scalar value (see Figure 9.8, *left*), that is one possible output value of the target function. This value will be chosen according to a criterion specific to the learning algorithm. Of course, since input features also will have continuous domains, some criterion for selecting the ones to be used for building the tests, and the specific values to be checked, ought to be added to the learning algorithm too.

When used for function approximation, the tests associated to the inner nodes can be geometrically interpreted as hyperplanes. Hence, each path in the tree identifies a region in the input space, which is defined by a set of hyperplanes. More specifically, because of the learning algorithms that have been chosen, after a tree has been built, it partitions the region of the input space containing all the learning examples into a finite set of closed regions. To each region a scalar value is then associated (the one that corresponds to the leaf at the bottom of the path). The *tree* is the *function approximator*, and the scalar value is the one returned by the approximator whenever it is applied to an input belonging to the region associated to it (see Figure 9.8). Thus, a RT approximates a function by means of a histogram, where the steps of the histogram are automatically computed by the algorithm.

Now, we will describe the algorithm for building RTs we have used. It has been designed by Breiman et al. (1984) and its name is CART.

Let x be an input vector, $y = f(x)$ the function to be approximated (target function), C_x the name of the leaf to the extension of which the input example belongs, $y = f^*(x)$ the output that will be produced; CART works as described in the following:

- Given a learning set made of pairs $\langle x_i, y_i \rangle$ and an error measure, which is usually the mean square error $e = (\Sigma_i^N (f^*(x_i) - y_i)^2)/N$,
- Start with a tree with only one node, classifying all samples in the same class, and let the label of this class be the average value for the target y (Figure 9.9a).
- While the desired precision or size are not achieved, repeat:
    - Choose a leaf, F, classifying incorrectly some samples (Figure 9.9b), *i.e.*, that places in the same region two values x_{i_1} and x_{i_2} for which the distance $\|y_{i_1} - y_{i_2}\|$ is greater than a fixed threshold.
    - Choose a feature j and a value k so that the classification error in the resulting tree is reduced.
    - Convert F into a node with the test $x^{(j)} \geqslant k$ and two new leaves F_T and F_F (where T stands for true and F for false) as children (Figure 9.9c).

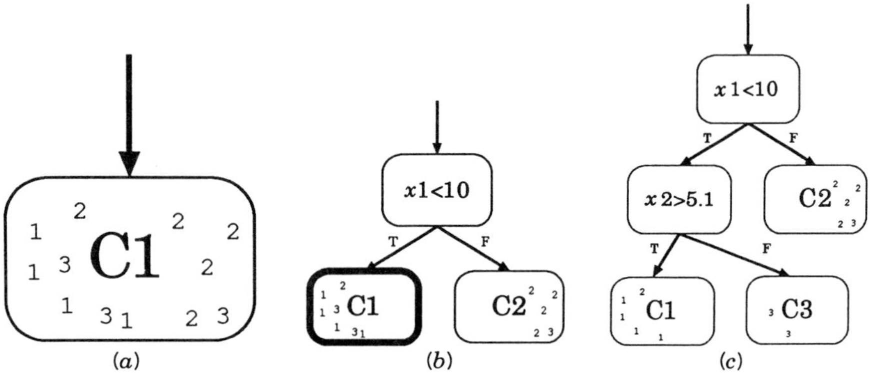

Figure 9.9. *How a regression tree is initialized (a) and built (b, c).*

- Label F_T and F_F with the average value for the y of the samples classified by each of them.

Thresholds on the size of the leaves (computed as the number of examples covered and their variance) are also used to stop the generation of nodes. To be noticed is that the generated tree might overfit the training examples generalizing poorly on unseen values.

Now, a refinement step can be executed in order to simplify the tree: the RT is explored bottom-up, suppressing nodes whose deletion does not increase the error on a second data set, called *pruning set*. In our experiments we used the classical error-complexity (post-)pruning; for a survey of various pruning algorithms, see Esposito et al. (1993).

3.1.1. Converting a Regression Tree into a Fuzzy Controller. A RT can be employed directly to build a controller, but it has some major disadvantages: it approximates the target function by a histogram, while smoothness of the output is desirable to avoid harmful mechanical solicitations; its consequents, that is, the output values, can be fine-tuned by means of some variant of backpropagation, but the input classification part cannot.

Because of these reasons, we worked to convert a RT into a fuzzy knowledge base that can undergo refinement using backpropagation. The algorithm we have developed turns each path in the tree into a fuzzy rule by applying the following steps:

- Identify the regions R_i in which the input is classified.
- For each dimension j of these regions, generate an input fuzzy set I_{ij}, encompassing the extension of the region along the jth dimension (if there is not already an identical set).

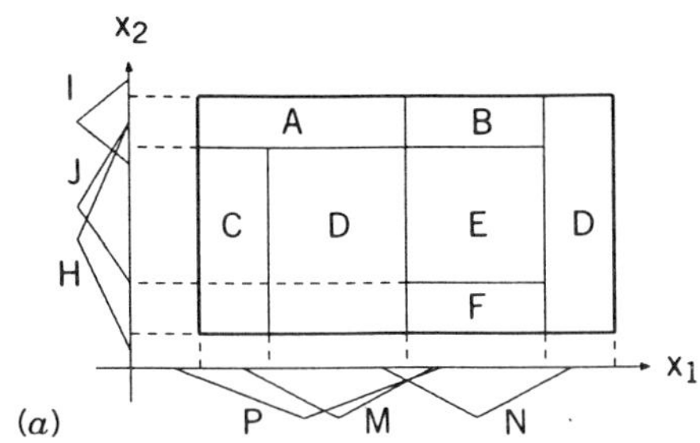

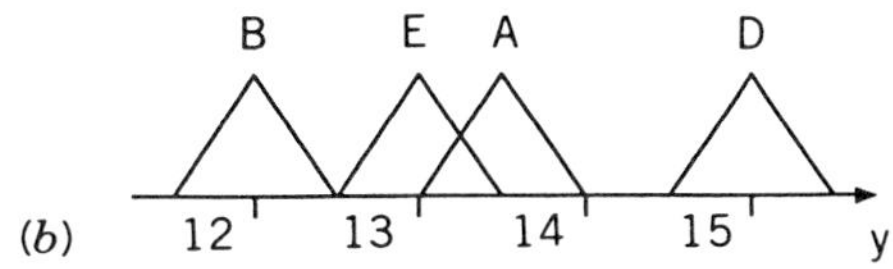

x_1 is $P \wedge x_2$ is $I \Rightarrow out$ is A

x_1 is $N \wedge x_2$ is $I \Rightarrow out$ is B

x_1 is $M \wedge x_2$ is $H \Rightarrow out$ is D

(c) x_1 is $N \wedge x_2$ is $J \Rightarrow out$ is E

Figure 9.10. *(a) Some of the fuzzy regions, (b) the output fuzzy sets in which the tree is converted, and (c) the resulting rules. Only four of them are shown for the sake of simplicity.*

- For every distinct output value, generate an output fuzzy set O_k centered on that value.
- For every region R_i, generate a rule combining all feature detectors I_{ij} defining R_i in the antecedent, and the output value associated with R_i in the consequent.

As an example, some of the fuzzy sets and rules given by the tree depicted in Figure 9.8 are shown in Figure 9.10. It is worth noting that the fuzzy regions identified by our algorithm partially overlap, allowing the desired degree of smoothness when passing from one region to another.

3.2. Compacting the Rule Set (Tangling)

The method presented above has a drawback: it tends to produce a big number of neurons in the first and third hidden layers, many of which might have similar activation functions. The problem is that, in this way, the complexity of the controller is increased. In particular, we notice that a greater number of neurons means a greater number of numerical parameters to tune, and, according to Computational Learning Theory, this implies using a wider learning set.

In order to reduce the complexity of the controller, similar fuzzy sets can be merged, where *merging* means that two or more neurons will be absorbed into a single one and the corresponding connections with the other layers will be transferred to the absorbing neuron.

Two fuzzy sets are considered mergeable if their ranges have a similarity degree S greater than an assigned value S_{min}, where the similarity is computed as:

$$S(I_{j,1}, I_{J,2}) = \frac{|I_{j,1} \cap I_{J,2}|}{max(|I_{j,1}|, |_{J,2}|)} \tag{3}$$

In (3) the notation $|I_{j,i}|$ denotes the width of a fuzzy set measured as the length of the segment in the feature space where the activation function has a value greater than 0.5.

The compaction of the rule base is performed by repeating the following steps separately first on the antecedent and then on the consequent fuzzy sets.

- Determine the similarity between all pairs of fuzzy sets.
- Merge the two most similar fuzzy sets provided that their similarity be greater than S_{min}.

As an example, with reference to the sets in Figure 9.10, the tangling algorithm can merge sets A and E (in the output series, Figure 9.10b), or P and M (in the input series, Figure 9.10a).

3.3. Gradient Descent

In our architecture, fine-tuning of the fuzzy sets is performed by means of the well-known error gradient descent method (Rumelhart et al., 1985; Rumelhart and McClelland, 1986).

The parameters of the network are modified in order to minimize a measure of the error on the training set: training samples $\langle \bar{x}, y \rangle$ are repeatedly presented to the network; for each of them, the controller computes the error $err = y - \hat{y}$ between the ideal value and the current output computed by equations (1–2), and then every weight $\{P_i\}$ is updated by the backpropagation formula:

$$P_i \leftarrow P_i + \Delta P_i; \qquad \Delta P_i = \beta(y - \hat{y}) \frac{\partial \hat{y}}{\partial P_i} \tag{4}$$

where $0 < \beta < 1$ is the *learning rate.*

As the same architecture is used both for supervised learning and for reinforcement learning (in the framework of the GARIC architecture) the rules for parameter updates can have a slightly different form, depending on the learning paradigm used. Here we will present only those for supervised learning.

3.3.1. Consequent Fuzzy Set Median Update. For output fuzzy set tuning, equation (4) reduces to:

$$\Delta \mathcal{Y}_s = \beta_C(y - \hat{y}) \frac{\partial \hat{y}}{\partial \mathcal{Y}_s} = \beta_C(y - \hat{y}) \frac{w_i}{\Sigma\, w_i}$$

To do this, it is necessary to share $\Delta\mathcal{Y}_s$ on the F_0 and F_1 parameters, memorized in the consequent of the rule:

$$\Delta F_{s0} = \begin{cases} \dfrac{\Delta\mathcal{Y}_s}{1 - w_s} & \text{if } w_s < \dfrac{1}{2} \\ \Delta\mathcal{Y}_s & \text{if } w_s = \dfrac{1}{2} \end{cases}; \qquad \Delta F_{s1} = \begin{cases} \dfrac{\Delta\mathcal{Y}_s}{w_s} & \text{if } w_s > \dfrac{1}{2} \\ \Delta F_{s0} & \text{if } w_s = \dfrac{1}{2} \end{cases}$$

3.3.2. Antecedent Gaussian Fuzzy Set Update. When bell-shaped memberships are used, the weights P_i to be changed are Gaussians center positions (C_i) and variances (σ_i).

Equation (4) is expanded in:

$$\Delta P_i = \beta_A(y - \hat{y}) \frac{\partial \hat{y}}{\partial P_i} = \beta_A(y - \hat{y}) \sum_c \frac{\partial \hat{y}}{\partial w_c} \frac{\partial w_c}{\partial \mu_i} \frac{\partial \mu_i}{\partial P_i}$$

Membership derivatives with respect to parameters σ and C are, respectively,

$$\frac{\partial \mu_i}{\partial \sigma_i} = \frac{2\mu_i(x - C_i)^2}{\sigma_i^3}; \qquad \frac{\partial \mu_i}{\partial C_i} = \frac{2\mu_i(x - C_i)}{\sigma_i^2}$$

4. THE CASE STUDY

The experimental testbed was represented by a robot KUKA-IR 361, equipped with an arm having six degrees of freedom, provided with force and torque sensors. As the robot is kinematically controlled, the control function had to predict two vectors $\bar{V}$ and $\bar{\omega}$, corresponding to the linear and angular velocities of the wrist, respectively.

The task to be performed was the *peg-into-hole*, that is, the robot was requested to insert a round peg into a vertical hole, having the same diameter, located on a plane surface. Since the robot is not provided with a vision system, it has to find the hole by keeping contact with the surface, using only the information from the force and torque sensors.

Note that the task could already be executed because a PID controller was available. The goal of the experiment was to learn to approximate the behavior of the PID controller using a set of examples of control actions. The hope is to increase, in future experiments, the performance of the controller, using reinforcement learning, until the PID controller will be outperformed.

The set of examples was obtained by recording the traces for the input sensors and of the corresponding speed values for seven insertion sequences executed starting from different initial conditions (different misalignments between the peg and the hole). Three of these sequences—the first, the second, and the fifth—were selected as a learning set. The remaining ones were used as a test set.

The same traces were used in Kaiser (1993), as well, for synthesizing controllers based on MLPs and TDNNs.

5. RESULTS

Both learning methods have been applied to the case study. The results obtained for the control signal V_z, the most critical for the successful execution of the task, are shown in Figure 9.11. In particular, Figure 9.11a shows how CART alone approximates the target function with an histogram; Figure 9.11b shows the output for the same sequence generated by the FC generated by converting the tree of CART into a set of fuzzy rules; then, Figure 9.11c shows the signal obtained after backpropagation was applied for 1,000 epochs.

Table 9.1 summarizes the results for V_x, V_y, and V_z, obtained using the method described in this chapter and Time Delay Neural Networks (TDNNs); results for TDNNs are cited from Kaiser (1993), where they proved to be the best approximator for this task, among a variety of networks based on the multilayer perceptron. The accuracy obtained by combining CART with the gradient descent, which is better than any other method we experimented, clearly emerges from this comparison. Finally, Table 9.2 shows how controller precision and complexity depend upon tangling and pruning for the FC generated by CART + backpropagation. Tangling was applied in three different ways: to input fuzzy sets only, to the output ones only, or to both; sets were merged when they overlapped by more than 75 percent. A tangled controller has fewer parameters to tune, and so the backpropagation converges faster while the accuracy is affected very little.

5.1. Test on Mackey-Glass Chaotic Time Series

Since the application of symbolic learning procedures is an important novelty and it gave very good results on robot traces, it was tested on another case study widely used in the literature in order to have a more extensive validation and comparison with other methods. The selected task was the prediction of a value 84 timesteps ahead in the Mackey-Glass chaotic time series, which is described by the following differential equation:

$$\dot{x} = \frac{0.2x(t-\tau)}{1 + x^{10}(t-\tau)} - 0.1x(t)$$

with $x(t < 0) = 0$, $x(0) = 1.2$ and $\tau = 17$. The task was to predict $x(t + 84)$ given $x(t - 24)$, $x(t - 18)$, $x(t - 12)$ and $x(t - 6)$.

The input features used by the approximators are the four values of the function at time, t, $t - 6$, $t - 12$, and $t - 18$. In order to correctly predict the function value 84 timesteps ahead it is necessary to capture the generative

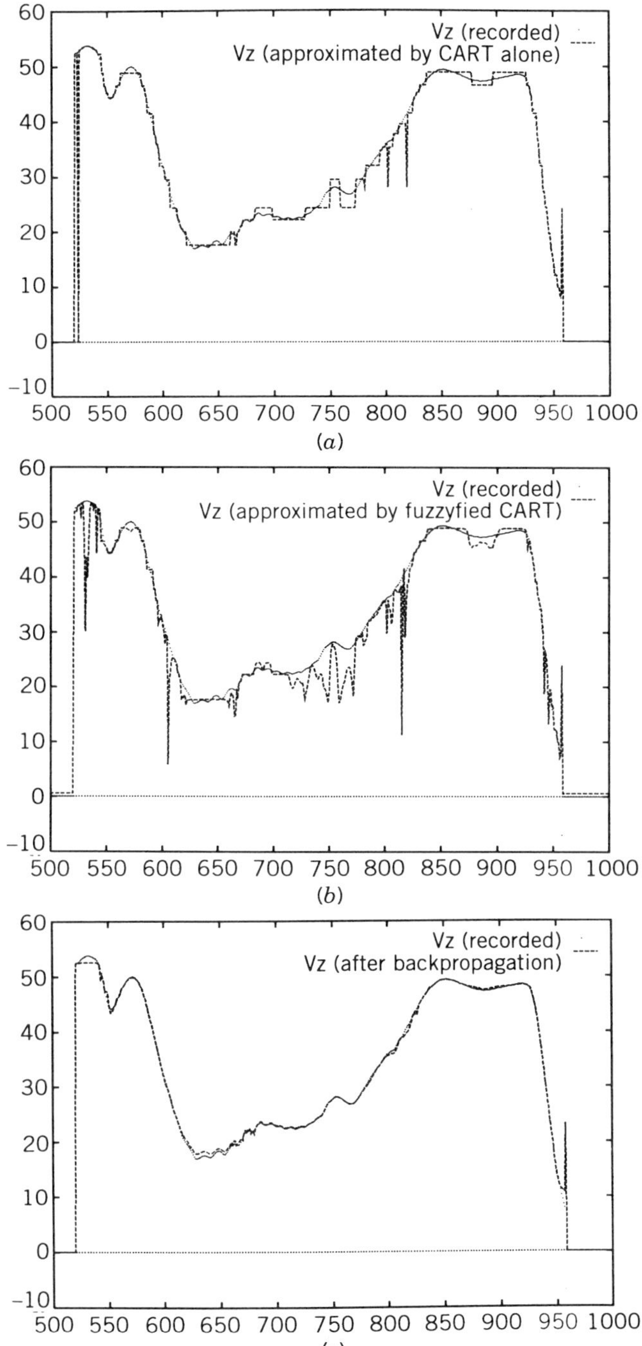

Figure 9.11 *Approximation of the control function V_z obtained using: (a) CART only; (b) after transforming the tree of CART into a fuzzy controller; (c) after training the fuzzy controller with backpropagation.*

TABLE 9.1. Statistical Comparative Results

Method	Track	Range	Err_{avg}	Standard Deviation
TDNN	V_x	11.41	2.25%	2.96%
	V_y	8.49	2.02%	2.38%
	V_z	56.10	1.80%	1.76%
CART alone	V_x	11.41	0.76%	1.05%
	V_y	8.49	1.02%	1.33%
	V_z	56.10	1.38%	5.01%
fuzzy CART	V_x	11.41	1.16%	2.51%
	V_y	8.49	1.77%	3.75%
	V_z	56.10	11.48%	20.92%
CART + backprop.	V_x	11.41	0.04%	0.17%
	V_y	8.49	0.05%	0.31%
	V_z	56.10	0.45%	2.75%

The columns Range and Err_{avg} report the maximum excursion of the signal and the average error respectively.

model of the phenomenon. This task was also used by Crowder (1990), Jang (1993), Jones et al. (1990), Lapedes and Farber (1987), Moody (1989), Moody and Darken (1989), and Sanger (1991), so results can be compared directly.

The experiment was organized in the following way. First, a sequence of 1,500 timesteps was generated. Then, the learning set was obtained by taking the first 1,000 whereas the remaining 500 were used as a test set.

The results are reported in Table 9.3. The Non-Dimensional Error Index (NDEI) is defined as the Root Mean Squared Error divided by the Standard Deviation of the target series.

TABLE 9.2. Effects of Tangling and Pruning on the Quality of the Signal V_z

Pruning	Tangling	Number of Feature Detectors	Number of Rules	Number of Consequents	Precision before Backprop.	Precision after Backprop.
NO	NO	116	50	50	2.46%	0.4%
NO	in.	65	50	50	2.49%	0.45%
NO	out.	116	50	12	5.32%	0.59%
NO	in. + out.	65	50	23	2.90%	0.99%
YES	NO	67	30	30	5.83%	1.40%
YES	in.	45	30	30	5.80%	1.34%
YES	out.	67	30	14	8.78%	0.92%
YES	in. + out.	49	30	14	8.80%	0.65%

TABLE 9.3. Comparative Results on the Mackey-Glass Chaotic Time Series Prediction

Method	Training Cases	Non-Dimensional Error Index
CART (alone)	800 + 200	0.36
$FC_{CART} + R$	1,000	0.037

Comparing the best results obtained in the literature in analogous experiments (Crowder, 1990; Jang, 1993), it appears that the predictor generated by CART shows excellent performances after training. In fact, the NDEI value of 0.036 reported by ANFIS, followed by the 0.05 of the MLP, are the only performances really comparable to it.

However, the big advantage of the methodology presented here is that the network has been synthesized *automatically* from example, whereas for both ANFIS and for the MLP it had been hand-coded.

The learning procedure based on modified k-means has not been tried here; nevertheless, by referring to the work by Moody and Darken (1989), we find a NDEI of 0.055, obtained with a training set of 10,000 examples and 1,000 hidden neutrons.

The controller generated by CART had 42 leaves in the tree, corresponding to a global number of 161 neurons in the first hidden layer.

6. CONCLUSIONS

In this chapter the problem of learning to approximate control functions from examples was addressed. As discussed in section 1, this is a key point in industrial robotic applications, which has been addressed in the framework of the project *B-Learn II*. Fuzzy controllers have been chosen as function approximators, owing to their structure, which allows a symbolic interpretation of the knowledge they encode, while preserving the adaptive capabilities of artificial neural networks.

An integrated learning methodology, based on regression trees (Breiman et al., 1984) and gradient descent, has been described, and we have shown that it can compete favorably with back-propagation networks and with alternative symbolic methodology on a real robotic task.

In the end, we believe that there are two main conclusions that can be drawn: the first is that fuzzy controllers are a good means for integrating symbolic and connectionist learning algorithms; the second is that this integration leads to definitively better results.

ACKNOWLEDGMENTS

The authors would like to thank the European Community for founding the B-Learn II project, and would like to thank M. Nuttin, M. Kaiser, and E. Blanzieri for their precious contributions. All scientific responsibility is assumed by the authors.

REFERENCES

Baroglio, C., Giordana, A., and Piola, R. (1994). Learning control functions for industrial robots. In *Workshop on Robotics, Machine Learning Conference, New Brunswick*. Rutgers University, NJ.

Barto, A., and Sutton, R. (1981). Landmark learning: An illustration of associative search. *Biol. Cybernet.* **42**, 1–8.

Barto, A., Sutton, R., and Watkins, C. (1990). Sequential decision problems and neural networks. In *Advances in Neural Information Processing System* (D. Touretzky, Vol. 2. Morgan Kaufmann, San Mateo, CA.

Berenji, H. (1992a). An architecture for designing fuzzy controllers with neural networks. *Int. J. Approximate Reason.* **6**(2), 267–292.

Berenji, H. (1992b). Fuzzy logic controllers. In *An Introduction to Fuzzy Logic Applications in Intelligent Systems* (R. Yager and L. Zadeh, eds.), pp. 69–96. Kluwer Academic Publishers, Dordrecht, The Netherlands.

Berenji, H., and Khedkar, P. (1992). Learning and tuning fuzzy controllers through reinforcements. *IEEE Trans. Neural Networks* **3**(5), 724–740.

Bonissone, P., and Chiang, K. (1993). Fuzzy logic controllers: from development to deployment. In *IEEE International Conference on Neural Networks*, pp. 610–619. IEEE Press, San Francisco, Vol. 2.

Breiman, L., Friedman, J., Ohlsen, R., and Stone, C. (1984). *Classification and Regression Trees.* Wadsworth & Brooks, Pacific Grove, CA.

Crowder, R. (1990). Predicting the mackey-glass time series with cascade-correlation learning. In *The 1990 Connectionist Models Summer School* (D. G. H. Touretzky and T. Sejnovsky, eds.), pp. 117–123. Carnegie-Mellon University, Pittsburgh, PA.

del R. Millán, J. (1994). Learning efficient reactive behavioral sequences from basic reflexes in a goal-directed autonomous robot. In *The Third International Conference on Simulation of Adaptive Behavior* (D. Cliff, P. Husbands, J. A. Meyer, and S. W. Wilson eds.), pp. 266–274. MIT Press, Cambridge MA.

Esposito, F., Malerba, D., and Semeraro, G. (1993). Decision tree pruning as a search in the state space. In *Proceedings of the Tenth European Conference on Machine Learning, Vienna, Austria* (P. Brazdil, ed.), *ECML-93*, pp. 165–184. Springer-Verlag, Berlin.

Hornik, K., Stinchcombe, M., and White, H. (1989). Multilayer feed-forward networks are universal approximators. *Neural Networks* **2**, 359–366.

Jang, J. (1993). ANFIS: Adaptive-Network-Based Fuzzy Inference System. *IEEE Trans. Syst., Man, Cybernet.* **23**(3), 665–687.

Jones, R., Lee, Y., Barnes, C., Flake, G., Lee, K., and Lewis, P. (1990). Function approximation and time series prediction with neural networks. In *IEEE International Joint Conference on Neural Networks*, pp. 649–665.

Kaiser, M. (1993). Evaluation of time-delay neural networks for control function approximation. In *The Workshop on Learning Robots, Torino*. (A. Giordana, ed.), pp. 68–78. University of Torino.

Lapedes, A., and Farber, R. (1987). *Nonlinear Signal Processing Using Neural Networks: Prediction and System Modeling*, Tech. Rep. LA-UR-87-2662. Los Alamos National Laboratory, Las Alamos, NM.

Mamdani, E., and Assilian, S. (1975). An experiment in linguistic synthesis with a fuzzy logic controller. *Int. J. Man-Mach. Stud.* **7**(1), 1–13.

Moody, J. (1989). Fast learning in multi-resolution hierarchies. In *Advances in Neural Information Processing* (D. Touretzky, ed). Morgan Kaufmann, San Mateo, CA.

Moody, J., and Darken, C. (1989). Fast learning in networks of locally tuned units. *Neural Comput.* **1**(2), 281–294.

Poggio, T., and Girosi, F. (1990). Networks for approximation and learning. *Proc. IEEE* **78**(9), 1481–1497.

Quinlan, J. (1986). Induction of decision trees. *Mach. Learn.* **81**, 81–106.

Raju, G., and Zhou, J. (1993). Adaptive hierarchical fuzzy controller. *IEEE Trans. Syst., Man, Cybernet.* **23**, 973–980.

Rumelhart, D., and McClelland, J. (1986). *Parallel Distributed Processing: Explorations in the Microstructure of Cognition*, Parts I and II. MIT Press, Cambridge, MA.

Rumelhart, D., Hinton, G., and Williams, R. (1985). *Learning Internal Representations by Error Propagation*, Tech. Rep. 8506. Institute for Cognitive Science, La Jolla University of California, San Diego.

Sammut, C., Hurst, S., Kedzier, D., and Michie, D. (1992). Learning to fly. In *Proceedings of the Ninth International Workshop on Machine Learning* (D. Sleeman and P. Edwards, eds.), pp. 385–393. Morgan Kaufmann, San Mateo, CA.

Sanger, T. (1991). A tree-structured adaptive network for function approximate in high-dimensional spaces. *IEEE Trans. Neural Networks* **2**(2), 285–293.

Specht, D. (1988). Probabilistic neural networks for classification mapping, or associative memory. In *IEEE International Conference on Neural Networks*, Vol. 1, pp. 525–532.

Specht, D. (1990). Probabilistic neural networks. *Neural Networks* **3**, 109–118.

Towell, G., Shavlik, J., and Noordwier, M. (1990). Refinement of approximate domain theories by knowledge-based neural networks. In *Proceedings of the Eighth National Conference on Artificial Intelligence*, (T. Dietterich and W. Swartout, eds.), pp. 861–866. AAAI Press/MIT Press.

Watkins, C., and Dayan, P. (1992). Technical note: Q-learning. *Mach. Learn.* **8**, 279–292.

Zadeh, L. (1965). Fuzzy sets. *Inf. Control* **8**, 338–353.

Zadeh, L. (1992). Knowledge representation in fuzzy logic. In *An Introduction to Fuzzy Logic Applications in Intelligent Systems* (R. Yager and L. Zadeh, eds.), Kluwer Academic Publishers, Dordrecht, The Netherlands.

CHAPTER 10

Some Developments in Statistical Credit Scoring

D. J. HAND
W. E. HENLEY

1. INTRODUCTION

Consumer credit is granted by a large variety of lending institutions, ranging from banks to retail outlets, and via many routes, such as formal loans or credit cards. The variety of institutions and routes reflects the rapid growth of the credit industry over the last thirty years. In the past, creditworthiness assessments were made on the basis of human judgment, but economic pressure resulting from increased demand and increased competition has led to the development of sophisticated statistical models to aid in the credit-granting decision. The term used to describe these more formal creditworthiness assessment methods is *credit scoring*. Such models use information about the applicant obtained from application forms and other sources to classify applicants into different risk categories. In principle, a continuum of risk categories exists (and even a multivariate risk continuum, if different kinds of risk are taken into account), but here we shall consider the simplest special case, and assume merely that each applicant is a *good* or *bad* risk. We shall also assume that each individual essentially belongs to one of these two classes, and does not change class over time. We will have available, in the traditional machine learning and statistical classification rule way, a *design* or *training* set, consisting of a sample of cases for which the true class is known and for which the application form and other characteristics are also known. Our aim is to use this set to construct a rule allowing the classification of future cases on the basis of their application form and other characteristics.

Machine Learning and Statistics, The Interface, Edited by G. Nakhaeizadeh and C. C. Taylor.
ISBN 0-471-14890-3

Each classification problem has its own particular properties and challenges, and credit scoring is no exception. Particular problems arising in a credit-scoring context are that the data are primarily categorical, that there can be a large number of potential variables to choose from, that the data sets can be very large, that there can be much missing data, that measurement error may be large, with some applicants setting out deliberately to deceive, that the underlying applicant population may be evolving (perhaps as the economic climate changes) so that the classification rule should also evolve, and that the common evaluation criterion of error rate may not be relevant. Further reviews of these issues will be found in Rosenberg and Gleit (1994), Henley (1995), and Hand and Henley (1996). The first of these presents an American perspective, the second a more detailed view of technical aspects of a subset of methods, and the third, an overall assessment of classification methods in credit scoring. In this chapter, after a brief review of earlier work, we concentrate on just three areas: a particular class of classification methods, which has only recently been deeply explored in the credit-scoring context, the lack of sensitivity of linear predictions to the values of the coefficients in the model, and the problem of 'reject inference'.

The performance of a classification rule may be measured in many different ways, but it is measured performance that is the pre-eminent concern. In particular, in the context of credit scoring, a tiny percentage difference in performance can mean a large difference in financial terms. In view of this, it is hardly surprising that credit-granting agencies have explored a wide class of classification methods. Section 2 reviews some of this work. This serves to set section 3 in context, where we describe our own investigation into the use of the k-nearest neighbor (k-nn) classification rule in credit scoring. This method has the attractive properties that it can model irregularities in the risk function over the measurement space, and that it can be used dynamically. That is, older elements of the design set can be tagged and gradually dropped from the database as new applicants arrive and are included, without requiring a recalculation of any decision surface. Of course, it is not without potential shortcomings: the categorical nature of the data may lead to many ties in the cases to be regarded as 'nearest'; one needs to define a suitable metric by which to measure proximity; and, a property that proved detrimental in some earlier work we carried out for Trustee Savings Bank (Hand, 1986) on applications for small loans, when there are many variables 'near' is really far. Nonetheless, as we show below, when properly applied in appropriate problems the method can prove attractive.

The second topic we examine is what is sometimes termed the *flat maximum effect*. This describes the fact that different linear models, with very different coefficients, may yield very similar predictive power. This matters if one's aim is to understand the relationships between the predictor variables and the response variable (credit risk, here) but is of no consequence, and may even be an advantage if one's aim is simply to predict an outcome. The latter, of course, obtains in the credit-scoring context.

The third area we consider in detail is that of *reject inference*. The *accepts* are those applicants who are granted a credit loan (that is, those the classifier classifies as 'good'), while the *rejects* are those who are not granted one (classified as 'bad'). Future behavior of the accepts will be known: unfortunately, not all of them will turn out to be good risks, since the classification rule is seldom perfect. Thus performance measures such as bad rate among the accepts can be estimated. But the future behavior of the rejects will not be known: the very fact that they have not been granted a loan means that they are not followed-up, so that their true class remains unknown. This, in turn, means that overall performance measures such as error rate cannot be estimated, at least not by standard methods. This opens up all sorts of interesting problems. For example: what is the best way of estimating the good/bad probability of an applicant in the reject region? And, a problem that has exercised many of the credit-granting agencies in the past: can the information that is available on the rejects (namely, the application form and other measurements) be used in conjunction with the information on the accepts (which includes their true classes) to produce improved classification rules? In this chapter we focus on the second of these problems. While here we set it in a credit-scoring context, which is the context in which we first encountered it, in Hand and Henley (1994) we discuss it in a medical screening context.

2. CLASSIFICATION RULES IN CREDIT SCORING

Although we have referred to an 'application form' above, things can be more complicated than this. For example, often an initial screening application form is used (perhaps printed as part of an advertisement in a newspaper or magazine), and only those who pass this stage provide 'full' information. In addition, further information may be available from national databases kept by credit-scoring agencies such as CCN and Equifax, which hold details of applicants' past repayment records. All of this suggests that often a multi-stage approach is needed—providing the opportunity for some interesting juggling with tuning parameters. In this chapter, however, we shall restrict ourselves to a single one-stage classification rule.

Although evaluations and assessments of most available classification techniques have been made in the credit-scoring industry, the techniques in widespread use tend to be the most simple. Given the power of modern computers, this is unlikely to be because of the complexity of modern sophisticated methods, or because of the time they take to reach a classification, which can usually be taken to be negligible (with clever programming, if necessary). The fact seems to be that, in the credit-scoring context, 'clever' classification methods seldom perform much better, if at all better, than simple ones. We suspect that there are two reasons for this. The first is that traditional credit-scoring methods (using, for example, simple linear or logistic models) have been around for some time: the developers of such approaches, who know their data well, have tuned them, included interaction terms, and so on, so that they are performing near the optimal for the problem in question. And the second is that the two classes in question—the 'goods' and the 'bads'—are not well separated. Given the qualifying remarks we made in section 1, about the reality being a (possibly multivariate) risk continuum, and a dynamic one at that, this is perhaps hardly surprising. Some classification rules try to take such things into account. For example, *behavioral scoring* monitors the behavior of a successful applicant over time, to detect a likely early default on repayment. However, such approaches are not relevant to the initial credit-granting decision and are beyond the scope of this chapter.

In the credit-scoring industry traditional classification rules are called *scorecards*. This arises from their structure of being a simple sum of values ascribed to the various categories of the variables. Often the particular values allocated to the categories are obtained from linear regression models. These might be by coding the categories as indicator variables or by adopting some scaling procedure to give the categories numerical values. A common coding system in credit-scoring (and in medicine) is the use of *weights of evidence*: the score for the jth level of the ith variable is defined as $w_{ij} = \ln(p_{ij}/q_{ij})$, where p_{ij} is the number of goods at level j of variable i divided by the total number of goods who respond to this variable, and q_{ij} is the corresponding proportion for bads. Typically in credit scoring the resulting numbers are rounded to be integers (which inevitably introduces approximations but leads to simpler explanations).

Measuring 'creditworthiness' by a simple weighted sum in this manner inevitably disposes one to think of it as a latent variable: a property of the applicant that cannot be directly measured, but which has to be measured indirectly via its relationships to the application form characteristics. This is a legitimate interpretation, though it does open deep philosophical questions. It seems to us that, if one is going to view the weighted sum in this way, the objective has changed from being merely one of trying to find an effective score (a pragmatic or operational approach) to being one of trying to model an underlying reality (a realist approach). If the latter perspective is being adopted then a more elaborate modeling strategy seems to us to be more appropriate, in which the various measured variables and their relationships, as well as their relationships with postulated latent variables (such as creditworthiness) are modeled together. An initial step in this direction is presented in Hand et al. (1996). During the course of that study, the measured application-form variables were found to contain more information about propensity to default than was contained in the weighted sum used by the bank funding the work. This supports the above contention that there is more to be modeled than is contained in a single weighted sum—as do the results of the nearest-neighbor analysis described in the next section. In this chapter, however, we restrict ourselves to a simple prediction of creditworthiness.

Apart from linear (and, less frequently, logistic) regression, other standard classification methods have been used. In particular, linear discriminant analysis has been applied, despite that fact that its theoretical basis might suggest it to be inappropriate: like regression, it requires the use of numerical scores such as weights of evidence, and it is based on first- and second-order moments. But it also assumes the covariance matrices of the good and bad risk groups to be equal, an assumption that is unlikely to be satisfied in practice. On the other hand, empirical studies suggest that, if the true decision surface is linear, then classical linear discriminant analysis often performs respectably. Myers and Forgy (1963) provide a comparison of such methods.

The predominantly categorical, high-dimensional data, with missing values and little-known distributional structure suggests that the fundamentally nonparametric approach of classification trees (recursive partitioning) might be suitable. Although we found the performance of such methods disappointing in our earlier work (Hand, 1986), other authors (for example, Srinivasan and Kim, 1987) have found that they can outperform other methods (but see the comments below about comparisons between methods).

The ability of genetic algorithms to search large spaces to find an optimum means that they have also been applied in the credit-scoring context (Fogarty and Ireson, 1994). Occasionally studies incorrectly describe such approaches as if genetic algorithms are themselves a classification tool. In fact, of course, they are merely an optimization algorithm, and they could be used to fit any class of model. Their strengths obviously become apparent when the model is complex—in situations where, for example, simple linear algebra cannot be used to find a solution.

A field with such evident commercial importance as credit scoring naturally attracts enthusiasts for 'novel' techniques, keen to demonstrate that their methods can outperform the traditional ones. Recursive partitioning and genetic algorithms might be put into this class, as might neural networks. In fact, however, no technique is completely novel and models similar to neural networks have been developed by the statistical community in other contexts (see Ripley, 1994). Such models are essentially flexible forms of decision surface. In the case of (feed-forward) neural networks, the flexibility is achieved by stacking sets of parallel transformations, each transformation consisting of a nonlinear transformation of a weighted sum of the outputs of the preceding set. The nonlinearity is essential to the flexibility—without it the whole reduces to a single weighted sum—but it requires the power of modern computers to enable parameter estimation in reasonable times. A comparison of various methods in the credit-scoring context, including neural networks, is described in Davis et al. (1992); but much recent exploratory work in this area remains unpublished.

This last remark raises the confidentiality aspect of much of the work. In a competitive commercial environment it might well be counterproductive to tell one's competitors how one has achieved superior performance, so there will be a natural reticence about publication. This could mean that major advances are being kept secret—though improved predictive ability would manifest itself in changing corporate practice and exceptional results and there is no evidence of this. In any case, there is a flow of personnel from one institution to another, so no overwhelmingly superior method would remain in the hands of one organization for long. Overall, judging from the comparative studies that have been published, it seems that none of the more recent developments in classification methodology have led to great improvements on more traditional methods.

Issues of confidentiality raise the related issues of legislative constraints. Some potentially discriminatory items of information cannot be used in a credit-scoring classification rule, as they breach laws relating to racial, sexual, or other discrimination. This does not affect the fundamentally statistical aspect of the problem, which is what is addressed in this chapter, but it does influence the level of performance which might be achieved. Some discussion of these issues is given in Hand and Henley (1996).

Henley (1995) gives a fairly comprehensive overview of comparisons of methods that have been published. Our general view is that sophisticated and sensitive use of any method will lead to performance near the optimum. Given the amount of work that has been undertaken with 'basic' statistical models, this suggests that the best hope for substantial improvement lies in improving the quality of the data—with finding new predictor variables that have high predictive power, for example.

There is a further aspect of comparative studies (of which there have now been a great many, see, for example, Michie et al. 1994; for an elaborate example) which is particularly relevant in the context of credit scoring. Classification techniques can be compared in two situations: (i) as used in standard 'commercial' applications, by someone who is competent but not a leading researcher in the methodology being used; and (ii) as used by someone who is actively involved in developing the methodology and will take advantage of subtleties of which the commercial user may be unaware (perhaps even developing new features to fit the characteristics of the data). It would not be surprising if comparisons undertaken in the two different ways led to different results. There is an analogy here to the 'explanatory' and 'pragmatic' distinction made in clinical trial research. The former describes studies aimed at understanding biological mechanism (for example, is one medicine better than another in some absolute sense? Is there a difference between medicines in their biological effect?), while the latter describes studies aimed at seeing if there are differences in effect when the medicines are used in practice (with poor compliance, missed doses, and so on). One consequence is that we might conclude that classification method A outperforms method B when the comparison is made by researcher X, but the converse holds when the comparison is made by researcher Y, not because of dishonesty, but merely because of different domains of expertise. A credit-granting agency considering the adoption of a new technology might be best served by trying out the novel method in-house themselves.

3. *k*-NEAREST NEIGHBOR METHODS

k-nearest-neighbor classification methods classify a new case on the basis of a majority vote amongst the k nearest design set elements to the new case, where proximity is measured in the space of the measured variables. The simplest approach (often called just the nearest-neighbor method), in which only the single nearest neighbor is used, has been used in a credit-scoring context by Fogarty and Ireson (1994). They obtained promising results, despite the simplicity of the basic method they adopted.

Clearly a fundamental aspect of a *k-nn* rule will be the distance metric which is adopted (see, for example, Fukunaga and Flick, 1984; Myles and Hand, 1990; Short and Fukunaga, 1982). Euclidean metrics and Mahalanobis distance measures are the most commonly applied. Mahalanobis distance between two points is defined as $(\mathbf{x}-\mathbf{y})'\Sigma^{-1}(\mathbf{x}-\mathbf{y})$, where Σ is the covariance matrix of the data (for example, the assumed common within class covariance matrix in discriminant analysis). This has been extended by replacing the matrix Σ by various matrices, including estimates which vary locally across the space. Henley and Hand (1996) propose replacing Σ by $(\mathbf{I}+D\mathbf{w}\mathbf{w}')$, where D is a parameter estimated from the data and $\mathbf{w}$ is a particular direction in the measurement space. Hence the distance between two points $\mathbf{x}$ and $\mathbf{y}$ is defined to be $(\mathbf{x}-\mathbf{y})'(\mathbf{I}+D\mathbf{w}\mathbf{w}')^{-1}(\mathbf{x}-\mathbf{y})$. The effect of the term including D is to modify the basic Euclidean metric so that differences in the direction $\mathbf{w}$ are emphasized. Ideally $\mathbf{w}$ would be taken to be a direction orthogonal to the equiprobability contours of belonging to class 1, at least if we assume these contours to be linear. In practice, of course, this direction has to be estimated, and we used the weights obtained from a linear regression of class on the predictor variables. (Logistic regression weights may be a more natural choice but, as noted above, experiments showed that they produced results barely distinguishable from those of the linear regression analyses.) With this choice of weights, the method approaches the simple *k-nn* method with the Euclidean metric as $D \to 0$ and approaches linear regression as $D \to \infty$.

To apply the method in practice k and D must be estimated.

The data used in Henley and Hand (1996) consisted of a sample selected from the entire population of applicants for credit from a large mail-order company. Especially noteworthy is the fact that this data set included true creditworthiness classifications of those who would normally have been rejected. This permitted evaluation of the rule on a data set not suffering from the bias problems described in section 5. (The presence of such data represents an enlightened scientific and commercially unusually sound attitude on the part of the organization involved. As we show below, in order to make valid deductions about all applicants, rather than merely the 'accepted' subclass, it is necessary to have a sample spanning the entire measurement space and not merely the region defining the accepts. That is, one must accept some of those thought to be poor risks, despite the fact that one might expect to lose money on them. The consequence, of course, is a more accurate classification rule, so that overall profit will be increased. Unfortunately, an all too common short-term attitude to profit makes the collection of such a global sample very rare.) Applicants who defaulted for three consecutive months were classified as 'bad' and the remaining applicants as 'good'. Sixteen categorical response variables were used, including such things as 'time on electoral role' (with six categories). The raw response categories for each of the variables were coded using weights of evidence. Using these scores we can calculate distances between a new applicant and those in the design set. The available data consisted of 19,186 applicants, with 54.6 percent bad risks. It is worth remarking here that this population was the result of an earlier selection process, the details of which are irrelevant here. This accounts for the high proportion of bad risks.

In the context of this data set, each accepted applicant incurs a cost (arising from printing and sending a catalog), so the creditors here specified beforehand that they wished to accept a certain percentage of the applicants. This meant that we were interested in the bad rate among those accepted for credit, rather than overall error rate. This criterion is somewhat unusual, so a few words about the consequences of using it are justified.

Without any bounds on the proportion to be accepted, the bad rate among the accepts that might be achieved ranges from 0 to 1. For example, by accepting only the very few of whom one is very confident one can make one's bad rate among the accepts arbitrarily close to 0. Conversely, if one accepted only those for whom one had most doubt, one could have a bad rate among the accepts close to 1. However, suppose that a specified proportion a is to be accepted and that a proportion b are known to be bad. Then, when $b < a$, the largest bad rate among the accepts one can achieve is b/a. On the other hand, the smallest bad rate among the accepts that one could achieve (if one was clever enough in devising one's classification rule) would be $\min\{0, 1-(1-b)/a\}$. Similar bounds apply to good rate among the accepts and rejects, bad rate amongst the rejects, and error rate. Note that these are only *bounds* on the achievable rates. How effective an instrument is in practice will depend on the data and the ingenuity of the person building the classifier. But they do serve to indicate whether or not a rule is good in the context of the data at hand. So, for example, a result of 36 percent bads among those accepted might superficially seem appalling. However, in the context of the data outlined above, with a bad rate of 54.6 percent and given that a proportion of 70 percent was to be accepted, the bad rate among the accepts must lie between 35.1 percent and 78.1 percent. Now 36 percent is seen to be virtually the best that can be achieved.

TABLE 10.1. Nearest Neighbor and Linear Regression Bad Rates for Five Samples

Sample	Linear Reg.	*k-nn*	*k* Value	*D* Value
1	43.87	43.96	1260	1.6
2	44.03	43.63	690	1.8
3	43.54	43.45	200	1.4
4	42.16	41.55	750	1.4
5	43.20	42.84	820	1.4

Using this criterion and a subsample of the available data as the design set, we used a bivariate optimization process to estimate k and D. Complications arise because bad rate is not a continuous function of either parameter, and we used various smoothing techniques to overcome this. Details are given in Henley (1995) and Henley and Hand (1996).

We conducted analyses on five samples, each obtained by randomly selecting a design set of 15,054 applicants from the available data, leaving a test set in each case of 4,132. We found that the optimal value of k chosen varied dramatically from sample to sample. However, the actual values of the criterion (bad rate among accepts) was very insensitive to k, showing broad flat valleys in all cases. Slight sampling induced irregularities in the shape of this valley presumably account for the range of optimal k. Of perhaps greater interest is the fact that the *k-nn* method performed slightly better than the conventional regression approach for a considerable range of values of k. Our results also showed that performance was insensitive to choice of D. Table 10.1 shows summary results for all five samples. The large range of the values of k is evident, ranging from 200 to 1,260. D does not show so dramatic a range. In four out of the five cases the nearest-neighbor method is slightly superior (has a lower bad rate) to the regression method. The differences are small (this is perhaps to be expected, as we commented above), but may be large enough to have commercial implications.

The nearest neighbor method described above has attractions for the credit-granting problem, such as its ease of updating as more classified cases become available and the ease with which it may be explained to potential users. But it also has properties that might be regarded as possible weaknesses. Classifying an applicant by this method requires considerably greater computational effort than does prediction from, for example, a linear or logistic regression model. But, in fact, nowadays the power of modern computers renders this effectively irrelevant. We used 486 PCs and SUN workstations and, with the data sets above, new cases were classified in less than 1.5 seconds for all values of k up to 3,000. There was no practical difference between simple linear methods and *k-nn* methods. Furthermore, we made no effort to program an efficient routine for searching through the multivariate measurement space to identify the nearest neighbors. Such methods exist, based on branch and bound strategies (for example, Hand, 1981). These involve preprocessing the data. Other preprocessing strategies can be used to reduce the numbers of design set elements which are retained (and, hence, which have to be examined in future classfications). For example, regions of measurement space densely populated with 'goods' and where there are very few or no 'bads', can be thinned out: the classification will be to class 'good' even with a sparser local data set. Conversely, isolated 'bads' in regions densely populated with 'goods' will never lead to a classification as bad and so can be removed from the design set without loss in performance. Such methods go under the names of condensed, reduced, and edited nearest neighbor, and are described in Hart (1968), Gates (1972), and Hand and Batchelor (1978).

A second property that might be regarded as problematic arises from the background nature of credit scoring. Current legislation requires that unsuccessful applicants be informed why they were rejected if they should ask. It is not sufficient to say, simply, that 'the classification method we are using estimated you to be a high-risk applicant'! For methods that base the classification on a weighted sum, one can identify those components of the sum that contribute most to the reject decision. It has been suggested that a similar approach could be used with the nearest-neighbor method described above: the weights in $\mathbf{w}$ serve to identify which dimensions are most important globally, and these might be reported as being the determining factors. Of course, this is not a strictly valid explanation, since the direction defined by the metric does not reflect local fluctuations in estimated risk (if it did, we could simply use regression and would not need to resort to nearest neighbor methods). Our view, however, is that this is not a problem specific to nearest-neighbor methods. It is, rather, one of asking the impossible. Classification is essentially a multivariate problem. The answer to the question 'why have you rejected me?' is simply that the applicant lies in a region of the space that has been estimated to have a high risk. The regression coefficients describe *conditional* weights, given the other variables included in the model. A small coefficient on some variable does not necessarily mean that that variable is unimportant in the classification process.

4. THE FLAT MAXIMUM EFFECT

In the previous section we explored the use of a nonparametric method as an alternative to standard linear approaches to formulating a credit-risk-scoring instrument. We produced evidence suggesting that the nearest neighbor method could yield improved classification rules. However, the improvement was not great. While a small improvement may have commercial implications, it seems likely that greater improvements could be obtained by changing the variables that are measured: linear methods look set to continue to be heavily used in the industry. They clearly have merits (such as their conceptual simplicity), which is influential beyond a mere measure of performance based on proportions misclassified. In this section we examine one particular property of linear rules. This property, variously described as the *curse of insensitivity* (Rapaport, 1975) or the *flat maximum effect* (von Winterfeldt and Edwards, 1982), is the fact that the predictive power of linear models is often insensitive to the precise values of the coefficients. Attention has been drawn to this effect in the context of credit scoring in Lovie and Lovie (1986) and Overstreet et al. (1992). It was forcefully brought home in Wainer (1976, 1978), where it was pointed out that, in a regression model with p variables, if standardized regression weights with values between 0.25 and 0.75 are replaced by the constant value of 0.5, then the loss in predictive accuracy is not greater than $p/48$. This provides strong justification for the common practice in credit scoring of rounding estimated regression coefficients to convenient integers.

This insensitivity is often regarded as a problem. But whether, in fact, it is such depends on the objective of the analysis. If the aim is model-building for the purpose of *understanding*, then what matters are the values of the regression coefficients. If this is the aim then being unable to distinguish effectively between substantially different models, which all happen to have similar predictive power, is, of course, a major problem. In contrast, if the aim is model-building in order simply to *predict* some criterion function, then the precise details of the model do not matter. The distinction again parallels the explanatory/pragmatic one noted above. And, in credit scoring, the objective is very much the latter: a hard-nosed aim of predicting default risk. (Whether or not this is the right thing to do is a different issue.)

Conditions for the existence of the flat maximum effect are provided in Lovie and Lovie (1986):

1. Choose the dominant predictor variables — usually those with the largest estimated standardised coefficients.
2. Choose the subset that maximizes the predictive power (such as R^2 in regression).
3. Choose mainly collinear predictors.
4. Align all predictors in the same direction with respect to the criterion variable.
5. Choose a dichotic or dichotomizable criterion variable.

Credit-scoring methods are seen to satisfy most of these. Typically (1) and (2) are satisfied by the variable selection process. Choice (4) is satisfied so as to yield 'interpretable' rules, which can be easily explained and justified. (We have had experiences of trying to introduce variables which improve predictive power, but which have been excluded on the grounds that they seemed counterintuitive.) Choice (5), of course, is the essence of the credit-scoring problem.

Two studies exploring the flat maximum effect in the credit scoring context are those of Stillwell et al. (1983) and Overstreet et al. (1992). The former describes a small study of credit scoring that shows, as Lovie and Lovie (1986) say: 'whether the loan officers had rank ordered the variables, assigned equal weights, or had attempted to produce more properly scaled responses, *made little or no difference to the outcome from the model*' (their italics). Overstreet et al. produce ten separate linear credit-scoring models for ten Southeastern U.S. credit unions (using stepwise regression) and compare them with a global model obtained using weighted averages of the coefficients of five of the systems. They summarize their conclusions by saying: 'In all cases, the generic model's performance was very close to that of the empirically derived models.'

5. REJECT INFERENCE

Reject inference is the term used for the process of inferring the good/bad status of applicants who lie in the reject region—those who have not been granted a loan. As noted above, there are several reasons why one might wish to to do this. One of them is the possibility that such inferred probabilities might be helpful in developing an improved classification rule. It is this that we explore in this section.

The obvious motivation behind using the rejects is that a new scorecard based only on the accepts will almost certainly be biased. The accepts represent a purposively selected subsample from the complete applicant population, and so, by definition (at least, if one assumes that the original scorecard was at all effective) the accepts differ from the rejects. Since the true class memberships of the rejects are unknown, the question is, can inferred class memberships (or, more generally, inferred probabilities of class membership) be used to reduce this bias?

This question can most easily be answered if we formulate the problem using the terminology of Little and Rubin (1987) for missing-data mechanisms. So, let x be an n by p variate random matrix containing the p items of application form information for the n applicants. Let c be a vector of the class memberships. This vector is incompletely observed, having unknown values for the cases corresponding to rejected applicants. We shall also use $c = (c_o, c_m)$, where c_o represents the classes for the accepts and c_m represents the (unobserved) classes for the rejects. Let b be a vector indicating whether a particular applicant has been accepted or rejected (and, so, telling us which components of c are unobserved). Let u, v, and w be parameter vectors for the distributions indicated in the joint distribution of the data given by

$$f(c, b, x; v, u, w) = f(b \mid c, x; u) f(c \mid x; v) f(x; w) \tag{1}$$

In this factorization $f(b \mid c, x; u)$ gives the probability that a particular pattern of missing values will occur, conditional on the true classes and the measurement vectors of the applicants, $f(x; w)$ gives the marginal distribution of the measurement vectors, and $f(c \mid x; v)$ gives the distribution of the class membership vector c, given the measurement vectors in x. It is this last distribution which is of particular interest to us since, using it, we can estimate the probability that a future observation vector will correspond to an applicant from each class. (Note that, since we will assume that observations on different applicants are independent, we have $f(c \mid x; v) = \Pi_i g(c_i \mid x_i; v)$, where g is the probability function for the class of a single applicant with measurement vector x_i.) So, we want to estimate v, the parameter vector for the distribution of class memberships, $f(c \mid x; v)$. The overall likelihood of the *observed* data is

$$\begin{aligned} L &= f(c_o, b, x; v, u, w) \\ &= \int f(c_o, c_m, x; v, w) f(b \mid c_o, c_m, x; u)\, dc_m \end{aligned} \tag{2}$$

The first factor inside the integral is the *data* model—it describes how the measurement vectors and classes arise. The second factor is the *non-response* model—it shows the probability that a class identifier is missing.

Little and Rubin describe three mechanisms for missing data that lead to different non-response models. Two of these are relevant for us.

The first is the *missing at random* case, for which $f(b \mid c_o, c_m, x; u) = f(b \mid x; u)$. In this situation the probability that a class-membership indicator is missing does not depend on the true classes (known or unknown). To put it another way, the probability that an applicant with application characteristics x is among the accepts does not depend on his or her true class. That is, the accept/reject decision was based solely on the information in x.

In this situation, we can factorize L as

$$L = \int f(c_o, c_m, x; v, w)\, dc_m f(b \mid x; u) = f(c_o \mid x; v) f(x; w) f(b \mid x; u) \tag{3}$$

We shall assume that the applicants are randomly sampled from the x distribution. (Hand and Henley, 1994, include a discussion of sampling conditional on x.) There are then two situations we must distinguish between: (a) when the parameter spaces have no intersection and (b) when v and w span intersecting spaces.

In the former case the likelihood for v is proportional simply to $f(c_o \mid x; v)$ and, since applicants are assumed to be independent, the information on the rejects does not assist in estimating v. Then, to make inferences of any kind over the reject region, it is necessary to extrapolate over the reject region. This will be done via some assumptions about the form of $f(c_o \mid x; v)$. This might be directly (e.g., assuming a logistic model) or it might be indirectly (e.g., assuming that the separate classes are each multivariate normal, along with some assumed prior probabilities).

In case (b), $f(x; w)$ does shed light on $f(c_o \mid x; v)$. The most familiar case of this is classical two class linear discriminant analysis (also discussed in Hand and Henley, 1994). This classification technique arises if one assumes multivariate normal distributions with identical covariance matrices (it also arises under more general conditions, but we need not be concerned with that here). In the usual situation one has a design set with known class memberships. This allows us to estimate the parameters in both w (namely, the class mean vectors, estimated by $\bar{x}_1$ and $\bar{x}_2$, and the common covariance matrix, estimated by $\hat{\Sigma}$), and v (which is a function of $\bar{x}_1, \bar{x}_2$, and $\hat{\Sigma}$) and, hence, derive the linear discriminant function. However, given that one is assuming that each class is multivariate normal, the overall distribution of x is a multivariate normal mixture. Consequently, one could estimate the parameters of the components from the sample distribution of x alone, using standard mixture decomposition methods (Everitt and Hand, 1981). That is, one need not have knowledge of the class memberships to estimate the parameters. In general, of course, $f(x; w)$ will only provide information about v if the two parameter spaces do overlap in the manner of w and v above. Put another way, this requires us to have information (beyond the sample) about the distributions of the separate class conditional distributions of the x's. Without such information—in effect, without making assumptions about the forms of these distributions—$f(x; w)$ does not provide any information about v.

To summarize, provided the accept/reject decision was based solely on x, then (i) without additional distributional assumptions, the rejects cannot help make an improved classification rule; and (ii), if assumptions are made then the rejects can help. In the context of credit scoring, where many of the variables are categorical, it will typically be difficult to produce realistic a priori distributional assumptions. Certainly the assumption of multivariate normality is untenable.

The second missing data mechanism described by Little and Rubin, which is of relevance to us, is the *non-ignorably missing case*. This is the case when no simplification of $f(b \mid c_o, c_m, x; u)$ is possible. Such a case arises when the initial selection of successful applicants is based on a set of variables y, which includes variables not in x, and when the complement of x in y contains useful information about risk. In this case Prob(good | accepted,x) will not, in general, equal Prob(good | rejected,x) and, hence, will also not equal Prob(good | complete sample,x). This means that accept sample is biased everywhere (at least potentially) in the space spanned by x, so that a scorecard based on the accepts will produce biased estimates of overall probability of being good everywhere in the measurement space. This situation is thus worse than the *missing at random* case: at least there the sample distribution in the accept region was an unbiased sample from the goods and bads. Just as in that situation, however, without making assumptions about the distributions involved, the rejects provide no further information about class membership.

An example is as follows (Henley, 1995; Hand and Henley, 1993–1994). Here y is based on several variables; that is, the initial accept/reject decision

TABLE 10.2. Proportion Good in Full and Accept Samples

Level	Full	Accept	Accept/Full
1	0.227	0.446	1.964
2	0.304	0.469	1.542
3	0.314	0.492	1.567
4	0.378	0.552	1.460
5	0.426	0.631	1.481
6	0.556	0.692	1.245

uses several application characteristics. In contrast, x, the variable to be used for the new classification rule, is the single variable 'Number of weeks since last County Court Judgement', recoded and scored as 1 to 6. The first column in the table of proportions in Table 10.2 shows the proportions of goods at each level of x estimated from the full sample (which we can estimate because, unusually, we know the true class details of the rejects, as explained above). The next column shows the proportions good as estimated from the accept sample. And the final columns shows the ratio of the preceding two columns. A gradual change with level of x is clear, as is the fact that for all categories of x (and hence for whichever set of categories we might choose as the accept reigion) the estimates based on the accept sample differ substantially from the estimates based on the full sample.

It follows from the above that the rejects, in themselves, contain no information which might be used to produce an improved classifier. However, it also follows (in the non-ignorable case, at least) that, since a new classifier based only on the accepts will generally be biased, there is scope for producing a classifier that does better than one based on only the accepts. Such improved performance would be obtained by luck or by using extra knowledge about the applicants (such as distributional assumptions). Such possibilities are discussed in Henley (1995) and Hand and Henley (1993–1994). It is our belief that it is this that underlies the occasional claims that reject inference can lead to improved classification rules, despite the fact that this is impossible using only the information available in x.

6. CONCLUSION

In this chapter we have briefly reviewed the application of formal statistical methods to the credit-scoring problem. As with any application domain, credit scoring has its own unique aspects. There are questions of precisely how the risk classes (if not a continuum) should be defined, and what the performance measure should be (profitability or default risk, for example) and also issues arising from such things as the multivariate categorical nature of the data and the possible desire for dynamic updating. Moreover, since the area has major commercial implications, it is not surprising that there has been substantial interest in finding the 'best' classifier, and that much investigative work has been carried out. All of this work has been moderated by the practical needs and constraints of credit scoring.

We discussed three issues. The first was the use of the nonparametric k-nearest neighbor classifier for credit scoring. Our earlier work had suggested that this, though superficially promising, had disadvantages. In the study we describe here, however, where, in particular, more attention was given to the choice of optimum metric, the method was shown to perform slightly better than the standard regression approaches in widespread use in the industry. The second issue we discussed was the flat maximum effect. This phenomenon is often perceived as problematic, but it is really only so when the aim is understanding the relationships underlying a model. When, in contrast, the aim is merely prediction, as is the case in credit scoring, the flat maximum effect may be seen as a useful property. Finally, we discussed reject inference and, in particular, whether the rejected applicants, for whom one does not know the true good/bad class, can assist one in building an improved classifier. The answer was negative—unless one is prepared to make additional assumptions about the distributions of the data. This will have consequences for the credit scoring industry, where reject inference is in widespread use.

ACKNOWLEDGMENT

W. E. Henley was sponsored in this work by a grant from Littlewoods Plc.

REFERENCES

Davis, R. H., Edelman, D. B., and Gammerman, A. J. (1992). Machine-learning algorithms for credit-card applications. *IMA J. Math. Appl. Bus. Ind.* **4**(1), 43–52.

Everitt, B. S., and Hand, D. J. (1981). *Finite Mixture Distributions.* Chapman & Hall, London.

Fogarty, T. C., and Ireson, N. S. (1994). Evolving Bayesian classifiers for credit control—a comparison with other machine learning methods. *IMA J. Math. Appl. Bus. Ind.* **5**, 63–75.

Fukunaga, K., and Flick, T. E. (1984). An optimal global nearest neighbour metric. *IEEE Trans. Pattern Anal. Mach. Intell.* **PAMI-1**, 25–37.

Gates, G. W. (1972). The reduced nearest neighbour rule. *IEEE Trans. Inf. Theory* **IT-18**, 431.

Hand, D. J. (1981). Branch and bound in statistical data analysis. *Statistician* **30**, 1–13.

Hand, D. J. (1986). *New Instruments for Identifying Good and Bad Credit Risks: A Feasibility Study.* Report prepared for Trustee Savings Bank.

Hand, D. J., and Batchelor, B. G. (1978). An edited nearest neighbour rule. *Information Sciences, (N.Y.)* **14**, 171–180.

Hand, D. J., and Henley, W. E. (1993–1994). Can reject inference ever work? *IMA J. Math. Appl. Bus. Ind.* **5**, 45–55.

Hand, D. J., and Henley, W. E. (1994). Inference about rejected cases in discriminant analysis. In *New Approaches in Classification and Data Analysis.* (E. Diday et al., eds.), pp. 292–299. Springer-Verlag, Berlin.

Hand, D. J., and Henley, W. E. (1996). Statistical classification methods in consumer credit scoring. *Submitted for publication.*

Hand, D. J., McConway, K. J., and Stanghellini, E. (1996). Graphical models of applicants for credit. *Submitted for publication.*

Hart, P. E. (1968). The condensed nearest neighbour rule. *IEEE Trans. Inf. Theory* **IT-14**, 515–516.

Henley, W. E. (1995). *Statistical aspects of credit scoring.* Ph.D. Thesis, Department of Statistics, The Open University, Milton Keynes, UK.

Henley, W. E., and Hand, D. J. (1996). A *k-nn* classifier for assessing consumer credit risk. *Statistician* **45**, 77–95.

Little, R. J. A., and Rubin, D. B. (1987). *Statistical Analysis with Missing Data.* Wiley, New York.

Lovie, A. D., and Lovie, P. (1986). The flat maximum effect and linear scoring models for prediction. *J. Forecast.* **5**, 159–168.

Michie, D., Spiegelhalter, D. J., and Taylor, C. C. (1994). *Machine Learning, Neural and Statistical Classification.* Ellis Horwood, New York.

Myers, J. H., and Forgy, E. Q. (1963). The development of numerical credit evaluation systems. *J. Am. Stat. Assoc.* **58**, 799–806.

Myles, J. P., and Hand, D. J. (1990). The multi-class metric problem in nearest neighbour discrimination rules. *Pattern Recognition* **23**, 1291–1297.

Overstreet, G. A., Bradley, E. L., Jr., and Kemp, R. S., Jr. (1992). The flat maximum effect and generic linear scoring models: A test. *IMA J. Math. Appl. Bus. Ind.* **4**, 97–109.

Rapoport, A. (1975). Research paradigms for studying dynamic decision behaviour. In *Utility, Probability, and Human Decision Making* (D. Wendt and C. A. J. Vlek, eds.). Reidel, Dordrecht, The Netherlands.

Ripley, B. D. (1994). Neural networks and related methods for classification (with discussion). *J. R. Stat. Soc. Ser. B* **56**, 409–456.

Rosenberg, E., and Gleit, A. (1994). Quantitative methods in credit management: A survey. *Oper. Res.* **42**, 589–613.

Short, R. D., and Fukunaga, K. (1982). The optimal distance measure for nearest neighbour classification. *IEEE Trans. Inf. Theory* **IT-27**, 622–627.

Srinivasan, V., and Kim, Y. H. (1987). Credit granting: A comparative analysis of classification procedures. *J. Finance* **92**, 665–681.

Stillwell, W. G., Barron, F. H., and Edwards, W. (1983). Evaluating credit applications: A validation of multiattribute utility weight elicitation techniques. *Organ. Behav. Hum. Perform.* **32**, 87–108.

von Winterfeldt, D., and Edwards, W. (1982). Costs and payoffs in perceptual research. *Psychol. Bull.* **83**, 213–217.

Wainer, H. (1976). Estimating coefficients in linear models: It don't make no nevermind. *Psychol. Bull.* **83**, 213–217.

Wainer, H. (1978). On the sensitivity of regression and regressors. *Psychol. Bull.* **85**, 267–273.

CHAPTER 11

Combination of Statistical and Other Learning Methods to Predict Financial Time Series

M. WESTPHAL
G. NAKHAEIZADEH

1. INTRODUCTION

The development of economic time series seems to be quite random. It is however possible that there is a chaotic process behind this random appearance. Indications for this already exist (Steurer, 1994).

The advantage of a chaotic process in contrast to total randomness is that the short-term development of such a process is forecastable. This is not possible for the long term, because it is necessary to know the chaotic function and the starting conditions. Even a minimal difference in the starting conditions can totally change the appearance of a chaotic process. "In the case of coin flipping, slightly different initial conditions give rise to quite different physical behavior of the coin, making its final outcome very difficult to predict" (Press, 1989, p. 5).

Many methods are developed or transferred from other fields, which can be applied to forecast the development of a chaotic time series, in particular the *k*-nearest neighbor method, Artificial Neural Networks, Model Trees, and the IBL-algorithms. One characteristic of the chaotic time series is the self-similarity. Structures exist in the time series that infinitely re-recur, sometimes on different levels or with different longitudes. The methods mentioned above attempt to identify such structures and employ them in the forecast.

In other fields of research attempts have been made to increase the power of single forecasting methods by combining them in one hybrid method. This chapter examines whether a hybrid technique is able to produce better results

Machine Learning and Statistics, The Interface, Edited by G. Nakhaeizadeh and C. C. Taylor.
ISBN 0-471-14890-3

than the single methods in forecasting the development of the DAX, the German stock price index.

First we will present the theory behind the single methods like *k*-nearest neighbor, IBL-algorithms, and M5. The theory needed for Neural Nets is spared because it is already well known. The second part of the chapter describes the empirical study of applying the methods to the predicting of the DAX.

In section 2 the *k*-nearest neighbors and the IBL-algorithms, the so-called case-based methods, are presented. In section 3 the model-based method M5 is explained. The description of the combination of the methods to gain a hybrid forecasting technique follows in section 4, and section 5 then deals with the empirical study. The last section is the summary.

2. SINGLE METHODS

2.1. *k*-Nearest Neighbor Method

The main idea behind the *k*-nearest neighbor (*k-nn*) method is predicting the target value of a new observation from known observations (case-base) made in the past. The new observation is compared with all elements of the case-base. The most similar past *k* observations are then chosen as references for the candidate. The similarity measure is often defined as the distance between new and old observations. The target values of the *k* references are then combined in some way to obtain the target value of the new observation. The *k-nn* method imitates a human ability, namely, reacting to a new situation with the aid of past experience. The *k-nn* method therefore belongs to the class of instance-based learning algorithms in the field of machine learning, because it employs instances rather than models. It can be also found in other fields of research. The most important field is pattern recognition, where it is mostly employed for classification. In statistics it is used as a smoothing technique to filter "white noise", and it is also applied to the fields like computational geometry and medicine. The *k-nn* belongs to the class of non-parametric methods. Assumptions about the distribution of the analyzed data is not necessary.

In practical applications of *k-nn* methods three questions arise:

1. What is the best value of *k*?
2. How is the distance obtained?
3. How are the target values of *k* references combined to obtain the target value of the new observation?

First, consider the last question: A simple solution in the case of a discrete-valued class is to take the value of the majority of the neighbors. For example, if 3 of 5 neighbors belong to class A then the candidate is put into class A (see Figure 11.1). Another possibility is to employ the qualified

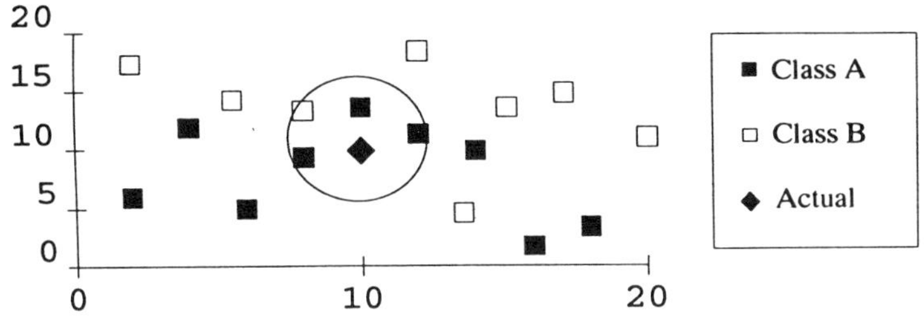

Figure 11.1. *Classification of the candidate (rhombus) according to four of its neighbors to class A.*

majority. The new observation is assigned to a class, when a minimum of k' neighbors belong to that class. This method is also called "nearest neighbor classification rule with a reject option" (Hellman, 1970, p. 237) or (k, k')-nearest neighbors.

If the target variable is continuous, the target value for the new observation can be estimated as the average, or it can be weighted according to the distances between the neighbors and the new observation. This weighted *k-nn* rule will be shown later in detail.

These examples are just a few of the most popular ways to obtain the target value for the new observation. In general, which solution is followed depends on the task to be performed and has to be decided by the user.

Important, but also difficult to answer, is the question of the value of k. In other words, how many neighbors should be taken to obtain the target value of the new observation? Similar to the previous case, no general rule exists for choosing the value of k. In the case of discrete-valued classes, a rule of thumb is to take an odd value for k so that a stalemate cannot occur.

To calculate the distance of two points a particular method is necessary. The most popular method is the L_p-distance. For the $\mathbb{R}^n$ defined as (Deussen et al., 1994, p. 83):

$$L_p(r, s) = \begin{cases} \left(\sum_{i=1}^{n} |r_i - s_i|^p \right)^{1/p}, & p \in [1, \infty) \\ \max\limits_{i=1,\ldots,n} |r_i - s_i|, & p = \infty \end{cases} \tag{1}$$

The vectors r and s contain the coordinates (attributes) of their data points. In $\mathbb{R}^3$ (i.e., $n = 3$) the coordinates could be width, height, and length. The L_1-distance is called the rectangular or Manhattan-distance. The distance is the shortest rectangular connection between two points. The L_2-distance is the most popular, called Euclidean-distance, which is a straight line between two points. Also well known is the L_∞-distance, called Maximum-distance, because the distance is the maximum of the differences of the vector components.

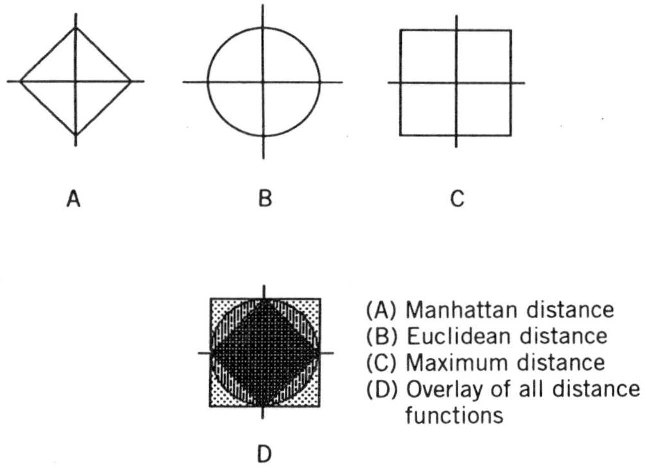

Figure 11.2. *Distances: The line shows the points with equal distance to the center. For B this results in a circle.*

Figure 11.2 shows the points with equal distance to the center depending on the methods used. The choice of the appropriate distance is also not pre-set for the *k-nn* method. Often the Euclidean distance is preferred for graphical problems or problems in two-dimensional space. It is recommended in three-dimensional space: "[...] it is sensible to take p to be Euclidean distance in applications where the independent variables are measurements of position in physical space; ..." (Clevland and Devlin, 1988, p. 597).

But not in every case are the results optimal. "[The result] confirms our expectation that the selection of an appropriate metric may significantly reduce the slope of the bias curve, resulting in more reliable estimates." (Fukunaga and Hummels, 1987, p. 210). For further discussion, see also Wettschereck (1995).

2.2. On the Training Data for *k*-Nearest Neighbors

Like other machine learning algorithms, the *k-nn* method is based on a representative set of training samples. So the training set should be carefully chosen. The quality of the weighted *k-nn* method depends on the volume of the training data. The number of training cases per class should not be taken as a constant figure, but, according to their part in the whole population. "Training sets should be fully representative of the populations from which they are drawn ..." (Davies, 1988, p. 130).

Attempts have been made to decrease the number of training cases necessary to classify the data effectively. In general this method is called edited *k*-nearest neighbors, and is similar to the IBL-algorithms described later in this chapter.

Here the case-base is pre-processed. This can be done by successive reduction of the instances, which belong to one class (see Chang, 1974), or by using two *k-nn* methods with decreasing value of k following each other, "... using the three-nearest neighbor rule followed by classification using the single-nearest neighbor rule ..." (Wilson, 1972, p. 299). The edited *k-nn* method is also used for smoothing a noisy case-base.

It is the aim of the edited *k-nn* method to filter those points from the case-base, which could be used as prototypes for their classes. According to Chang (1974), in the successive reduction a point is chosen randomly from the set of all known points, set B. The nearest neighbor p to this point is sought in set A, the set of the prototypes. If p belongs to the same class as the chosen point, then a new point p^* with the average of the attributes of both points is created and is assigned to set A replacing p. If the chosen point belongs to an other class than p, then the point itself is assigned to set A. A new point is continuously chosen from B until B is empty.

The *k-nn* method afterwards uses the set A of the prototypes as the case-base. The value of k should be reduced to one; that one neighbor is sufficient for classification was the aim of this algorithm. (See Figure 11.3.) Wilson proposed a different editing method. He used a three-nearest-neighbors rule followed by a single-nearest neighbor. The case-base is pre-processed by a *k-nn* rule. After this a 1-*nn* rule is sufficient for the classification with the help of the resulting case-base.

First the k neighbors of every Point X_i from the original case-base are obtained. The class θ is the class to which the majority of the k neighbors belong. The point X_i is removed from the case-base if it is an element of this class θ. The resultant reduced case-base is used for the classification by the 1-*nn* rule.

The edited *k-nn* method holds a risk, see also the last example: "... when pruning training sets there is considerable scope for upsetting the balance of class populations and introducing a bias due to altered a priori probabilities" (Davies, 1988, p. 130). In other words, the training data (case-base) is not representative for the original class any more.

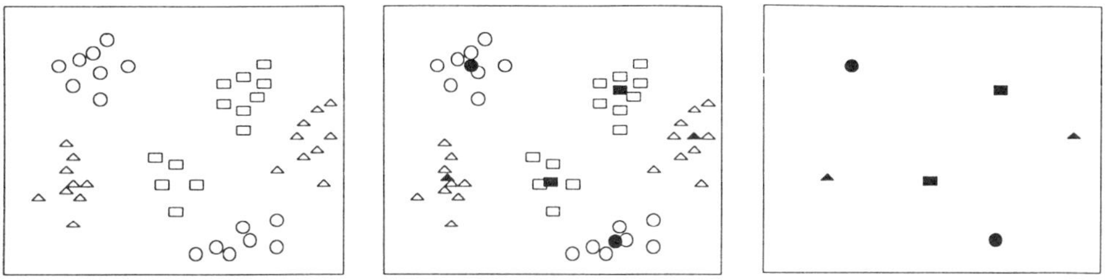

Figure 11.3. *Example of editing a case-base following Chang (1974). The filled points represent the obtained prototypes.*

2.3. Runtime Behavior in Practical Applications

The following shows the runtime behavior of popular algorithms for the *k-nn* method. The runtime behavior is an important factor in practical applications. An optimal algorithm can be useless, if it is too time consuming.

The distance calculation is the basis for finding the k neighbors of a candidate p. "This problem can be solved in time $O(n)$ by comparing p with all points ..." (Müller, 1989, p. 60).

Friedman et al. (1975) developed an algorithm for the *k-nn* method, employing the projection of the d-dimensional points onto the d axes. This algorithm needs $O(n^{1-(1/d)})$ for the distance calculation.

Yunck (1976) proposed an algorithm, which also employs the projection onto the d axes but works iteratively. The runtime behavior of this algorithm is $O(1)$. A single pre-processing of the data in $O(n)$ is necessary. This algorithm will be described in section 2.4 and builds the base *k-nn* algorithm we used to perform our empirical study.

Friedman et al. (1977) used a *k-d* Tree also with $O(1)$, but the pre-processing was decreased to $O(kn \log n)$.

Kim and Park (1985) developed the fastest algorithm so far. It combines the projection of attribute values with a search tree and needs a runtime of $O(1)$. Its speed is better described by the following statement: "... only 4.6 distance calculations, on the average, were required to find a nearest neighbor of a test sample among 10000 training samples ..." (Kim and Park, 1985, p. 371). This algorithm reaches nearly the theoretical optimal value for the *k-nn* method: "$O(\log N)$ comparisons are necessary to find a nearest neighbor of a point ... in any dimension" (Preparata and Shamos, 1990, p. 193).

2.4. The Algorithm of Yunck

This algorithm is one of the fastest algorithms and is also easier to implement than the algorithm by Kim and Park. The algorithm of Yunck is based on the projection of the data points onto all axes. Every single coordinate of a d-dimensional point is projected onto one of the d axes. Then an interval of side length l is put around the candidate y. This interval is also projected onto the axes. The side length l of the interval is then changed until the exact k of the other points lie inside this interval. This is where the algorithm ends. The interval of length l forms a Hypercube C (Figure 11.4). This implies that the algorithm mainly uses the maximum-distance for the distance calculations necessary in the *k-nn* rule. The points are numbered from $1, \ldots, N$, the axes are indexed with $i = 1, \ldots, d$. For the management of the data, two 1-dimensional arrays are necessary:

1. The array A_i, whose components $A_i(j)$ contain the axis i coordinate of point j.
2. The array I_i, whose components $I_i(j)$ contain the number of the point to which this coordinate $A_i(j)$ belongs.

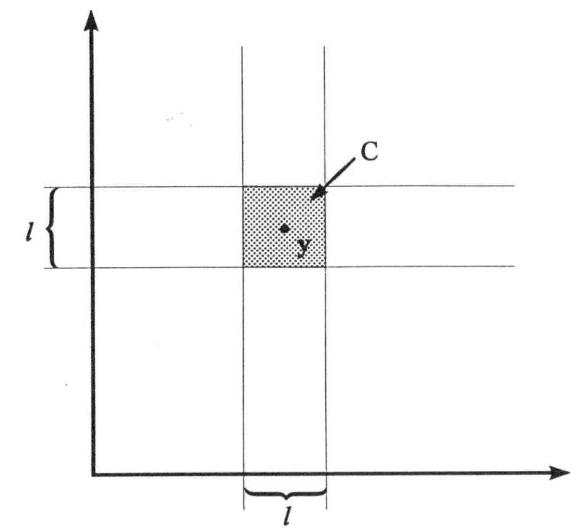

Figure 11.4. *Projection of the hypercube.*

This structure is necessary, because the coordinates must be separately sorted for each axis. For instance: For the points $a = (3, 5)$ and $b = (8, 2)$ the values for the arrays as shown in Table 11.1 are obtained. Two pointers for each axes, which point to the beginning and the end of the interval l, are needed. The pointers are called $P1_i$ and $P2_i$. The array $Z(\,)$ is used to register the identified neighbors. Before the algorithm can start, A_i is sorted once. For every axis i the pointer $P1_i$ is set to the smallest A_i, for which $A_i \geqslant y_i - l/2$ is valid and pointer $P2_i$ is set to the greatest A_i for which $A_i \leqslant y_i + l/2$ is true. The array $Z(\,)$ is initialized by 1 (see Figure 11.5).

After the initialization each value of $Z(I_i)$ is shifted left one bit for the I_i, which $A_i(j)$ lies in the interval between $P1_i$ and $P2_i$. This is equal to a multiplication by two. The value of the $Z(\,)$ shows how many of the coordinates of the corresponding point is inside the projection of the hypercube C. The amount L of $Z(\,)$'s whose value is equal to 2^d gives the number of neighbors identified for the distance $l/2$ from the candidate.

The l is decreased if $L > k$ and the $Z(\,)$, whose coordinates $A_i(j)$ does not lie in the hypercube C any more, are shifted back once. The l is increased if $L < k$ and the $Z(\,)$ are increased for the coordinates $A_i(j)$, which are new in

TABLE 11.1. Example for the Used Arrays

j	A_1	I_1	A_2	I_2
1	3	a	2	b
2	8	b	5	a

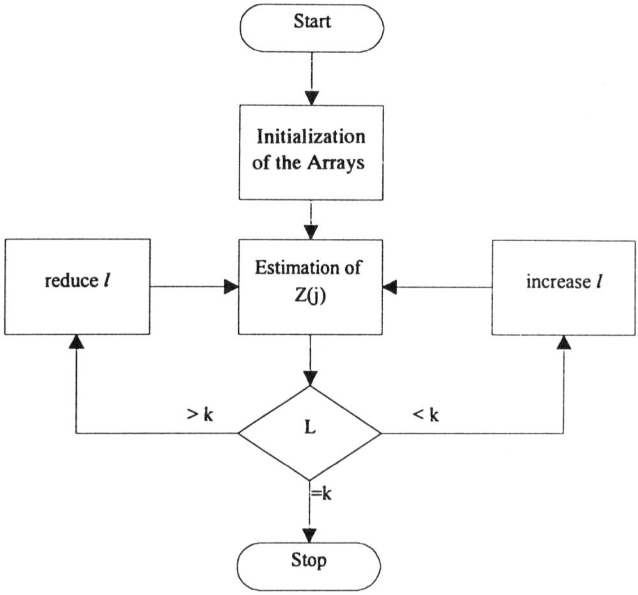

Figure 11.5. *Flow chart for the algorithm of Yunck.*

hypercube C. This is done iteratively until $L = k$. At this point the algorithm stops. In the implementation we used for the empirical part of this chapter, we have changed the algorithm of Yunck slightly. This modified version is described below (Figure 11.6).

In Yunck's algorithm it is possible to go too far, to identify more neighbors than preset by k. In such a case l is corrected again. The algorithm of Yunck swings around the optimal l before actually settling on l. Therefore l should be removed from the algorithm. The k-neighbors will be obtained only with the help of the array I_i. Two of the variables are to be changed: L will be the set of identified neighbors not just their number. And $Z(\,)$ will be initialized by zero. As seen above I_i are sorted according to A_i and the A_i are sorted by value.

First, the candidate y is positioned in the A_i. Now a step is taken from the y_i in positive and negative direction. But this step is not taken according to the A_i as by Yunck, it is taken according to the I_i. The start is $\mathrm{P1}_i = y_i$ and $\mathrm{P2}_i = y_i$.

$\mathrm{P1}_i$ is increased by one and $\mathrm{P2}_i$ is decreased by one. So the position in I_i, which will be further examined, is set. $Z(I_i(\mathrm{P1}_i))$ and $Z(I_i(\mathrm{P2}_i))$ are increased by one. Is $Z(I_i(\,)) = d$ true for one of the $Z(I_i(\,))$ then a neighbor is inserted into set L. If L then has less than k elements then $\mathrm{P1}_i$ is further increased and $\mathrm{P2}_i$ is further decreased, until L has exact k elements. Finally the indeces of the neighbors can be taken from L.

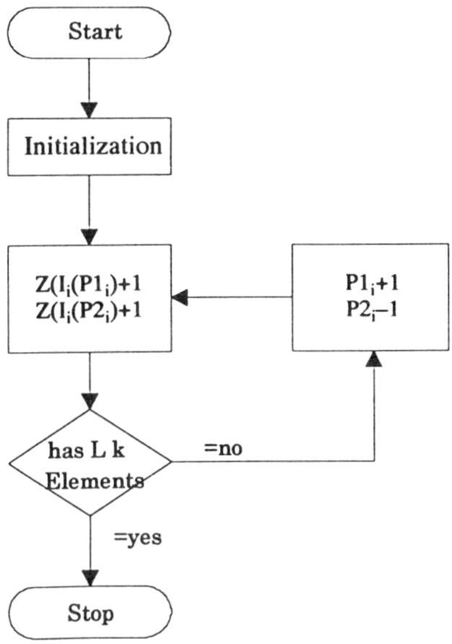

Figure 11.6. *Flow chart of the variation of Yunck's algorithm.*

This variation of Yunck's algorithm leads to the following advantages:

- The search for the $A_i \geqslant y_i - l/2$ and the $A_i \leqslant y_i + l/2$ is avoided, because $P1_i$ and $P2_i$ point to a position in I_i.
- The disadvantage of the iteration of the l is avoided. An identification of more neighbors than needed does not occur.
- In general, the problem of determining a suitable start value for l and a suitable step size ("It is rather vaguely stated that l should be modified 'appropriately'" (Yunck, 1976, p. 343) for the change of l is avoided. For $P1_i$ and $P2_i$ are successively increased in I_i. Unnecessary steps where a change of l brings no change to the array $Z(\,)$ are prevented.

2.5. The IBL-Algorithms

In the field of machine learning the *k-nn* method belongs to the group of "Instance-Based-Learning" (IBL) methods. These methods are applied to classification problems. They form a case-base for the whole population by incremental, supervised learning. The IBL-algorithms are also based on the hypotheses that similar instances belong to the same class (Aha, 1992).

The IBL-Algorithms can be compared with the edited *k-nn* methods, because their case-base is also, only a reduced selection of the whole popula-

TABLE 11.2. Example of the Ambiguity of Incremental Normalization in the IBL-Algorithms

	Original Values		Normalized Values	
Point No.	x	y	x	y
1	2.0	1000	1.0	1.0
2	1.0	950	0.5	0.95
3	1.5	850	0.75	0.85
4	30.0	920	1.0	0.92
5	2.0	1000	0.067	1.0

tion. The editing is done by an incremental algorithm. The case-base is formed in a successive manner.

All IBL-Algorithms use only the Euclidean distance and they normalize each coordinate of the instances (as demanded by Cleveland and Devlin 1988). This normalization will guarantee that the single coordinate is equally weighted in the distance calculation. The coordinates of the points are divided by the maximum value of their axis. So all values for each axes lie in the interval between zero and one.

But in IBL-Algorithms the normalization is a continuous procedure. The attributes of the points from the case-base are normalized according to the maximum value so far encountered. This is the consequence of the incremental behavior of the algorithm. So a new data point can have attributes which are a new maximum for one of the axis. "Renormalization is required whenever a new highest and lowest value is presented" (Aha et al., 1991, p. 64). Aha and colleagues recognized this problem but they mentioned it only in a footnote without further reflection.

The normalization technique used by Aha and colleagues (1991) should be considered critically. The aim is to weight all the attributes equally, but if the basis for the normalization is changed during the procedure, then an equal weighting cannot be guaranteed. A numerical example which demonstrates this fact is given in Table 11.2.

In Table 11.2 the real and the normalized values for data points with the two attributes x and y are shown. For the data point number 4 the renormalization becomes necessary, because the attribute x encounters a new maximum value. Then by observing point 5 the ambiguity becomes obvious. Although point 5 has the same attribute values as point 1, its attributes are no longer equally weighted. The attributes of point 1 were equally weighted in distance calculations but for point 5 the attribute x seems to be less important. This example shows that this kind of normalization does not achieve its aim.

2.6. The IBL-Algorithms in Detail

The algorithm IB1 is equal to a normal 1-*nn* method. The difference Aha mentions is that the case-base is formed incrementally. In our opinion this is not a real difference because in the end all data points are put into the case-base. No selection is carried out.

In contrast to the classical *k-nn* method, IB2 uses only special points, called prototypes, for its case-base. Aha and co-workers call this case-base "Conception Description". IB2 therefore belongs to the class of edited *k-nn* methods.

IB2 also works incrementally. It starts with an empty case-base and each of the new points is compared with its nearest neighbor from the case-base. If the class of both points is not the same then the new point is added to the case-base. In other words every point which can not be classified by the existing case-base is added as a reference.

In Figure 11.7 the resultant case-bases of the two algorithms, IB2 and the edited *k-nn* rule by Chang (1974) are presented. Both methods form their case-base incrementally or successively. As Figure 11.7 shows, in IB2 a single solution is irreversible.

The algorithm by Chang, in contrast, adapts its prototypes whenever new information is encountered. This results automatically in the visible order seen in Figure 11.7. This is further the reason why Chang's case-base contains fewer prototypes than the one for IB2, because if a special case is encountered it is not added to the case-base, but the existing prototypes are adjusted where possible. The structure of the case-base does not depend in Chang's algorithm on the sequence in which the instances are processed. For IB2 the case-base is different from sequence to sequence for the same population.

IB3 extends IB2 so that the IBL-algorithm is not so sensitive toward noisy data any longer. It tries to reduce the number of exceptions in the case-base. IB3 achieves this by adding an additional "classification record" counting the correct classifications by each reference point. Noisy points can be so identified

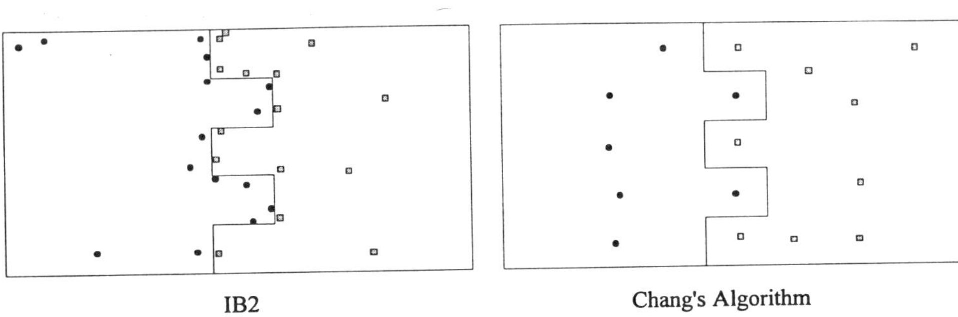

Figure 11.7. *Comparison of the resultant case-base of IB2 and the Algorithm by Chang. (Figures are adapted from figures by Aha et al. 1991, p. 49, and Chang (1974), p. 317).*

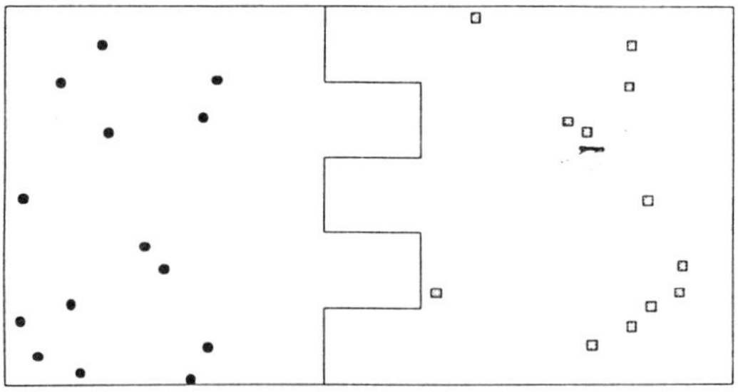

IB3

Figure 11.8. *Distribution of the prototypes in the case-base of IB3 (see Aha et al. 1991, p. 57).*

in the case-base because the number of correct classifications is small. These points are removed from the case-base afterwards. It should be mentioned that Wilson (1972) had already worked on the problem of classifying noisy data by a *k-nn* method. His algorithm already implements a noise reduction.

The classification record in IB3 holds the proportion between correct classifications with the help of the point y belonging to it and the frequency of the class of point y, compared with the whole population. The reference to the size of the classes, in contrast to the total number n of the known data points, is necessary, because the quality of a point could otherwise be thought of as bad only because the points of its class are rare.

IB3 requires a greater effort than the algorithms so far. After every single classification the classification records have to be updated. As Figure 11.8 shows, none of the prototypes lie near the border of the two classes. Only the points which can be used as a prototype for a great number of points remain in the case-base. This means that, points in the center of a class are preferred by the algorithm. "IB3 ... tends to use instances further from the actual concept boundaries than does IB2 for classification decisions" (Aha et al., 1991, p. 59). The resulting disadvantage is stated by Aha et al. themselves: "... it is unable to distinguish noise from disjuncts described by few instances" (Aha et al., 1991, p. 59).

IB3 performs a strong reduction of the noise compared to Wilson's method. In Wilson's algorithm the filtering can be finely adjusted by the choice of k for the first *k-nn* method and the effort required is less. The further developed algorithms IB4 and IB5 differ by the equal weighting of all attributes (coordinate axis) of the points. But the attributes are still normalized. An additional weight is included for each attribute in the distance calculation. These attribute weights are altered during the learning process.

The change to the Euclidean distance is:

$$\sqrt{\sum_{a=1}^{d} c_a^2 (x_a - y_a)^2} \tag{2}$$

where d is the number of the attributes of the points, x and y are two points and c_a is the weight for attribute a.

The idea behind the attribute weight is the same as that behind the classification record. The weight c_a for attribute a consists of the proportion of weight of correct classifications and the weight of all classifications:

$$c_a = \max\left(\frac{\text{CumulativeWeight}_a}{\text{WeightNormalizer}_a} - 0,5; 0\right) \tag{3}$$

The c_a weight shows the proportion between "hits" and "trials".

The weight for correct classifications is estimated as:

$$\text{CumulativeWeight}_a = \text{CumulativeWeight}_a + ((1 - \lambda_x)(1 - |x_a - y_a|)) \tag{4}$$

if x was classified correctly. The classification record of x is called λ. Otherwise if the classification of x was wrong, the following term is added to the $\text{CumulativeWeight}_a$:

$$((1 - \lambda_x)|x_a - y_a|) \tag{5}$$

To get a reference value as a basis the $\text{WeightNormalizer}_a$ is increased each time with $(1 - \lambda_x)$ independently of the success with classification. IB4 does not just learn the prototypes, but it also learns the relevance of each attribute. The difference between IB5 and IB4 is the learning process for the weights. IB5 also allows missing values, which is not the case in the other IBL-algorithms.

It was shown that the IBL-algorithms are similar to the *k-nn* methods in the field of pattern recognition. In some parts they are even inferior in comparison to the *k-nn* rules. Only IB4 and IB5 have something new. Both concentrate not just on the raw data points but they try to identify the important attributes depending on the classification, [IB1–IB3 are described in Aha et al. (1991) and IB4 and IB5 are described in Aha (1992)].

3. MODEL-BASED METHODS

3.1. General Remarks

The model-based methods which we used were an artificial Neural Net and M5, a model tree algorithm, introduced by Quinlan (1992). The Neural Net was a standard three-layer feedforward backpropagation net with the sigmoid function as the activation function. Due to the fact that this algorithm is well known, we will describe in the following only the algorithm M5, which combines the idea of regression trees with classical regression analysis. M5 works in a similar way to CART, an algorithm by Breiman et al. (1984). "Regression trees ... are based on a similar divide-and-conquer strategy, but have values rather than linear models at the leaves" (Quinlan, 1993, p. 237). A model tree divides a set of known data points (the training set) into parts. A linear regression model is adapted for each of the subsets. The model tree belongs to the group of binary trees. In each node of the tree the subset is divided into two parts, according to a discriminator. In the terminal nodes a regression model for the remaining subset is saved.

3.2. The Algorithm M5 in Detail

M5 starts with the root-node. The root-node contains a regression-model for the whole training set T. This set T is divided into two subsets, for the left and the right son of the root-node. (See Figure 11.9.) A discriminator marks the border between the two subsets. It is the result of a test. The set T can be divided into i possible pairs of subsets. For M5 the standard deviation (SD) of the T_i is taken as a measure of error. This error should be minimized by the regression tree of the actual node. The expected reduction of the error (Δerror), if the set T_i is used for one of the sons, can be written as

$$\Delta\text{error} = \text{SD}(T) - \sum_i \frac{|T_i|}{|T|} \times \text{SD}(T_i) \tag{6}$$

After estimating all possible Δerrors M5 chooses the partition with the maximum Δerror. This is done recursively for all sons of the actual node of the

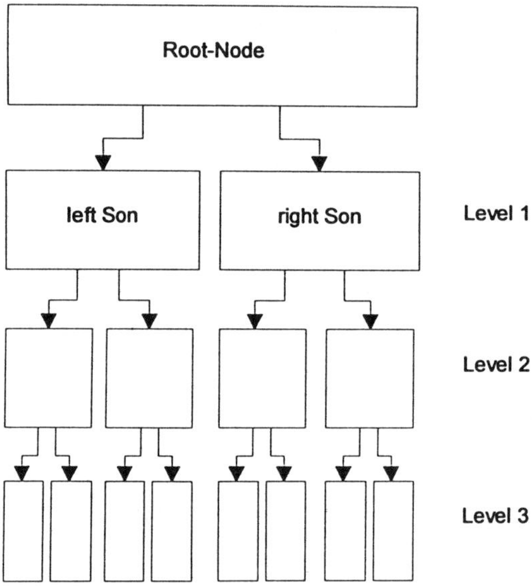

Figure 11.9. *Structure of a model tree. Each level encloses the whole training-set, but with a more detailed partition.*

model tree. "The major innovations of M5 come into play after the initial tree has been grown" (Quinlan 1992, p. 344). These innovations of M5 are:

- It is robust toward unseen data points;
- The models are linearly developed;
- Simplification of the models by elimination of regression attributes;
- Pruning of the constructed model tree; and
- A smoothing technique following Pregibon.

The use of linear regression models in the nodes of the tree rather than the average values brings the advantage of robustness towards unseen cases that have not been in the training set. This is also the major advantage compared to regression trees: "... regression trees will never give a predicted value lying outside the range observed in the training cases, whereas model trees can extrapolate" (Quinlan, 1992, p. 348). Furthermore the error of the models is corrected to increase the robustness toward unseen points.

Quinlan believes that M5 underestimates the error of the models for unseen cases. The error is therefore adjusted by a multiplication with $(n + v)/(n - v)$, where n is the number (equal to $|T_i|$) of data points which were used to form the model, and v is the number of parameters in the model. This adjusting factor means that models with a lot of parameters formed by fewer points are punished. Their estimated error is increased.

The simplification of the models achieved through parameter elimination should decrease the estimated error. The mentioned adjusting factor is important for the simplification. The fewer parameters a model has, the less is the adjustment for the estimated error. So it is possible that an increase of the error can be compensated by droping one of the parameters. M5 is therefore allowed to remove a parameter from a model if it has less influence on the model performance. It is possible that M5 leaves just a constant instead of a model in one of the nodes.

The linearity in the development of the model is achieved by using only attributes of a subtree that have already been used in higher levels of the tree. In this way the models in a subtree are concentrating on the same information and become comparable to each other. This quality is used in the process of "pruning." Each node aside from the terminal nodes possesses a linear model and a subtree. M5 compares the estimated error of the node and its subtrees. If the error of the model in the node is less than all the errors in the subtree, the subtree is pruned. The examined node then becomes a terminal node.

Quinlan added a smoothing technique to M5. It is observed that the quality of a forecast with a tree-based algorithm could be improved by applying a smoothing technique. The idea is that the prediction of a model in a terminal node is influenced by the models higher in the subtree. The prediction is therefore sent back through the subtree to the root. The process starts in the obtained terminal node. The father is named S, the chosen son is named S_i.

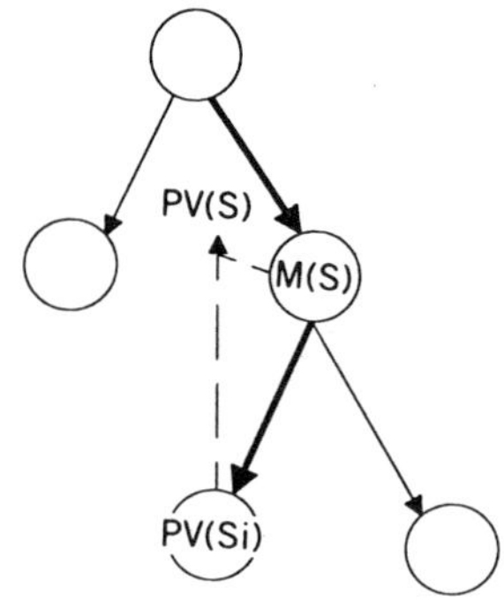

Figure 11.10. *Process of smoothing.*

This means that for the first step S_i is the terminal node. The number of data points in the subset of the son-node is n_i. PV(S_i) is the prediction of the son-node and M(S) is the estimated value of the model in the father-node. (The thick line in Figure 11.10 is the search-path in the model tree.) The smoothed prediction PV(S) for the father-node reads:

$$\mathrm{PV}(S) = \frac{n_i \times \mathrm{PV}(S_i) + k \times M(S)}{n_i + k} \tag{7}$$

where k is a smoothing constant chosen by the user. Equation (7) is obviously similar to an exponential smoothing. "Smoothing has most effect when leaf models are constructed from few training cases and do not agree with models higher in the tree" (Quinlan, 1993, p. 238).

As a rule of thumb, Quinlan (1992, 1993) recommends to set $k = 15$. It has to be known that the series of data examined by Quinlan enclosed maximally 563 points. With this, $k = 15$ is similar to a smoothing factor of 0.9 for exponential smoothing.

In our opinion the choice for k could become problematic, if k is greater than n_i in the terminal nodes. In such a case the prediction PV is discriminated from the beginning and the model value M is always preferred. If this is accepted, then the terminal node can be pruned without a lack in the accuracy of the prediction.

4. COMBINATION OF THE TECHNIQUES

Different methods for the combination of prediction methods exists. Below, three of them will be described. Quinlan (1993) applies a strategy based on the combination of an instance-based and a model-based learning technique. The instance-based technique is not able to predict a value outside of the range of the values of the target variable given in the training set. Contrary to the instance-based learning, a model-based technique has this ability. Quinlan uses this advantage to improve the results of instance-based algorithms.

In this approach, the target value of an unseen point U is to be predicted. The prototype P is identified by the instance-based rule and leads to prediction $V(P)$. The prediction of the model-based method is called $M(\,)$. Quinlan suggests the modification of $V(P)$ as

$$V(P) - (M(P) - M(U)) \tag{8}$$

By this modification, the predition of the instance-based method is adjusted by the difference between the prediction of the model-based method for the unseen case U and prototype P. In other words, the unknown error of the instance-based method is estimated with the help of the known error of the model-based method for the two points U and P. Then the prediction of the instance-based method is adjusted with this estimated error. The real values of the prediction of the model-based method are not noticed. Only the gradient of the model according to the points P and U is employed for a linearly correction of the prediction by the instance-based method. An alternative strategy for the combination of two or more predictions is trivially building their average. The idea is that the forecast error of each technique can be balanced leading to a better performance. Another possibility is using the prediction made by one technique as additional input for an alternative prediction. For example, the prediction of the k-nearest neighbors can be considered as an additional attribute to train a Neural Net.

5. EMPIRICAL RESULTS

5.1. General Remarks

To examine the performance of the single and combined approaches discussed in the previous sections, we have predicted the future development of DAX. Besides the technical indicators which will be described later in this section, we have used the following time series as input:

- the Dow Jones index;
- the Nikkei index;
- the exchange rate between U.S. $ and Deutsche Mark; and
- the German floating rate.

The DAX is the target that is to be predicted. The period from 01.01.1990 until 01.01.1993 was taken as the training set. The remaining data of the year 1993 formed the test set. We did not use the absolute daily values of the time series,

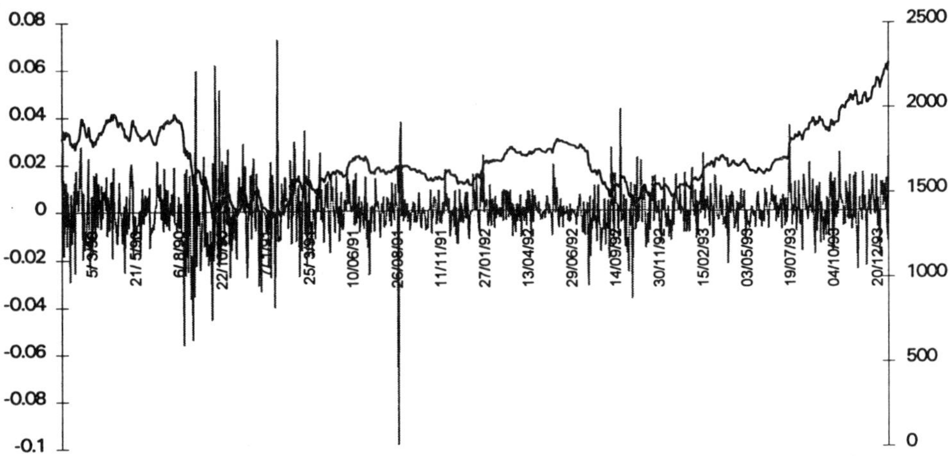

Figure 11.11. *The DAX and its daily logarithmic changes; black is the DAX with the corresponding right axis, grey is the line of the daily changes with corresponding left axis. A value of 0.04 on the left axis means a change of the DAX of +4%.*

but the daily changes. Furthermore, all the time series were converted to logarithmic differences of two following days. The converted DAX for the period 01.01.1990 until 01.01.1994 is shown in Figure 11.11.

A disadvantage of this method is that the strength of the linear relations between the time series are weakened. This can be seen from the correlation coefficients. The correlation coefficient for the time series plotted in Figure 11.11 between DAX and the exchange rate of the dollar is 0.53, the correlation between the logarithmic changes between the two time series is, however, 0.01.

The correct forecast of the direction of a change in price is more important. Because if the direction is wrong a correct height of the change is worth nothing anymore. The number of input series was increased to improve the performance of the neural net. The best way to improve the performance was to present not only single values to the net, but also short intervals of the time series, for example, three days. Technical indicators were used as additional input.

Applying technical indicators, which are used as additional input attributes, brings two advantages: First, not so many real time series are necessary for the test, because the technical indicators are estimated from the original series. Second, the forcasting method gains access to analyze tools common in the practice of financial markets, even if these tools are not scientifically established. The "technical analysis," in contrast to the "fundamental analysis," does not attempt to explain the level or the development of a series. It tries instead to identify early symptoms for changes in the market. All applied time series are given in Table 11.3.

TABLE 11.3. Time Series Used for Prediction

Fundamental	Technical Indicators
DOW	RSI 5
US $/DM	RSI 14
Nikkei	Moving Average 12–26 days
Umlaufrendite	Momentum 1
DAX	Momentum 5
	Momentum 10
	OBOS 5
	OBOS 10
	OBOS 15
	OBOS 20
	Deviation for 2 days
	Deviation for 3 days
	Deviation for 4 days

In Table 11.3 the variable "Umlaufrendite" means long-term average interest rate, RSI 5 and RSI 14 mean Relative Strength Index estimated for 5 and 14 days, respectively. OBOS stands for Over Bought Over Sold or "Williams R%".

The logarithmic daily changes were used as input for the net. By considering the values of four following days, 43 input attributes are used in total. The

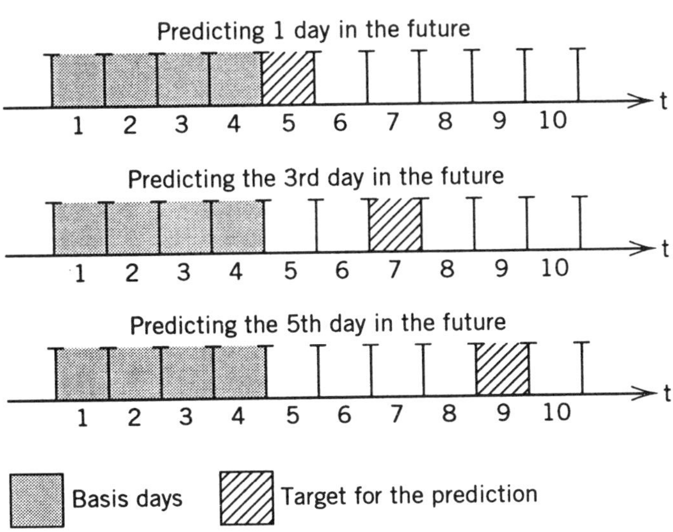

Figure 11.12. *Four days build the basis for the prediction of the 5th, 7th, and 9th day.*

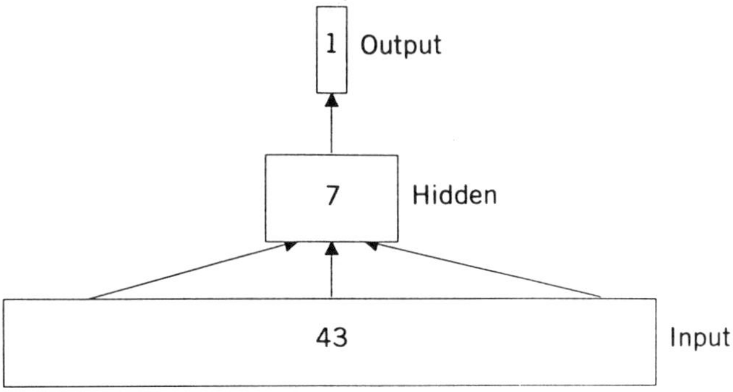

Figure 11.13. *Topology of the neural net.*

development of the time series in four days was employed to forecast the value of the DAX on the fifth day (Figure 11.12). The topology used for the Neural Network consisted of one output, seven nodes in the hidden-layer and 43 input-nodes (Figure 11.13).

The softward used to estimate the classical k-nearest-neighbor method is based on our own optimization of Yunck's algorithm, described in section 2.4. The *IBL*-algorithms are made available by David Aha, and *M5* by Ross Quinlan. The algorithms were trained with the data from the years 1990 until 1992 and the data of the year 1993 was used for the prediction. The results were also compared with the naive prediction; here the prediction is the same value as today. Besides Mean Square Error (MSE), we used the accuracy rate as evaluation measure where the prediction was reduced to the statements the DAX will "rise" or "fall."

5.2. Prediction Results

Table 11.4 shows the results for the DAX-prediction of the next day. It can be seen from this table that IB3 is the best version of IBL-algorithms. It is interesting that IB4 was outperformed by the less developed IB3. IB4 should be able, in contrast to IB3, to identify, automatically, the attributes which are important for the prediction. IB3 achieved his accuracy-rate with only 74 prototypes. The case-base was strongly reduced. As we will see the earliest IBL-algorithms performed mostly better than the high developed ones, in this task. The best performance of the k-nearest neighbor method was achieved with $k = 5$. The technical indicators had mostly a high significance. Especially including of the deviations was advantageous. Figures 11.14 and 11.15 show the predictions of the neural net and M5.

The results of the DAX for the third day is presented in Table 11.5. It can be seen that here the neural net performed slightly better (from 55% to 57%) when a second hidden-layer containing two nodes was included. IB1 was the

TABLE 11.4. Results for the DAX-Prediction of the Next Day

	Neural Net	M5	IBL	*k-nn*	Naive Prediction
MSE	0.0106	0.0074	*	0.0090	0.0111
Accuracy-rate	66.80%	69.53%	54.50%	53.13%	51.95%

*The *IBL*-software does not allow estimation of this value.

best of the IBL-algorithms and it is interesting to see that IB3 was this time the worst, with an accuracy-rate of only 34%. For the classical *k-nn* method the choice of $k = 3$ was the best. Overall the accuracy-rate decreased compared with the prediction of the next-day-DAX. The very good performance of M5 was not repeated.

Considering the prediction results for the change of the DAX at the fifth day in the future presented in Table 11.6, we can see that again IB1 was the best of the IBL-algorithms and again IB3 was the worst with 34.5 percent. The classical *k-nn* method performed best with $k = 6$. For prediction of the level of the DAX five days in the future, again the data of four days builds the basis. This time not the change of the DAX at that day was predicted but the change of the level of the DAX compared to the current day. Figure 11.16 shows, the time series seems smoother than the one before. The prediction results of this case are presented in Table 11.7.

As it can be seen from Table 11.7, the accuracy-rate is overall higher than that of the prediction of the changes at the fifth day. But this time also the

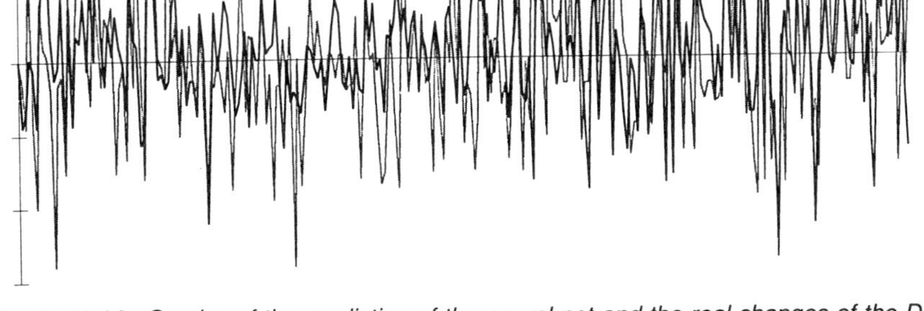

Figure 11.14. *Overlay of the prediction of the neural net and the real changes of the DAX.*

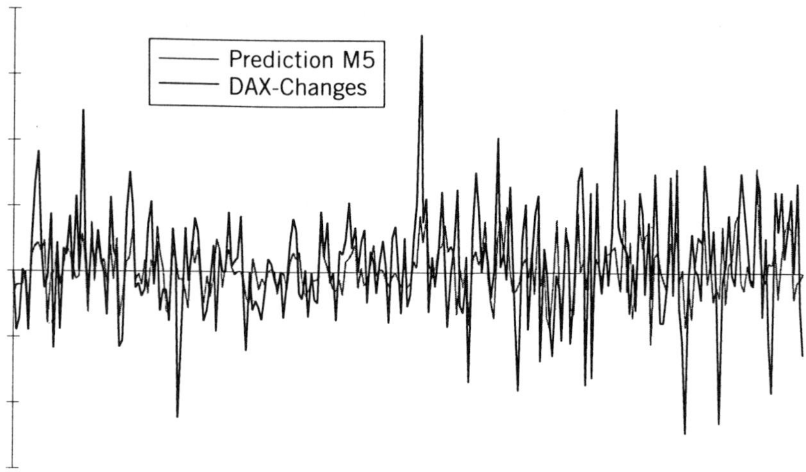

Figure 11.15. *Overlay of the prediction of M5 and the real changes of the DAX.*

accuracy-rate of the naive prediction is much higher, depending on the smoother curve. None of the forecasting-methods achieved the performance of the naive prediction. It is interesting to see that this time the *k-nn* method outperformed the others, with a choice of again $k = 5$. The best of the IBL-algorithms was IB2 with 48.2%, followed by IB3 with 47%, and the worst was IB4 with 43.5%.

TABLE 11.5. Results Achieved by the Different Methods for the DAX-Prediction for the Third Day (see Figure 11.12)

	Neural Net	M5	IBL	*k-nn*	Naive Prediction
MSE	0.0101	0.0087	*	0.0092	0.0111
Accuracy-rate	57.03%	47.27%	45.80%	54.30%	51.56%

*The *IBL*-software does not allow estimation of this value.

TABLE 11.6. Results for the Fifth Day Prediction (see Figure 11.12)

	Neural Net	M5	IBL	*k-nn*	Naive Prediction
MSE	0.0103	0.0087	*	0.0090	0.0113
Accuracy-rate	55.08%	46.09%	46.00%	52.73%	51.56%

*The *IBL*-software does not allow estimation of this value.

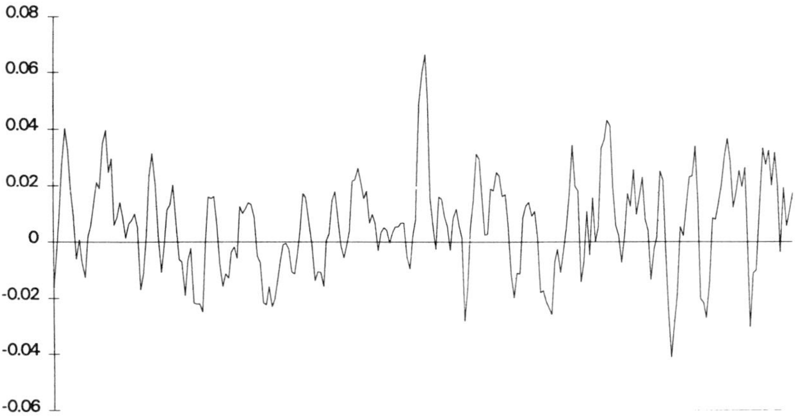

Figure 11.16. *Change in the level of the DAX during five days.*

TABLE 11.7. Results for the DAX-Prediction Five Days Ahead

	Neural Net	M5	IBL	*k-nn*	Naive Prediction
MSE	0.0228	0.0188	*	0.0189	0.0121
Accuracy-rate	55.86%	49.22%	48.20%	56.25%	78.91%

*The *IBL*-software does not allow estimation of this value.

5.3. Discussion of the Results of the Single Methods

In the last subsection the results of the predictions of the single methods were briefly presented. Now we want to comment on these results. Considering the figures that show the development of the target variables, we can determine that the prediction task has not been an easy one. Specially, we can see that the time series of the daily changes of the DAX was highly volatile and seems to be random. However, the results are partially very good. The employed methods seem to be able to identify hidden structures in the data. As mentioned before, an important difference between the methods is, that in contrast to the *k-nn* and IBL-algorithms the Neural Net and M5 are able to estimate values also outside the range of the target variable during the training process.

Using M5, the predictions were, however, not so volatile as the ones of the Neural Net, as can be seen in Figures 11.14 and 11.15. Furthermore, using the default version of M5 showed a tendency to predict a constant value for the whole time. It was not the case, using other values for the parameters of M5.

In performing most of the predictions by M5, we used all possible attributes rather than only the attributes involved in parent models. So far M5 employs just linear-regression models. An interesting variant would be to use also non-linear regression models. The costs of estimation might rise but perhaps the performance of M5 could be increased, too.

Dealing with the used Neural Net, it becomes obvious that the choose of the net input is of greater importance than the optimal architecture and complexity of the net. The pre-choice of the input time series highly determines the performance of the prediction.

Concerning *k-nn*, the parameters of the applied algorithms had to be adjusted individually for each task to achieve the best performance. This was not the case for the IBL-algorithms. The versions IB1 and IB2 mostly outperformed their further developed versions, namely, IB3 and IB4. In theory, IB3 should be better, because of the filtering of noisy data, and IB4 should have been the best, because it learns not only the best prototypes but also the most important attributes for the task, confirmed by Aha et al. (1991 and 1992). The results of the IBL-algorithms were, generally, inferior, compared with the results of the other methods.

5.4. Prediction of the DAX with the Combined Methods

Now we present the results of the combined methods. As mentioned before, the suggested combination using Quinlan's approach uses a model-based method to adjust the prediction result of an instance-based algorithm. The function for the resulting prediction is:

$$V(P) - (M(P) - M(U)) \tag{8}$$

It means, we need two predictions of the model-based method, $M(P)$ and $M(U)$ and the prediction $V(P)$ of the instance-based method. Table 11.8 contains the results. The used Neural Net algorithm was developed at the University of Karlsruhe (Germany) within the project NEXUS. We can see

TABLE 11.8. Results of the Combination of the Neural Net with the *k-nn* Method Using Quinlan's Approach

Sort of Prediction	Neural Net		*k*-nearest neighbor		Combination	
	Accuracy	MSE	Accuracy	MSE	Accuracy	MSE
1 Day	66.80%	0.0106	53.13%	0.0090	64.80%	0.0138
3. Day	57.03%	0.0101	54.30%	0.0092	49.69%	0.0135
5. Day	55.08%	0.0103	52.73%	0.0090	52.34%	0.0106
in 5 Days	55.86%	0.0228	56.25%	0.0189	51.56%	0.0254

TABLE 11.9. Results of the Combination per Average, the Best Method Is That with the Highest Accuracy-Rate

Sort of Prediction	NN & *k-nn*		NN & M5		M5 & *k-nn*		NN & M5 & *k-nn*		Best of the Single Methods	
	Accuracy	MSE	Accuracy	MSE	Accuracy	MSE	Accuracy	MSE	Accuracy	MSE
1 Day	64.45%	0.0082	70.70%	0.0076	63.67%	0.0077	69.92%	0.0074	69.53%	0.0074
3. Day	53.91%	0.0089	54.30%	0.0089	54.69%	0.0087	50.78%	0.0086	57.03%	0.0101
5. Day	51.95%	0.0088	54.30%	0.0088	48.83%	0.0087	49.22%	0.0086	55.08%	0.0103
in 5 Days	59.77%	0.0185	52.73%	0.0186	53.13%	0.0179	55.08%	0.0177	56.25%	0.0189

from Table 11.8 that neither the accuracy-rate nor the MSE could be improved by Quinlan's approach. The results of the Neural Net alone are better in each case. An alternative method of combination is to use the average of two predictions from different methods. This average is then the new prediction.

Table 11.9 shows the results of the average of the predictions of the involved forecasting methods. These results can be compared with the results of the best single methods. In two of four cases the accuracy rate of the single methods is surpassed by the combination. But more interesting are the values of the MSE. Nearly all the combinations have an MSE lower than that of the single methods.

Table 11.10 compares the first column of Table 11.9 with the results of Quinlan's approach. It is obvious from this table that, for this domain, the results of Quinlan's approach in general, are not better than the results of a simple combination per average. Uniquely two of the accuracy rates are slightly better by using the Quinlan's approach. Another alternative method, which was discussed already, is the use of the prediction of one method as additional input attribute for the Neural Net or M5. The results of the combination with the classical *k-nn* rule are presented in Table 11.11.

The results of this sort of combination are encouraging. The accuracy-rates of the values at the fifth day in five days are the best of all. The results of the MSE are also very good. The value of "T", which is used as an additional evaluation measure, indicates the performance of the method compared with the "naive prediction," where a value of $T < 1$ is better than the naive prediction. The values of T are in general, also very good. The prediction of the level of the DAX in five days by the combination with the Neural Net is the only one which is surpassed by the "naive prediction." This case is also the only one where the combination with M5 performs better than the one with the Neural Net. An analysis of the significance of the inputs of the Neural Nets shows that the predictions of the *k-nn* method is an important input attribute for the Neural Net model. Furthermore the past values of the DAX and the Nikkei were important as well.

TABLE 11.10. Comparison of Quinlan's Combination with the Combination per Average

Sort of Prediction	NN & *k-nn* Accuracy	NN & *k-nn* MSE	Quinlan Accuracy	Quinlan MSE
1 Day	64.45%	0.0082	64.80%	0.0138
3. Day	53.91%	0.0089	49.69%	0.0135
5. Day	51.95%	0.0088	52.34%	0.0106
in 5 Days	59.77%	0.0185	51.56%	0.0254

TABLE 11.11. Neural Net and M5 Get the Prediction of the *k-nn* Rule as Additional Input

Sort of Prediction	NN with input from *k-nn*			M5 with input from *k-nn*		
	Accuracy	MSE	*T**	Accuracy	MSE	*T*
1 Day	65.23%	0.0088	0.7965	64.06%	0.0076	0.6829
3. Day	55.47%	0.0107	0.9686	46.09%	0.0089	0.7879
5. Day	59.38%	0.0090	0.7987	45.70%	0.0089	0.7918
in 5 Days	66.80%	0.0205	1.7012	82.03%	0.0121	0.9902

*With T is Theil's coefficient:

$$T = \sqrt{\frac{\Sigma_t (x_t - \bar{x}_t)^2}{\Sigma_t (x_t - x_{t-1})^2}}$$

with the prediction $\bar{x}_t$.

6. CONCLUSIONS

Four different forecasting techniques were examined in this chapter to predict the future development of DAX, two model-based methods (Neural Net and M5) and two instance-based methods (*k-nearest neighbor* and *IBL*). None of the methods really dominated the others, only the IBL-algorithms performed generally weaker. After the estimation of the prediction quality of the single methods, we examined whether the results could be improved by a combination of the techniques. Three combination concepts were tested. Only one of the combinations seems to be really superior compared with the single methods. It is the combination technique where the prediction of an instance-based method is used as additional input for a model-based method.

In this study, the combination proposed by Quinlan (1993) is hardly superior compared with a simple average of some predictions. The model tree M5 alone, was also not able to satisfy here. Differentiated predictions were only possible by disabling some of the smoothing and prunning techniques of the algorithm. The used Neural Net showed an over all satisfactory performance. The disadvantage of this method is that it is much more time consuming during the training phase. The k-nearest-neighbor rule produced a reliable nearly constant MSE, but the accuracy-rate was mostly slight higher than that of the "naive prediction." The IBL-algorithms showed contradictory performance compared with the theoretical expectations. Especially, the further developed algorithms IB3 and IB4 performed weaker than expected.

Put together, the results of this work encourage applying methods from the field of machine learning on forecasting tasks. Their main advantage is that the user needs not pre-determine the dependencies of the data. Therefore these methods can be used for automatic analysis of economic time series, and to aid human experts. The methods have different strengths and weaknesses, so that a combination can be advantageous.

ACKNOWLEDGMENTS

The authors would like to thank David Aha and Ross Quinlan for providing *IBL*- and *M5* software.

REFERENCES

Aha, D. W. (1992). Tolerating noisy, irrelevant and novel attributes in instance-based learning algorithms. *Int. J. Man-Mach. Stud.* **36**, 267–287.

Aha, D. W., Kibler, D., and Albert, M. K. (1991). Instance-based learning algorithms. *Mach. Learn.* **6**, 37–66.

Breiman, L., Friedman, J. H., Olshen, R. A., and Stone, C. J. (1984). *Classification and Regression Trees.* Chapman & Hall, New York and London.

Chang, C. L. (1974). Finding prototypes for nearest neighbor classifiers. *IEEE Trans. Comput.* **C-23**(11), 1179–1184 (reprinted in Dasarathy, 1991).

Cleveland, W. S., and Devlin, S. J. (1988). Locally weighted regression: An approach to regression analysis by local fitting. *J. Amer. Stat. Assoc.* **83**(403), 596–610.

Cover, T. M., and Hart, P. E. (1967). Nearest neighbor pattern classification. *IEEE Trans. Inf. Theory* **IT-13**, 21–27 (reprinted in Dasarathy, 1991).

Dasarathy, B. V., ed. (1991). *NN Pattern Classification Techniques.* IEEE Computer Society Press, Los Alamitos, CA.

Davies, E. R. (1988). Training sets and a priori probabilities with the nearest neighbour method of pattern recognition. *Pattern Recognition Lett.* **8**, 11–13 (reprinted in Dasarathy, 1991).

Deussen, O., Kreeb, M., and Schmitt, A. (1994). *Graphisch geometrische Algorithmen*, Lect. Pap. Institut für Betriebs- und Dialogsysteme Abteilung Dialogsysteme und graphische Datenverarbeitung, Universität Karlsruhe, Germany.

Farmer, J. D., and Sidorowich, J. J. (1988). *Exploiting Chaos to Predict the Future and Reduce Noise.* Los Alamos National Laboratory, Los Alamos, NM.

Fix, E., and Hodges, J. L. (1951). *Discriminatory Analysis: Nonparametric Distribution: Consistency Properties*, Proj. 21-49-004, Rep. No. 4. USAF School of Aviation Medicine, Randolph Field, TX (reprinted in Dasarathy, 1991).

Fix, E., and Hodges, J. L. (1952). *Discriminatory Analysis: Nonparametric Discrimination: Small Sample Performance*, Proj. 21-49-004, Rep. No. 11. USAF School of Aviation Medicine, Randolph Field, TX (reprinted in Dasarathy, 1991).

Friedman, J. H., Baskett, F., and Shustek, L. J. (1975). An algorithm for finding nearest neighbor. *IEEE Trans. Comput.* **C-24**, 1000–1006.

Friedman, J. H., Bentley, J. L., and Finkel, R. A. (1977). An algorithm for finding best matches in logarithmic expected time. *ACM Trans. Math. Software* **3**(3), 209–226.

Fukunaga, K., and Hummels, D. M. (1987). Bias of nearest neighbor error estimates. *IEEE Trans. Pattern Anal. Mach. Intell.* **PAMI-9**(1), 103–112 (reprinted in Dasarathy, 1991).

Graf, J. (1990). Klassische statistische Methoden zur kurzfristigen Prognose des Aktienmarkts und deren Vergleich mit den Methoden der künstlichen Intelligenz.

Diplomarbeit, Institut für Statistik und Wirtschaftstheorie, Universität Karlsruhe, Germany.

Graf, J. (1993). *Technische Analyse.* SGZ-Bank Karlsruhe, Germany.

Hellman, M. E. (1970). The nearest neighbor classification rule with a reject option. *IEEE Trans. Syst. Sci. Cybernet.* **SSC-6**(3), 179–185 (reprinted in Dasarathy, 1991).

Keller, J. M., Gray, M. R., and Givens, J. A., Jr. (1984). A fuzzy *k*-nearest neighbor algorithm. *IEEE Trans. Syst. Man, Cybernet.* **SMC-15**(4), 580–585 (reprinted in Dasarathy, 1991).

Kim, B. S., and Park, S. B. (1985). A fast *k* nearest neighbor finding algorithm based on the ordered partition. *IEEE Trans. Pattern Anal. Mach. Intell.* **PAMI-8**(6), 761–766 (reprinted in Dasarathy, 1991).

Müller, H. (1989). *Computer Graphics II. Computational Geometry and Image Generation*, 5th ed. Fakultät für Informatik, Institut für Betriebs- und Dialogsysteme, Universität Karlsruhe, Germany.

Preparata, F. P., and Shamos, M. I. (1990). *Computational Geometry and Introduction.* Springer-Verlag, New York and Berlin.

Press, S. J. (1989). *Bayesian Statistics: Principles, Models and Applications.* Wiley, New York, and Chichester.

Quinlan, J. R. (1992). Learning with continuous classes. In *Proceedings of the Fifth Australian Joint Conference on Artificial Intelligence*, pp. 343–348. World Scientific Press, Singapore.

Quinlan, J. R. (1993). Combining instance-based and model-based learning. In *Proceedings of the Tenth International Conference on Machine Learning*, Amherst, MA, pp. 236–243. Morgan Kaufmann, San Mateo, CA.

Steurer, E. (1994). *Exchange Rate Prediction by using Symbolic Machine Learning.* Discuss. Pap. Daimler Benz AG, Research and Technology, Ulm Germany.

Wettschereck, D. (1995). *Choosing the Distance Function for k-Nearest Neighbor Algorithm*, Discuss. Pap. GMD, Sankt Augustin, Germany.

Wilson, D. L. (1972). Asymptotic properties of nearest neighbor rules using edited data. *IEEE Trans. Syst., Man, Cybernet.* **SMC-2**(3), 408–421 (reprinted in Dasarathy, 1991).

Yunck, T. P. (1976). A technique to identify nearest neighbors. *IEEE Trans. Syst., Man, Cybernet.* **SMC-6**(10), 678–683 (reprinted in Dasarathy, 1991).

CHAPTER 12

Combining Statistical Techniques and Search Heuristics to Perform Effective Feature Selection

M. RICHELDI
M. ROSSOTTO

1. INTRODUCTION

Both in the scientific and in the industrial worlds there is a growing realization and expectation that data, intelligently analyzed and correctly presented, will be a valuable resource to be used for a competitive advantage. Knowledge and data discovery in databases, or data mining, is a new area of research that integrates techniques originated in several fields, such as data visualization, inferential statistics, machine learning, and database management, to develop techniques for intelligent data analysis and to fill the growing gap between data generation and data understanding (Matheus et al., 1993; Piatetsky-Shapiro et al., 1994).

Many knowledge discovery algorithms perform well on relatively simple and familiar domains (Rendell et al., 1989). Recently, attention has been shifted toward hard domains, which are characterized by high noise level and strong interactions between data attributes, also called features (Rendell and Ragavan, 1993).

Some studies (Caruana and Freitag, 1994; Piatetsky-Shapiro, 1991; Rendell and Ragavan, 1993) show that addressing hard domains is easier when machine-learning algorithms are supported by statistical techniques that can analyze and grasp dominant relationships between the data. This result lead to the definition of an effective methodology for data mining, depicted in Figure 12.1. In such a methodology (Piatetsky-Shapiro and Matheus, 1994; Richeldi, 1993), the core process of knowledge and data discovery is carried out in three

Machine Learning and Statistics, The Interface, Edited by G. Nakhaeizadeh and C. C. Taylor.
ISBN 0-471-14890-3

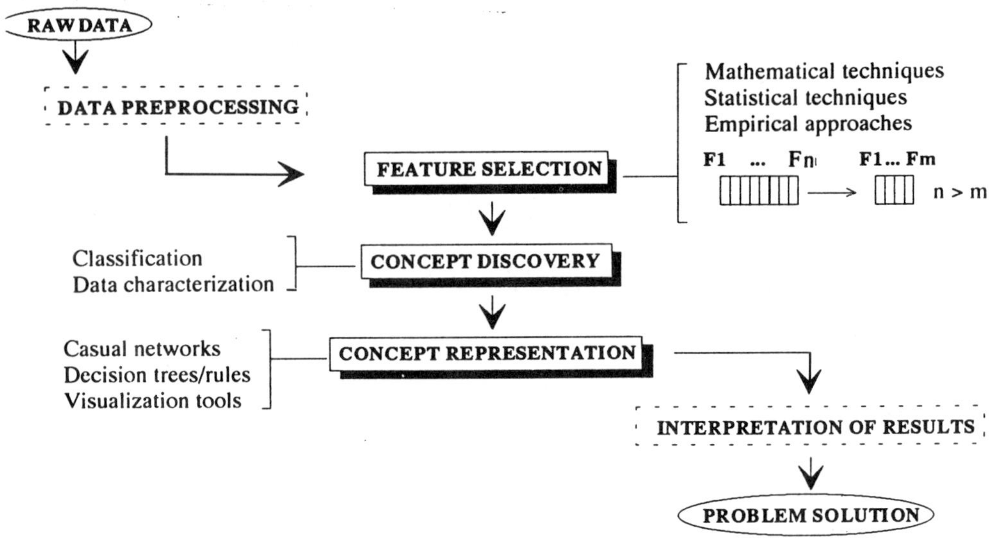

Figure 12.1. *An effective methodology for the data mining task.*

successive steps (depicted in the figure as shadowed boxes), called the Feature Selection phase, Concept Discovery phase, and Concept Representation phase, respectively.

- *The Feature Selection Phase.* When analyzing real-world data, concept learners have to cope with abundance and redundancy of features. The representation of data often requires many features, only a few of which may be related to the target concept. Irrelevant features degrade the performance of concept learners, both in speed and in predictive accuracy. The problem of choosing the minimal set of features that properly describes the target concept is named *feature selection* (Kira and Rendell, 1992a; b). In the feature-selection phase, relationships between features are investigated. As a result, irrelevant features can be eliminated and the remaining features ranked according to their relevance to the concept description.
- *The Concept Discovery Phase.* In this phase, interesting and useful new patterns are identified and extracted from data. The best results are obtained when the discovery task takes the knowledge gained in the feature-selection phase into account. Possible goals of the concept discovery phase are class identification, concept description,[1] and outlier detection (Rendell and Cho, 1990; Quinlan, 1983).

[1]Concept description provides the user an abstract or intensional description that summarizes interesting qualities about the classes of data.

- *The Concept Representation Phase.* This phase is aimed at presenting the results of the discovery task in a simple and concise notation, thus making easier their interpretation and understanding. The question of providing an adequate representation to results is at least as important as the choice of learning algorithms appropriate to the given problem.

The focus of this chapter is on the feature-selection phase of the data-mining methodology. Section 2 introduces the general problem of feature selection. It also presents the two major approaches to data dimensionality reduction, that is, the filter model, in which features are selected independently of the learning algorithm used in the concept discovery phase, and the wrapper model, in which the learning algorithm is used to select the features that maximize its generalization performance. Examples of techniques that perform data reduction according to the filter model are principal component analysis, factor analysis, and cluster analysis. These statistical techniques are reviewed and analyzed in section 3. Section 4 describes ADHOC (Automatic Discoverer of High-Order Correlations), a new algorithm that combines the advantages of both the filter and the wrapper model. ADHOC addresses the feature-selection problem in two steps.

In the first step, which is independent of the learning algorithm, it discovers the major dimensions underlying the data by investigating direct and indirect relationships between features. The algorithm provides a new technique for discovering the deep structure of data that overcomes some of the problems that affect both factor and cluster analysis.

In the second step, ADHOC takes the learning algorithm into account by incorporating it into the feature-evaluation process. A branch-and-bound procedure is used to search a feature subset that generalizes well with the learning algorithm. The search is not blind; the algorithm selects at least one feature among those that reflect every dimension of the problem. Analysis of two real-world data sets were carried out to evaluate the feature selection using ADHOC; section 5 summarizes and discussed experimental results. Section 6 presents the conclusions.

2. THE FEATURE-SELECTION ISSUE

Any object or situation that can be classified by a learning algorithm has to be expressed as a set of features, or properties. The first step in any learning process, performed either by a machine or by a human being, is to figure out whether discriminatory features do exist and, if any, consider the problem of what discriminatory features to select and how to extract them. It is quite intuitive that the number of features required successfully to perform a given learning task depends on their discriminatory qualities. However, the problem of feature selection is often complicated by the fact that the most important features are not necessarily easily measurable or, in many cases, their measurement is inhibited by economical considerations.

Consider, for example, the issue of estimating software development effort. Software quality has been defined as being on time, within budget, and meeting user needs (Jones, 1991). As a consequence, project schedules and costs are considered to be important quality factors. Development effort, or cost, may be estimated by characterizing and measuring those aspects of programs (features) that seems to affect cost. These features are called *software metrics* (Fenton, 1992). Examples of software metrics are the number of delivered source lines; the required software reliability; execution time constraints, and so on. A large number of metrics are usually computed, but not all of them necessarily carry much discrimination information. For instance, the "lines of code count," "statement count," "total number of lines excluding comments" metrics measure the same dimension of a software project. Every single metric conveys sufficient information about the length of the code delivered in the project; using the three metrics altogether is worthless. Moreover, too many metrics may even complicate the subsequent classification scheme, because they contain insufficient discriminatory information. Excess features affect generalization:[2] noise in the training sample is magnified by adding more noisy features. The task of separating signal from noise gets harder as the number of noisy features increases. The more noisy features, the less likely the learning algorithm will be able to identify properly those features that lead to good generalization, given a fixed training sample.

As a second example, consider the problem of oil prospecting: an oil industry wishes to ascertain whether a given geographical region contains or lacks an acceptable quantity of oil. In this case, oil prospecting may be regarded as a two-class classification problem. The most obvious approach to oil prospecting would be drilling oil wells throughout the region until either oil is found, or the number of dry or semidry oil wells approaches a quantity large enough to indicate that the region is, in fact, oil-less, for all practical purposes. These measurements will yield the most relevant features for this particular problem. But the high cost of oil drilling (Hayes-Roth et al., 1983) prevents us from taking this solution into account. Thus, oil industry scientists figured out a set of features that are more economical to obtain, although conveying less information. These features are seismic events obtained from low-frequency reflection waves artificially created on the earth's surface at several points throughout the region in question. In this manner a local map of the earth's inner crust is obtained, and the region can be classified as having or lacking the potential of holding oil. In this problem, a necessary compromise in feature-selection results in a classification process that is considerably far from optimum. This trade-off between feature selection and predictive accuracy is present in most classification tasks of practical significance.

[2]In this context, the training sample consists of the projects being examined.

The above observations support the evidence that feature selection plays a central role in learning and in the data analysis process in general. As manual feature selection is very complex, much effort has been spent to automate this task. The advantages of automatic selection are quite intuitive:

- It gives the chance of identifying as many combinations of potentially useful features as possible and then let the system select the best one;
- It provides a method to ascertain the effect of different feature combinations on predictive accuracy; and
- It allows the learning task for taking into account different set of features according to changes in the problem specification or in the training set.

There exist two major approaches to feature selection. The first one, called the *Filter Model* (John et al., 1994; Caruana and Freitag, 1994; Kohavi and Sommerfield, 1995), accomplishes feature selection independently of the performance of the learning algorithm used in the concept-discovery phase. Optimal feature selection is achieved by maximizing or minimizing a criterion function. The disadvantage of this method is that it ignores the effects of the selected features on the generalization performance of the learning algorithm. The FOCUS (Almuallin and Dietterich, 1991) and Relief (Kira and Rendell, 1992a; b) algorithms are instances of the filter model. Statistical techniques such as principal component analysis, factor analysis, and cluster analysis fit the filter model in the data-reduction framework.

The second approach, named the *Wrapper Model*, is performance-dependent. Its effectiveness is directly related to the performance of the learning algorithm, usually in terms of predictive accuracy and minimal use of input features. Algorithms that induce decision trees, such as CART (Breimen et al., 1984) and C4.5 (Quinlan, 1993), are examples of the wrapper model. Indeed, they perform dynamical feature selection by employing some theoretical measure, such as the entropy function. John et al., (1994) describe an interesting implementation of the wrapper model that uses cross-validation to evaluate the performance of the inductive algorithm and greedy search heuristics to identify useful feature subsets.

3. STATISTICAL TECHNIQUES FOR DATA REDUCTION

Human beings use physical features as the basis for classification. Indeed, sensory organs are trained to recognize these kind of features accurately. Since the capabilities of human sensory organs are too difficult to imitate in most practical situations (Editor, 1980), the effectiveness of physical features in the learning process is sharply reduced. On the other hand, machines can extract and select mathematical features of data, which a human being has difficulty in determining without mechanical aid. Examples of these type of features are statistical means, correlation coefficients, eigenvalues, and eigenvectors of covariance matrices, and other invariant properties. Inferential statistics provides several techniques for data reduction, in general, and feature selection, in particular. Some of these methods are still under development and refinement. This section provides a brief survey of three of the most employed data-reduction techniques: factor analysis, principal component analysis, and cluster analysis.

3.1. Principal Component Analysis

Principal Component Analysis (PCA) (Fabbris, 1991; Monge and Cappella, 1980) is a statistical technique used to reduce the number of the observed features. The central idea of PCA is to transform a set of features of dimension n, consisting of a large number of interrelated features, into a new set of features of a reduced dimension m, called the principal components (PCs), which are uncorrelated, and which are ordered so that the first few retain most of the variation present in all of the original data.

Consider, for example, a common problem in the social sciences, that is, the description of a community of people. A large number of features can be used to describe a community: population, mobility, average family income, degree of industrialization, commercial activity, extent of home ownership, and so forth. For instance, Jonassen and Peres (1960) examined 82 features to describe 88 communities in Ohio.

PCA allow one to replace the large and unwieldy set of observed community features by a smaller set of summary features, which preserves most of the information of the larger set. In particular, this method forms linear combinations of the original features. Any combination is called a PC. The first PC accounts for the largest amount of variance in the sample. The second PC accounts for the next largest amount of variance and is uncorrelated with the first. Successive components explain progressively smaller portions of the total sample variance, and all are uncorrelated with each other. PCs that explain a great portion of total variance retain most of the information provided by the original features.

Table 12.1 shows the amount of variance accounted for by PCs for a subset of the features used by Jonassen and Peres (1960). The total variance in the sample is simply the sum of the variance of each feature. The greater the portion of total variance explained by the PCs, the better the PCs account for all the correlations that exist between the observed features, to within sampling error. If one computes as many PCs as there are features, the total variance can be explained, but no simplification or summarization occurs. All one has gained is to transform a set of correlated features into a set of uncorrelated features. The basis of the PCA is that not all of the PCs will be kept, that is, only the first several are to be used as summary features for the whole set. Table 12.1 contains statistics for each PC; columns that list the absolute value, the percentage, and the cumulative percentage of the total variance explained by each PC are depicted. Table 12.1 shows that 73 percent of the total variance

TABLE 12.1. Results of the PCA on a Set of Community Features

Feature	Variance		
	Absolute Value	Percent (%)	Cumulative Percent (%)
POPUL_STAB	6.1	44.53	44.53
NEWSPAP_CIRC	2.0	14.60	59.13
FEMALE_LAB	1.9	13.87	73.00
FARMERS	0.7	5.11	78.11
PRO_CAPIT_SALES	0.7	5.11	83.22
COMM_ACT	0.6	4.38	87.60
INDUS_INDEX	0.5	3.65	91.25
HEALTH_INDEX	0.3	2.19	93.44
AID_TO_CHILD	0.2	1.46	94.90
COMM_EFF	0.2	1.46	96.36
DWELLING	0.2	1.46	97.82
MIGRATION	0.1	0.73	98.55
UNEMPLOYMENT	0.1	0.73	99.28
MENTAL_ILLNESS	0.1	0.73	100.0

is attributable to the first three PCs, referred to as population stability (popul_stab), newspaper circulation (newspap_circ), and female labor (female_lab). The remaining 11 components together account for only the 28 percent of the variance. Thus the three PCs may be adequate to represent the data.

Several procedures have been proposed for determining the best number of PCs to summarize the data, but none is a good solution for every domain. The experience and intuition of the analyst is required to make the right choice.

Another major pitfall of PCA is that PCs cannot be named in case of many features. As a consequence, PCs do not give the analyst an idea of the qualitative characteristics that they describe, and reveal nothing of value.

3.2. Factor Analysis

Factor analysis is a statistical technique aimed at identifying a relatively small number of factors that reflect the relationships between a set of features (Monge and Cappella, 1980; Sadocchi, 1980). For example, features such as scores on a set of aptitude tests can be expressed in terms of a set of factors that represent verbal skills, mathematical aptitude and perceptual speed. Features that represent human behaviors can be summarized by a linear combination of factors that measure cultural background, ambitions, and religious beliefs. Often, correlations between behavior traits are all so low that they suggest that there is no structure to personality, that is, that behaviors are mostly independent. But this is not true. Indeed, there are quite a number of ways of behaving to achieve the same purpose; any behavior can be influenced, in part, by environmental conditions, randomness in the person's response, and so on. We presume that, over the traits which represent momentary behavior, there exists a more stable motivational structure that represents the person's basic purposes and desires in life. The goal of factor analysis is to determine a small, easy-to-understand, set of factors that expresses this motivational structure.

The basic assumption of factor analysis is that correlations among variables result from their sharing the factors. For example, correlations between the community features might be due to factors such as the welfare of the community (Welfare), the population stability (Stability), and the amount of urbanization (Urbanism). The mathematical model for factor analysis is similar to a multiple-regression equation. Each original feature is expressed as a linear combination of factors. For example, the commercial activity index might be expressed as

$$\text{comm\_act} = c_1 * \text{Urbanism} + c_2 * \text{Welfare} + c_3 * \text{Stability} + \text{Ecomm\_act}$$

Factors are not observed features as in multiple regression, but labels for a group of features that characterize these concepts. Urbanism, Welfare, and Stability are called *common factors*, because all features are expressed as function of them. Ecomm_act is called a *unique factor*, and represents the part of commercial activity that cannot be expressed by the common factors. Unique factors are assumed to be uncorrelated with each other and with the common factors.

It is possible that all of the features contribute to a factor, but this conveys little summarization. Factor analysis is useful when the observed relationships between features are explained using as few factors as possible, and each feature contributes to explain only one factor. Furthermore, factors should be meaningful and easy to interpret. Unfortunately, factor interpretation is not always possible. Strange factors may be an artifact of failing to distinguish between common and unique factor variance or due to dimensions of the phenomenon under consideration that have not been captured in the sample data.

Many statistical rules have been proposed to determine the number of factors that provide an easy and adequate representation of the data. However, these methods perform poorly in many cases. Skilled factor analysts, therefore, conduct multiple experiments by evaluating the adequacy of factor sets of increasing cardinality. It is basically up to the person doing the study to assess which solution fits his purpose best. Intuition and experience are important factors for the analyst to make the best choice.

3.3. Cluster Analysis

Cluster analysis (Editor, 1980; Monge and Cappella, 1980; Sadocchi, 1980) places objects into groups or clusters suggested by the data, not defined a priori, such that objects in a given cluster tend to be similar to each other in some sense, and objects in different clusters tend to be dissimilar. The goal is obtaining a symbolic description of the problem and from this an identification procedure. In image processing, cluster analysis is used to classify objects scattered in an image according to color, shape, texture, and other geometrical properties. In marketing, cluster analysis is used to identify people with the same buying habits. By examining their characteristics, it would then be possible to target future marketing strategies more efficiently.

Cluster analysis can be used to find homogeneous groups of features as well. The set of observed features is partitioned into groups of features in such a way that features in a given cluster are thought to be measures of the same underlying concept or phenomenon. When cluster analysis applies to features, the absolute value of the correlation between features is used to measure their similarity. The resulting partition is such that members of the same cluster are all fairly correlated with each other and behave similarly in their correlation with other features.

The groups of features generated by a clustering algorithm might be the same as those established by the factor analysis. This is not a chance occurrence. Although factor analysis has an underlying model while cluster analysis is more ad hoc, both identify groups of related features. But cluster components are usually easier to interpret than principal components or factors. Furthermore, cluster analysis can be performed on different levels of clustering. Comparison between clusters of different levels can define the level of clustering that sufficiently describes the sample variation.

4. ADHOC

We next introduce some considerations about the techniques presented in section 3.

First, statistical techniques devoted to data reduction are well-established procedures that are effective in many domains. But there exists a great deal of different algorithms that are based on specific assumptions and capture some particular aspect of the data (Monge and Cappella, 1980). Choosing the right one for a given problem is not a simple task for a non-specialist. Still, the outcome of these algorithms is rarely easy to interpret; the likelihood of grasping the true dimensions of the phenomenon and the reliability of these inferences strongly depends upon the experience and skill of the data analyst.

Second, statistical techniques are based on the analysis of direct correlations between the features to perform data summarization. Their goal is to obtain factors that help to explain the correlation matrix of the features. The underlying assumption is that, if the correlation between the features is low, it is unlikely that they share common factors.

But correlation may not provide a reliable measure of the association between two features, for it does not take other features into account. Spurious correlations may occur, that is, two features are correlated because both are related to a third feature which is the true predictor (Kendall and Stuart, 1979). For example, suppose a survey shows that the crime rate of a country is correlated to the immigration rate. Although it may appear that most of the immigrants are involved in criminal activities, a further analysis will show that both crime and immigration rates are correlated with the industrialization index. The association between criminality and immigration is due to the presence of the latter feature.

It may also happen that one obtains a weak correlation, despite the fact that intuition suggests that a strong relationship should exist. In this case, it is likely that other features hide the expected association. For example, it may be that A is not correlated with B, because A is negatively related to C, which is positively related to B. Assume that a socioeconomical survey reveals almost no correlation between pollution and the density of green areas. The startling result may be due to the combination of two factors: the positive correlation between pollution and industrialization and the negative one between industrialization and green area density. Industrialization masks the inverse relationship between pollution and density of green areas.

Finally, PCA and factor analysis apply to numerical features only, and standard clustering algorithms modified to handle symbolic features lose some of their effectiveness.

We can draw the following conclusion: current statistical techniques may not represent an optimal solution to the feature selection issue in the data-mining framework.

The algorithm described in the rest of this section, called *Automatic Discoverer of Higher-Order Correlations* (ADHOC), provides a different approach to feature selection that overcomes some of the problems mentioned above. ADHOC accomplishes feature selection in two steps.

The first step is concerned with reducing the dimensionality of the problem and is independent of the learning algorithm used in the concept-discovery phase. Groups, or clusters, of features that relate equally to each other are identified by investigating both direct and indirect relationships. Related features are classified into different clusters according to the type and the strength of their mutual association. The algorithm produces a hierarchy of clusters; each level provides a description of the concepts underlying the data from a different level of abstraction. Then, the cluster that is evaluated as the most significant is returned to the user.

In the second step, a branch-and-bound search is used to select one or more features from each cluster. The evaluation function of the feature subsets is the generalizations performance of a learning algorithm selected by the user. This step is therefore dependent on the learning algorithm; it returns a small set of features that provide good performance.

ADHOC deals with both numerical and symbolic features and handles missing values. Most important, it is simple to use and yields results easy to interpret; thus, it may be appealing for those who have to analyze a large amount of data but lack strong statistical background.

4.1. Discovering Higher-Order Correlations

ADHOC identifies groups of features that are equivalent measures of some factor. It can be regarded, therefore, as a clustering technique. However, the mechanism underlying the formation of clusters is very different from the one mentioned in section 3.3. Cluster analysis assumes correlation as a measure of feature similarity. If correlations are either imposed artificially or masked by other features, or the data is so noisy that all the correlations assume similar values, the cluster analysis model will fail to fit the data. In this case, indirect relationships between features need to be investigated, since direct associations, which are measured by correlation, do not convey enough information to discover the deep structure of the data.

The ADHOC search for indirect association is based on the concept of feature *profile*. The profile of a feature F denotes which features F is related to and which ones F is not related to. For example, let A, B, C, D, E, and F be six features that characterize a given data set. Also, let 0.2, 0.1, −0.8, 0.3, and 0.9, be estimates of the direct relationships between F and A, B, C, D, and E, respectively.[3] F's profile is defined as the vector ⟨0.2, 0.1, −0.8, 0.3, 0.9, 1.0⟩.[4]

Features that have similar profiles are related, for they are related to the same group of features and are not related to the rest of the features. This is quite intuitive, since features which measure the same concept must share similar relationship with features that contribute to the definition of the other concepts. If the converse were true, two concepts would be related in two contrasting ways at the same time, a very unlikely situation in nature. For example, let ⟨0.25, 0.1, −0.75, 1.0, 0.9, 0.3⟩ be the profile of feature D. The direct association between features F and D, 0.3, is weak, but the similarity of their relationships with features A, B, C, and E is too strong to be ascribed as mere chance. It is very likely that F and D correlate more than 0.3 and the true correlation has been blurred by noise or imperfect sampling. Comparing feature profiles may yield a more reliable estimate of true association than a direct measure of association, such as correlation, provided the cardinality of the feature profile is not too small (at least 4).[5]

Since, in the simplest case, components of the profile vector express correlations, comparing feature profiles may be viewed as correlating correlations. The result of the comparison has been named *2nd-order correlation* in Doyle (1992) to stress the double application of correlation. Accordingly, standard Pearson's correlation coefficient is named *1st-order correlation*.

4.2. Description of the Algorithm

ADHOC accomplishes feature selection in two successive steps, as depicted in Figure 12.2. The first step identifies the underlying dimension of the phenom-

[3]By applying Pearson's product–moment correlation coefficient, for example.

[4]Of course, the correlation between *F* and itself is 1.0.

[5]The cardinality of a feature profile is given by the number of observed features.

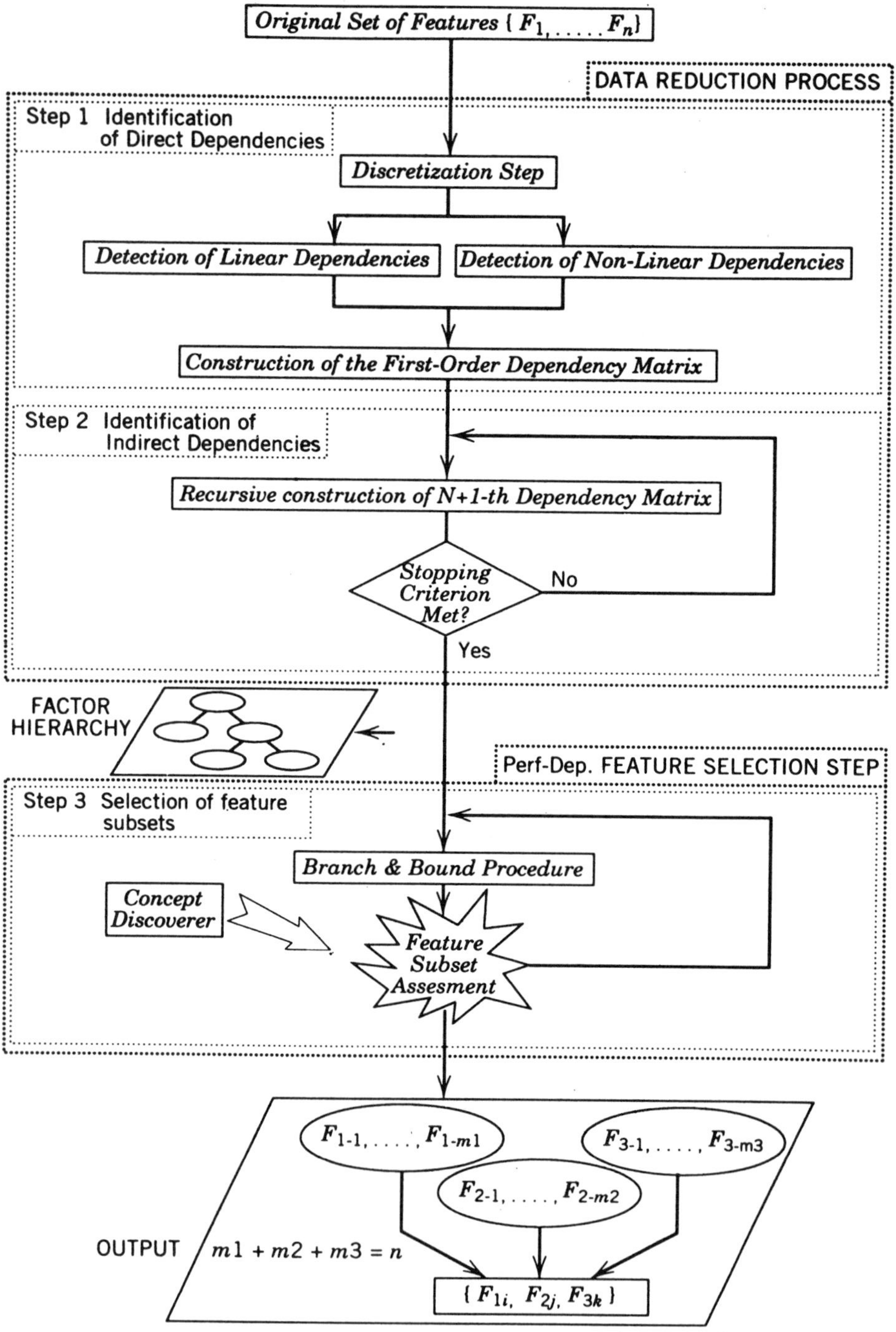

Figure 12.2. *The architecture of ADHOC.*

enon described by the sample data. It consists of two major phases. The former phase investigates direct, namely first-order, association between the observed features. The latter phase seeks for association at different level of indirection, that is, for higher-order dependencies between features. Since, in the simplest case, dependencies may occur as correlations, the algorithm has been called ADHOC, which stands for Automatic Discover of Higher-Order Correlations. As a result, features are grouped into a set of factors according to the typology of mutual relationship that has been detected.

The second step seeks a subset of features that make the learning algorithm used in the concept discovery achieve a good generalization performance. The algorithm picks up a limited number of features from every factor to form feature subsets that reflect all the problem dimensions. This heuristic has two major advantages: it increases the likelihood of selecting good performing features, and consistently diminishes the complexity of search. The three phases of the algorithm are briefly described next.

4.2.1. Phase 1—Identification of Direct Interactions between Features. Input to the algorithm is a training set of feature-valued examples. In the first phase, ADHOC investigates linear and nonlinear relationships between features by employing non-parametric statistics (Agresti, 1990; Kendall and Stuart, 1979). The algorithm selects the most appropriate statistics for any given pair of features automatically. Numeric features are automatically discretized (Richeldi and Rossotto, 1995a; b) if they need to be compared with symbolic features. Measurements of the dependencies between any pair of features are collected in a matrix called the *1st-order dependency matrix*.

4.2.2. Phase 2—Identification of Indirect Interactions between Features. In the second phase, feature profiles are compared with investigate indirect dependencies. A statistical test, named *Rsim*, was designed to estimate profile similarity. Let $x = \langle x_1, \ldots, x_n \rangle$ and $y = \langle y_1, \ldots, y_n \rangle$ be two feature profiles. *Rsim* is computed as follows:

$$\text{Rsim} = \sum_{i=1}^{n} \left(\frac{(x_i - \mu_x) * (y_i - \mu_y) * \text{Sim}(x_i, y_i)}{n * \sigma_x * \sigma_y} \right),$$

where μ_x, σ_x, μ_y, σ_y, are the expected and standard deviation of profiles x and y, respectively. *Rsim*, which spans in the range $[-1, 1]$, is a correlation measure between profiles. The term $\text{Sim}(x_i, y_i)$ is a non-euclidean metric which estimates similarity of profile components. Profile values are weighted proportionally to their absolute value, that is, similarity between large values is rewarded more than similarity between small values. Moreover, the sign of profile components is compared to account, and provide a weight, for different kinds (positive or negative) of association between features. Thus, *Rsim* scores high when applied to profile pairs that are strongly correlated in the most important (largest) components.

Higher-order correlations between features are computed by recursive application of the *Rsim* statistics. Nth-order correlations result in a matrix called the *nth-order dependency matrix*. By examining the nth-order dependency matrix, one can determine the strength of the relationship between features, and group those features that appear to be related. The analysis can then be repeated on each group of features in turn.

Four different types of clusters of features can be generated by the algorithm. They are called *Positive_Concept*, *Negative_Concept*, *Unrelated_features*, and *Undefined*, respectively, and reflect the different typology and strength of dependencies that may exist between a set of features. Features that share very similar profiles, i.e., that appear to contribute to the same dimension of the phenomenon, are grouped into a cluster of type Positive_Concept. Features related by a negative association to other features are assigned to a Negative_Concept-type cluster. Features that appear not to influence or be influenced significantly by the rest of the features form the Unrelated_features cluster. The Undefined cluster holds all of the remaining features which cannot be assigned to one of the other three types of clusters. Subsets of type Positive_Concept, Unrelated_features, and Undefined can be further partitioned, with the aim of refining the classification, as for the Positive_Concept clusters, or identifying relationships that could be masked by other features, as for the Undefined clusters.[6] ADHOC returns a hierarchy of feature clusters which would resemble the hierarchy of concepts that characterize the observed phenomenon. A test of homogeneity of content is then applied to every cluster to determine a good factorization of features.

4.2.3. Phase 3—Identification of Factor Prototypes.

Once significative groups of features have been generated, a branch-and-bound technique is used to select the smallest number of features from each group. The criterion function for the branch-and-bound is the generalization performance of the concept discoverer, which may be a decision tree inducer like C4.5 (Quinlan, 1993) or CAL5 (Muller and Wysotzki, 1994), or a multivariate regression model (Ryrkit, 1987). To fairly estimate the prediction accuracy of the knowledge discoverer, a cross-validation technique is employed. The original data set is broken into a training set and test set. The learning algorithm is given a subset of the observed features, one or many from every cluster generated in phase 2. The error rate on the test set is measured by k-fold cross-validation, with k specified by the user, and the generalization performance of the learning algorithm is obtained by averaging error rates over all the k runs. In the case a multivariate regression model is constructed by incorporating the feature subset, the usual goodness-of-fit statistics (Byrkit, 1987) are employed to validate the model and yield an estimate of its prediction

[6] Strong related features often hide association between the rest of the features. Masked associations may be discovered by dropping out features that share a significant relationship.

accuracy. When the branch-and-bound stopping criterion is met, the feature subset that performed best is returned.

5. EXPERIMENTAL RESULTS

In order to evaluate the feature selection using ADHOC, we performed the analysis of two real-world domains. The first domain is the estimation of software development effort. ADHOC was applied to the COCOMO software project data set (Bohem, 1981), to investigate the relative accuracy of software cost estimators. Credit scoring is the second domain. A data set of credit applicants (Brazil et al., 1994) was analyzed to determine which features the credit manager takes into consideration to assess creditworthiness. Experimental results are described in this section.

5.1. Analysis of the COCOMO Dataset

Many plausible metrics, or cost drivers, have been proposed in the literature to produce a reliable estimation of software development effort. Despite much research having been done, results are very controversial. However, recent studies showed that a reduced set of cost drivers can help estimate development effort equally well. In particular (Capacci and Stamelos, 1993; Subramanian and Breslawski, 1993) report interesting results obtained by performing factor analysis on the COCOMO data set (Bohem, 1981). We applied ADHOC on the COCOMO data set as well, with the twofold goal of identifying the major factors that contribute to project cost and selecting a reduced number of cost drivers that best synthesize these factors. The results are summarized next.

The COCOMO data set is a collection of data on 63 software projects, including business, scientific, process control, and other applications. Projects are characterized by 17 features, or cost drivers, that have been suggested to influence software costs. The first feature, KDSI-Kilo Delivered Source Instructions, provides the project size in code lines. The second feature, MODE, subdivides software projects into three modes, namely organic, semidetached and embedded, according to their size and complexity. Projects of the organic type are small in size, have relaxed delivery requirements, and are developed in a stable in-house environment. Projects of the embedded type are large, operate with tight constraints and have a high degree of hardware, software, and customer interface complexities. Projects of the semidetached mode fall between the other two. The rest of the features are grouped in four categories: software product features (RELY-Required Software Reliability, DATA-Database size, CPLX-Product Complexity), computer features (TIME-Execution Time Constraint, STOR-Main Storage Constraint, VIRT-Virtual Machine Volatility, TURN-Computer Turnaround Time), personnel features (ACAP-Analyst Capability, AEXP-Applications Experience, PCAP-Programmer Capability, VEXP-Virtual Machine Experience, LEXP-Programming Language

Experience), and project features (MODP-Modern Programming Practices, TOOL-Use of Software Tools, SCED-Required Development Schedule). Feature values are expressed as ratings, ranging from very_low to extra_high. The lower the value for features RELY, DATA, CPLX, TIME, STOR, VIRT, and TURN, the lesser the effort required for the project. Conversely, the lower the value for features ACAP, PCAP, AEXP, VEXP, LEXP, MOPT, TOOL, and SCED, the higher the effort required for the project. The true cost for each project in the database is also given, that is, the development effort, expressed as the number of man-months required to complete the project.

The first-order dependency matrix (not reported for lack of space) showed high correlations (>0.7) between the following pairs of features: RELY-TIME, TIME-STOR, and VEXP-LEXP. Many other features are slightly correlated, but no significant non-linear correlations are identified.

The first level of the factor hierarchy is obtained after nine steps; that is, differences between the 8th- and the 9th-order dependencies matrices are so negligible that the stopping criterion is met. As a result, observed features are partitioned into four groups. The recursive higher-order construction process is then applied to each of this group in turn. Only one group, the largest one, is further split into two sub-groups after 18 steps. ADHOC took 15 seconds on a DEC 3100 to complete the analysis. The resulting cluster hierarchy is depicted in Figure 12.3.

ADHOC identified five factors, called Project Size, Developer Skills, Project Complexity, Developer Experience, and Supports to Development, respectively. The first factor consists of KDSI and DATA, the features that express the size of the software project. As a consequence, we named the factor Project Size. The second factor consists of ACAP and PCAP. As ACAP and PCAP

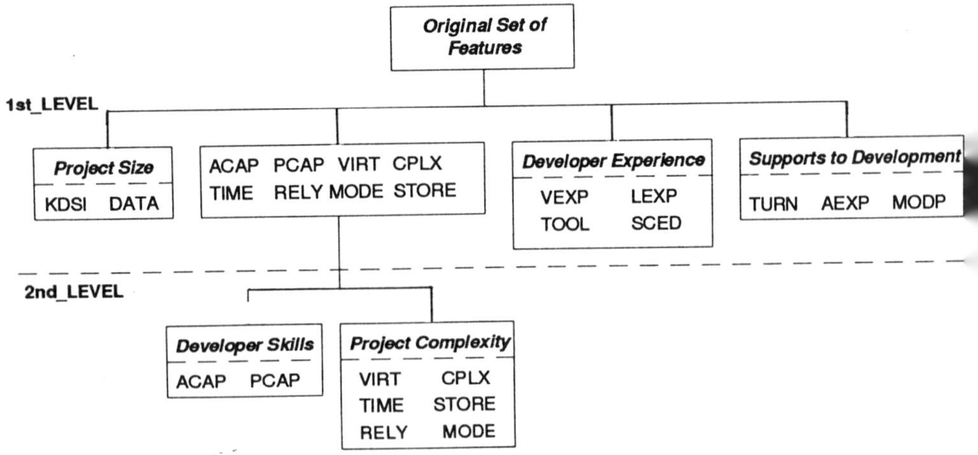

Figure 12.3. *Results from applying ADHOC's first phase to COCOMO cost drivers.*

measure the capability of the development team, we called the factor Developer Skills. The third factor consists of VIRT, CPLX, TIME, STORE, RELY, and MODE. Each one of these features represents a different aspect that contributes to the project cost. It is quite intuitive that any of the following occurrences:

- Increase in product complexity (CPLX) or in the required software reliability (RELY);
- Higher execution (TIME) and main storage (STORE) constraints;
- Less stable virtual machine environment (VIRT); and
- Increase in project size, complexity, and need for innovation (MODE),

result in an increase of the software development effort, and vice versa. This factor, therefore, was called Product Complexity. The fourth factor consists of VEXP, LEXP, TOOL, and SCED. VEXP and LEXP reflect the experience in the virtual machine and in the language, respectively. As good experience is required to make use of sophisticated software tools and to complete the project within schedule constraints, TOOL and SCED can be regarded as indirect measures of the development team experience. Thus, the fourth factor was named Developer Experience.[7] The last factor consists of TURN, AEXP, and MODP. These features represent different methods to cope with software development complexity, that is, supports to software development. Indeed,

- a higher level of application experience (AEXP);
- the use of modern programming practices (MODP); and
- on-line coding and debugging (TURN) instead of batch processing:

may result in shorter coding times, thus decreasing the project effort. Consequently, we named this factor Supports to Development.

ADHOC generated a reasonable clustering of the COCOMO cost drivers. The resulting groups have an intuitive, simple explanation and seem to capture the true factors that affect the software development effort. However, to assess the validity of our result, we constructed an analytical model for predicting the cost of a software project and compared it to the ones reported in Subramanian and Breslawski (1993). Phase 3 of ADHOC is applied to pick up one feature from each of the groups generated in phase 2. The selected features are used to build a multivariate regression model that predicts the effort required for a project on the basis of the cost driver values. Since cost-estimation models often include non-linear relationships between the cost drivers and the project effort, transformations F_i^2 and $\log F_i$ of the feature F_i are considered rather than the feature itself. Shown below is the model that best fits the COCOMO

[7]Note that the algorithm correctly captures the difference between developer experience and skills. Unfortunately, experience does not always convert into skills, and vice versa.

data set:

$$\log Effort = -1.426 + 0.960 * \log KDSI + 2.533 * PCAP$$
$$- 1.111 * \log MODE + 1.545 * TURN^2 + 8.255 * \log LEXP$$

$R^2 = 0.92$; $F = 149.9$; $p = 0.00001$. This model was obtained after 100 iterations on a DEC 3100 that took about five minutes.

Subramanian and Breslawski (1993) reports the results of performing a factor analysis on the COCOMO data set. The resulting factors differ from the ones obtained by ADHOC. The authors of this study apply stepwise selection to generate two multivariate regression models, called model A and model B, for estimating project effort. Model A contains six features, while model B only four. Table 12.2 compares the predictive accuracies of the two models with the one obtained through ADHOC.

$$MRE = \frac{\Sigma_i(|E_i - \hat{E}_i|)/E_i}{n}$$

is the mean magnitude of relative error; E_i is the actual effort and $\hat{E}_i$ is the predicted effort. *Pred(L)* $= K/N$ is the percentage of projects with acceptable relative error, that is, the percentage of projects in the given set of projects whose MRE < L; RMS is the relative root-mean-square error and represents the relative square root of the mean value of the error minimized by the regression model;

$$\text{RMS} = \frac{\sqrt{\Sigma_i(E_i - E_i)^2/n}}{\Sigma_i E_i/n}$$

As can be seen, the model constructed by ADHOC, which contains five features, performs equally well to model A (6 features) and better than the more parsimonious model B (4 features). The comparison result confirms that ADHOC feature selection is effective and efficient as well. Indeed, the complete analysis of the COCOMO data set took a few minutes. More importantly. the same result would be achieved almost without effort by a user not acquainted with the statistical techniques for feature selection.

TABLE 12.2. Comparison Results for the COCOMO Credit Data Set

Model	No. of Features	MRE	PRED(0.2)	RMS
A	6	0.0967	88.89%	0.1030
B	4	0.1287	80.95%	0.1042
ADHOC	5	0.0960	88.85%	0.1000

5.2. Analysis of the Credit Data Set

A credit-scoring system estimates the risk associated with each application for credit. Banks, engineering companies and service providers need to assess the risk of granting a credit to improve their marketing and debt-recovery procedures, in order to decrease the probability of losing business. Credit-scoring procedures yield a numerical score that represents an objective measure of the credit risk. Our analysis involves a data set taken at a German bank from Brazdil et al. (1994). It contains data of about 1,000 customers who applied for a credit. The credit manager of the bank classified applications into two classes: bad risk (debt hard to retrieve) and good risk (debt easy to retrieve). The goal of a data-mining analysis would be to replace the human expert assessment of risk by an algorithmic procedure (Carter and Catlett, 1987).

The data set contains seven numeric and thirteen symbolic features. The features include: Status of existing checking account (Stat_acc); Duration in month (Duration); Credit history (Cred_hist); Purpose for credit request (Purpose); Credit amount (Cred_am); Saving account and/or bounds (Saving); Length of employment (Length_emp); Installment rate in percentage of disposable income (Install_rate); Personal status and sex (Status); Other debtors/guarantors (Other_debt); Length of time at present resident (Length_res); Property (Property); Age (Age); Other installment plans (Other_inst); Housing (Housing); Number of existing credits at this bank (Num_cred); Number of people being liable to provide maintenance for (Num_man); Telephone (phone); and Foreign Worker (For_work).

The first phase of ADHOC identified many weak correlations between several features but no relevant non-linear association. After 42 steps, the recursive process of higher-order correlation matrix construction terminated by producing the feature clustering hierarchy depicted in Figure 12.4.

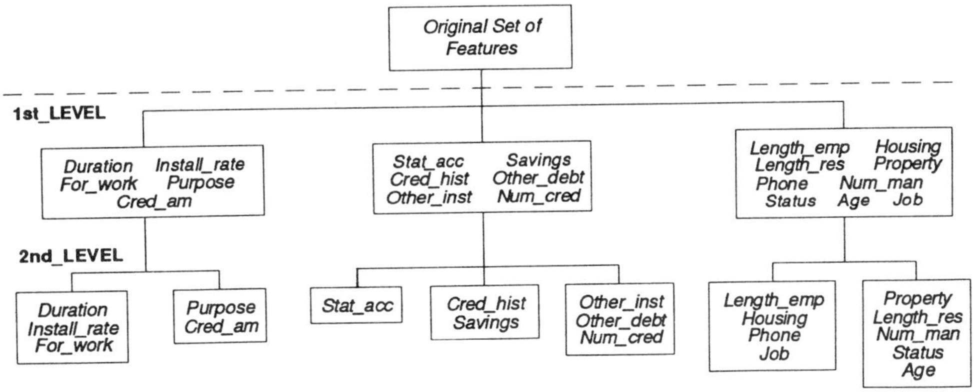

Figure 12.4. *Results from applying ADHOC's first phase to the Credit Data Set.*

The upper level of the hierarchy identifies three major factors that explain the interactions between the observed features. We called them Customer's Current Situation, Description of the Credit, and Customer's Outline, respectively. Customer Current Situation consists of Status of account. Credit history, Saving account/bounds, Other debtors, Other installment plans, and Number of credits at this band. These features provide a snapshot of the current economic status of the applicant and of his current, hypothetical "degree of reliability," expressed by the Credit history. The second factor consists of Duration, Installment rate in percentage, Purpose, Credit amount, and Foreign worker. These features provide a description of the requested credit. The citizenship of the applicant (Foreign worker) contributes to define the credit risk: intuitively, granting a credit to a foreign worker is hypothetically more dangerous than granting one to a local worker. Therefore, we called this factor Description of the Credit. The third factor underlies the rest of the features. It was named Customer Description, as it includes features that describe applicant's characteristics (age, job, and so on) with special regard to his income and properties.

It is worth noting that the algorithm captures the difference that lies between features which reflect the current, dynamic economic situation of the customer and features that describe factors that are unlikely to change in a short time, like the possession of real estate and the job position. The lower level of the clustering hierarchy shown in Figure 12.4 refines the three factors by generating seven groups. Providing an interpretation of these groups is not as easy and intuitive as for the three original factors; thus, we refrain from attempting unnatural explanations. However, we show next that this latter clustering is as interesting as the former one.

In the third phase, ADHOC identifies factor prototypes by employing a forward stepwise selection procedure. We forced the algorithm to pick up only one feature from each cluster, in order to achieve for high data reduction. Generalization performance was selected as the best measure for assessing feature subsets obtained through branch-and-bound search. A decision tree algorithm, namely C4.5 (Quinlan, 1993), was used to classify the credit data set by accounting solely for the selected features. We employed tenfold cross-validation; in every run, the training and test set contained 90 percent and 10 percent of the original examples, respectively. By averaging the error rate on the test set over the ten runs, the generalization performance of the feature subset is obtained. ADHOC returns the features that score the highest generalization performance. We run ADHOC to determine prototypes for both the groups of factors obtained in phase 2. We then run C4.5 on the credit data set by making use of the entire set of original features.[8] Table 12.3 shows comparison results. A two-sided t-test at the 5 percent level confirmed that both the reduced sets of features showed a significant improvement in perform-

[8]With the same conditions (10-fold cross validation; 90/10 the ratio between training and test set cardinalities) and settings as in the previous experiment.

TABLE 12.3. Comparison Results for the German Credit Data Set

Features in the Subset	Average Number of Tree Nodes	Average Number of Leaves	Average Error on the Test Set (%)
{Duration, Cred_hist, Property}	21.0	13.9	27.4
{Duration, Cred_am, Stat_acc, Savings, Other_inst, Housing, Num_man}	120.9	67.4	27.9
All	154.0	94.3	30.1

ance with respect to the entire set of original features. This result supports the claim that abundance of features may mislead the learning algorithm and degrade its performance. Note also that the tree generated by running C4.5 on the group of three features contains about 15 percent of the leaves of the tree obtained by employing the whole set of features. The first tree is far simpler and easier to interpret than the second one. The rationale lies in the fact that feature selection, by reducing the complexity of the domain, allows concept learners to yield more concise and significant results.

6. CONCLUSION

Abundance and redundancy of features thwart the knowledge-discovery task, for irrelevant features degrade the performance of a learning algorithm both in speed and in predictive accuracy. The first step in a learning process is to figure out whether features which can discriminate between the target concepts exist and, if any, consider the problem of what discriminatory features to select and how to extract them.

In this chapter, we addressed the problem of selecting a small subset of features that are necessary to describe the target concept and provide good generalization performance. First, we presented and reviewed some statistical approaches to the problem of reducing the dimensionality of the data. Then, we described a new algorithm, called ADHOC, which combines statistical techniques and search heuristics to accomplish effective and efficient feature selection. ADHOC employs a branch-and-bound technique to search for a small subset of features, which allows a learning algorithm to achieve high prediction accuracy. To reduce the complexity of the search task, direct and indirect relationships between features are investigated to identify groups of features that reflect the main dimensions underlying the data. We reported results of experiments conducted on two interesting real-world domains, namely, the problem of estimating software development effort and credit scoring.

REFERENCES

Agresti, A. (1990). *Categorical Data Analysis.* Wiley, New York.

Almuallin, H., and Dietterich, T. G. (1991). Learning with many irrelevant features. In *Proceedings of the Ninth National Conference on Artificial Intelligence,* pp. 547–552. MIT Press, Cambridge, MA.

Bohem, B. W. (1981) *Software Engineering Economics.* Prentice-Hall, Englewood Cliffs, NJ.

Brazdil, P., Gama, J., and Henery, B. (1994). Characterizing the applicability of classification algorithms using meta-level learning. In *Proceedings of the 7th European Conference on Machine Learning,* pp. 83–102. Springer-Verlag, Berlin.

Breimen, L., Friedman, J. H., Oshen, R. A., and Stone, C. J. (1984). *Classification and Regression Trees.* Wadsworth, Pacific Grove, CA.

Byrkit, D. (1987). *Statistics Today.* Benjamin Cummings, Menlo Park, CA.

Capacci, C., and Stamelos, I. (1993). Constructing software cost models. *Proc. Achieving Quality in Software-93, Venice, Italy.*

Carter, C., and Catlett, J. (1987). Assessing credit card applications using machine learning. *IEEE Expert,* Fall Issue, pp. 71–79.

Caruana, R., and Freitag, D. (1994). Greedy attribute selection. In *Proceedings of the 11th International Conference on Machine Learning,* pp. 28–36. Morgan Kaufmann, San Mateo, CA.

Doyle, J. (1992). Mcc-multiple correlation clustering. *Int. J. Man-Mach. Stud.* **37**, 751–765.

Editor, K. S. F. (1980). *Pattern Recognition.* Springer-Verlag, New York.

Fabbris, L. (1991). *Analisi explorativa di dati multidimensionali.* CLEUP, Padova.

Fenton, N. E. (1992). *Software Metrics.* Chapman & Hall, London.

Hayes-Roth, F., Waterman, D. A., and Lenat, D. B. (1983). *Building Expert Systems.* Addison-Wesley, Reading, MA.

John, G. H., Kohavi, R., and Pfleger, K. (1994). Irrelevant features and the subset selection problem. In *Proceedings of the 11th International Conference on Machine Learning.*

Jonassen, C. T., and Press, S. H. (1960). *Interrelationships of Dimensions of Community Systems.* Ohio State Univ. Press, Columbus.

Jones, C. (1991). *Applied Software Measurement: Assuring Productivity and Quality.* McGraw-Hill, New York.

Kendall, M., and Stuart, A. (1979). *The Advanced Theory of Statistics.* Macmillan, New York.

Kira, K., and Rendell, L. (1992a). The feature selection problem: Traditional methods and a new algorithm. In *Proceedings of the Tenth National Conference on Artificial Intelligence,* pp. 129–134. MIT Press, Cambridge, MA.

Kira, K., and Rendell, L. (1992b). A practical approach to feature selection. In *Proceedings of the Ninth International Conference on Machine Learning* (D. Deeman and P. Edwards, eds.). Morgan Kaufmann, San Mateo, CA.

Kohavi, R., and Sommerfield, D. (1995). Feature subset selection using the wrapper method: Overfitting and dynamic search space topology. In *Proceedings of the First International Conference of Knowledge Discovery and Data Mining.*

Matheus, C. J., Chan, P. K., and Piatetsky-Shapiro, G. (1993). Systems for knowledge discovery in databases. *IEEE Trans. Knowl. Data Eng.* **5**(6).

Monge, P., and Cappella, J. (1980). *Multivariate Techniques in Human Communication Research.* Academic Press, New York.

Müller, W., and Wysotzki, F. (1994). A splitting algorithm based on a statistical approach in the decision tree algorithm cal5. In *Proceedings of the Workshop on Machine Learning and Statistics*, ECML-94.

Piatetsky-Shapiro, G. (1991). Discovery, analysis, and presentation of strong rules. In *Knowledge Discovery in Databases* (G. Piatetsky-Shapiro and W. Frawley, eds.). MIT Press, Cambridge, MA.

Piatetsky-Shapiro, G., and Matheus, C. J. (1994). Knowledge discovery workbench for exploring business databases. *Int. J. Intell. Syst.* **7**, 657–686.

Piatetsky-Shapiro, G., Matheus, C., Smyth, P., and Uthurusamy, R. (1994). Kdd-93: Progress and challenges in knowledge discovery in databases. *AI Mag.*

Quinlan, J. R. (1983). Learning efficient classification procedures and their application to chess end games. In *Machine Learning: An Intelligence Approach*, 463–482.

Quinlan, J. R. (1993). *C4.5 Programs for Machine Learning.* Morgan Kaufmann, San Mateo, CA.

Rendell, L., and Cho, H. (1990). Empirical learning as a function of concept character. *Mach. Learn.* **5**, 267–298.

Rendell, L., and Ragavan, H. (1993). Improving the design of induction methods by analyzing algorithm functionality and data-based concept complexity. In *Proceedings of the 13th International Joint Conference on Artificial Intelligence.*

Rendell, L., Cho, H., and Seshu, R. (1989). Improving the design of similarity-based rule-learning systems. *Int. J. Expert Syst.* **2**, 97–133.

Richeldi, M. (1993). *A New Methodology for the Automatic Analysis and Characterization of Large Databases*, Tech. Rep. CSELT (Centro Studi E Laboratori Telecomunicazioni S.p.A.).

Richeldi, M., and Rossotto, M. (1995). Class-driven statistical discretization of continuous attributes. In *Proceedings of the Eighth European Conference of Machine Learning*, Heraklion, Crete, Greece.

Sadocchi, S. (1980). *Manuale di Statistica Multivariata.* Franco Angeli Editore, Milano.

Subramanian, G. H., and Breslawski, S. (1993). Dimensionality reduction in software development effort estimation. *J. Syst. Software* **21**, 187–196.

CHAPTER 13

Comparison of Three Inductive Numerical Law Discovery Systems

M. MOULET

1. INTRODUCTION

In data analysis, one major task is to induce a numerical law from the input experimental data. There are two extreme situations: in the first one, a lot of prior information is available: the context of the data, the laws previously discovered, or some other constraints hold concerning the form of the law (for instance, only polynomials are accepted because they are easy to use), and so on: in summary the universe is quite well understood. In this case, a simple solution is first to define a list of possible models (or forms) of the law, to apply regression on them, and to select the best law according to some classical statistical criterion. In the opposite case, where the problem is totally new and unknown, the general solution consists in first defining the variables or descriptors, choosing the measuring devices, setting up hypotheses on the law and on its model and recursively testing these hypotheses and setting up new ones, like in any research task (Bertrand, 1986).

Numerical law discovery systems try to tackle the second problem: they do not seek the goal of automating the entire discovery cycle, but they aim to assist the user by discovering the form of the law. Among these systems, a first approach, that we call *heuristic*, deals with finding heuristics able to guide this search efficiently, and tries to simulate human experts solving similar problems. These systems are called BACON-like systems (Langley et al., 1986) because the initial heuristics were proposed by Langley (1986) and implemented in the system BACON. These simple heuristics help iteratively to build the model and

Machine Learning and Statistics, The Interface, Edited by G. Nakhaeizadeh and C. C. Taylor.
ISBN 0-471-14890-3

are able to reduce significantly the size of the search space. In BACON.3 the initial variables are combined by the operators $\times, /$ and by applying regression on the simple model $y = ax + b$, while ABACUS (Falkenhainer and Michalski, 1986, 1990) combines variables with the four arithmetical operators $+, -, \times, /$ and without using regression. Our system ARC.2 (Moulet, 1993, 1994) is a direct descendant of BACON.3 (Langley, 1979) and of ABACUS. It integrates heuristically polynomial regression in an iterative search. In this first family of systems, the main principle is to discover the law step by step, like a statistician usually does.

The second and more recent approach does not discover the law by decomposing it, but explores exhaustively a solution space bounded by some parameters. This approach, which we call *exhaustive*, has no cognitive explanation since it cannot be practicable by hand but, on the contrary, relies on brute computational force and assumes that the law is more probably simple and rapidly accessible. It is based on statistical regression, whose role is to approximate a data plot by the best instance of a given model (or form). Regression computes the coefficients of a given model such that they minimize the error between the value of the dependent variable and its value predicted by the law. For instance, a common model is the polynomial form: $y = \beta_0 + \beta_1 x_1 + \beta_2 x_2 + \cdots + \beta_p x_p$, where x is the independent variable, y is the dependent variable, and β_i are the coefficients of the model. A law is an instantiation of the model. Two main systems belong to this family: *Equation Finder* (also called EF) (Zembowicz and Zytkow, 1991, 1992) looks for the best adequate solution among many possible relations between only two variables x and y, and LAGRANGE (Dzeroski and Todorovski, 1993) applies linear regression on all products of the initial variables and detects each satisfactory relation. These two systems use regression in a classical way, applying it to different complete models, using a statistical criterion in order to select the best law.

These two approaches, the *heuristic* one and the *exhaustive* one, consider the discovery task in an inductive way, that is, only from the given data, with little additional knowledge, and they discover the model and the law at the same time. A third approach is based, like the second approach, on regression, but solves the problem of model selection by previously defining a list of prototypes, which can be considered as some background knowledge and defines the search space. E* (Schaffer, 1990) considers only two variables and six simple models of the form $y = kx^n$ where $n \in \{-2, -1, -0.5, 0.5, 1, 2\}$. KEPLER (Wu and Wang, 1991), the successor of *Discover*.2 (Wu, 1988), considers many variables, linear models but also non linear models, like $x^{r1} * y^{r2}$ which, however, need the help of the user. In KEPLER, the user can also propose additional models. In this third approach, the goal is not to discover laws in an open space, but to propose an efficient search in a small space.

Whenever the user has a precise idea of the model of the law, the best solution is to select between the systems that can test this model. On the other hand, if he does not have any idea on the model of the law, he must choose a search system. In this case, the user needs to consider systems belonging to the two previously presented inductive approaches. Since we consider that the user needs more information about the systems when he lacks information on the model, we have restricted our comparison to inductive approaches, through the comparison of three systems; on the one hand, ARC.2 is representative of the *heuristic* approach, on the other hand, *Equation Finder* (EF) and LAGRANGE are two main representatives of the *exhaustive* approach. We compare these systems according to a list of characteristics we present in section 2. Along with a comparison of these systems, we aim at giving some information which could help a user to choose the system most adapted to his application.

Equation Finder, LAGRANGE and ARC.2 are successively described in sections 3, 4, and 5. A synthesis of the theoretical comparison is given in section 6. In addition to this theoretical comparison, a part of this study will be devoted to an empirical comparison of the three systems (section 7). This comparison will be separated in two parts: first, the performances of the three systems will be compared on various polynomial forms, which are the possible models all three systems can discover, secondly the performances of only both LAGRANGE and ARC.2 will be tested on some multivariate laws.

2. COMPARISON CRITERIA

Although the two inductive approaches have a similar goal, they may be opposed for many reasons. Instead of proposing a list of goals for quantitative discovery as did Falkenhainer and Michalski (1990), we propose to compare the three systems EF, LAGRANGE and ARC.2, according to some characteristics which could help a user selecting the most adequate system for his task. We want also to test the use of each system's facility since their role is rather, in our sense, to help a human user rather than to be used in a completely automatic process. This facility can be measured by testing first how inputs are easy to select or to get and secondly how results are easy to interpret. The following list of characteristics is clearly not exhaustive but is important both from the theoretical and the practical point of view:

1. *The inputs.* The inputs define the applicability degree of the system as well as it is easy to use. The systems can be compared both on the type of data they can deal with, and on the facility to get these data:
   - The nature of the variables (or attributes): numeric (or continuous) or symbolic (or discrete);
   - The uncertainties of the data. The capacity of a system to take into account this knowledge generally influences its robustness to noise;
   - The available knowledge: in experimental science, the dimensions of the variables are in general known, and can help a system to verify the

relevance of a law. Another type of information often used in statistics is which variables are dependent (i.e., function of the others) and which ones are independent. When dealing for instance with chemical reactions, the volume and other measures of the substances produced in a reaction represent the dependent variables: they can be observed but not controlled. Many statistical methods focus on the prediction of one variable; and

- The parameters of the system. We have not included in the list of comparison criteria the ergonomy of the systems, since all three systems use very simple interfaces: calling Lagrange simply consists in a line command of the form "lg-options file-name". EF and ARC.2 both propose very simple menus without graphical interface. In EF, some graphical functions like plotting data are implemented but their portability is very limited (Allegro Commonlisp on DEC-stations). Therefore, we consider that from the ergonomy point of view, the three systems are comparable.

  However, we can estimate their use interface by counting the number of parameters, and the facility of setting them. This facility depends both on the parameter influence on the discovery process, and on the user ability to understand their role.

2. *The algorithm. Equation Finder* and LAGRANGE are based on a statistical evaluation criterion, while ARC.2 is based on many heuristics. One main point concerns the criteria deciding the validity of a solution.
3. *The reachable laws.* For this purpose, we first consider the list of operators used to create the models, then we define the form of the reachable functions. We will consider in particular polynomials, since they are very common in many sciences. One could recommend a system only if it is able to discover at least polynomials.
4. *The complexity.* Evaluating the complexity of each system helps to estimate the performance, mainly the execution time.
   - The complexity is classically a function of some characteristics like the number of examples, the number of variables, but also the parameters' values. When the complexity is a direct function of the parameters' values, the user needs to fix them carefully in order to prevent a long execution time. With this aim, either the parameters are set depending on the resources or the user must have some information or some additional hypotheses on the law. In the second case, parameters constitute another source of knowledge. Thereby, the parameters role must be studied when comparing such systems with other systems which do not use this knowledge.
   - An interesting point is, when discovering polynomials, the relation between the system complexity and the polynomial degree. As a reference, regression complexity is awkward to the polynomial degree. One could expect a discovery system not to be less efficient than this

simple statistical tool, and therefore, not to be too sensitive to polynomial degree.

- Some other elements can influence the performance of the discovery systems, like the general conditions under which the data has been collected or the possibility to detect irrelevant variables.

3. EQUATION FINDER (EF)

The main characteristic of *Equation Finder* (Zembowicz and Zytkow, 1991, 1992), in comparison with the two other systems under study, is that it looks for only two-dimensional functions, like BACON.1 and E* (Schaffer, 1990). While BACON.1 discovers the model in an iterative and heuristic way and E* applies regression on a list of prototypes, EF automatically explores a space of models until it finds a satisfactory one.

3.1. Input

a. Nature of the Data. EF applies to only two numerical variables (x, y). Its input is a list of triples (x_i, y_i, σ_i) where σ_i is the uncertainty of the dependent variable y_i. EF assumes that noise can appear in the dependent variable but not in the independent one which is assumed to be controlled. This is due to the inherent limit of efficient statistical regression used by EF, where only the dependent variable errors are used for computing the resulting coefficients errors. However, dealing with uncertainty only on the dependent variable cannot always be justified since in general the independent and the dependent variables are both measured, even if the first ones are controlled while the second ones are only observed. Consequently, there is no reason why only the measuring device for the independent variable would present an error, and not the one used for measuring the resulting dependent variable.

b. Available Knowledge. The only additional knowledge is the designation of the dependent and independent variables, and the list of operators.

c. Parameters. Two parameters control the search process: d represents the maximum polynomial degree and t is the maximum transformation search depth or the maximum number of iterations. By default, $d = 3$ and $t = 1$, which are the values chosen in our empirical comparison.

3.2. Algorithm

First, EF automatically generates the list of all polynomials of the form $y = P(x)$ up to the given degree limit d. If none of these models satisfies the given statistical criterion, x and y are transformed in order to create new terms

x' (function of only x) and y' (function of only y or both x and y). By default, there are four operators on one variable (but the user can define additional operators): logarithm, exponential, square root and inverse and there are two operators on two variables: multiplication and division. In this way, by default, the first generation of new terms are: $\log x$, e^x, $\sqrt{x}$, x^{-1}, x^2, (5 new terms x') and xy, $\frac{x}{y}$, $\frac{y}{x}$, $\log y$, e^y, $\sqrt{y}$, y^{-1}, y^2, (8 new terms y'). Then, EF creates for each possible couple (x', y') all polynomials of degree from 1 up to d. That gives a list of models of the form $y' = P(x')$ to fit the data. All these models are in turn evaluated according to a statistical criterion. If none satisfies the criterion, the same process is repeated: it creates a new list of models from a new set of transformations, and so on until a solution is found or the externally defined maximum number t of transformations is reached.

The statistical criterion is as follows: first, the solution minimizes the value of χ^2, that is, the sum of squares of deviations of data points to the model (which is minimized by regression), secondly EF assigns to the best fit to each model the probability Q that this χ^2 value has not been reached by chance and a model is *acceptable* if Q is higher than some minimum value Q_{min} (Zembowicz and Zytkow, 1992).

Even if no model has been found acceptable, the user can ask for the three best models, as well as their χ^2 value and Q value and he can thus select a law by himself.

3.3. Scope

Theoretically, EF can explore the infinite space of models one can create with two variables and the given operators. The reachable laws have therefore the following form:

$$y' = P(x'),$$

where y' is a combination of x and y, x' is a function of x and P is a polynomial of degree limited to d. For instance, when the degree is limited to 1, it creates models such as: $y = a + bx$; $y = ae^{b/x}$; $y = ax + b \log x/x; \ldots$

3.4. Complexity

The search space is bounded by the parameters d and t and by the number of operators used. Let n be the number of possible operators applicable to one variable (by default $n = 4$) and let us consider multiplication and division as the only operators between two variables. We can roughly compute EF's complexity by recurrence, by separating the terms involving only x from those involving y. Let X_i be the number of new terms x' created at the level i. X_i is the result of the application of n operators, multiplication and division to the previous levels. For instance, in the first three levels, X_i has the following complexity:

$X_0 = 1$ (the only term is x), $X_1 = n + 1$, the new terms are: $\log x$, e^x, $\sqrt{x}$, x^{-1}, x^2. Therefore, $\mathcal{O}(X_1) = n$, $\mathcal{O}(X_2) = nX_1 + 3/2X_1(X_1 - 1)$ (by applying n operators to X_1 and by applying multiplication and division to all pairs of the X_1 previous terms). Therefore, $\mathcal{O}(X_2) = n^2$. At this point, one can only estimate the complexity and not the exact value, since EF detects the equivalences and does not create twice the same model. Recursively, the complexity of X_i can be estimated to $\mathcal{O}(n^{2^{i-1}})$.

Similarly, let Y_i be the number of new terms y' created at the level i. The Y_i new terms are the result of the n operators applied on the Y_{i-1} previous terms, and the result of the products between the Y_{i-1} previous terms y' and x':

$$Y_0 = 1,\ Y_1 = n + 1 \text{ then } \mathcal{O}(Y_1) = n,\ Y_2 = nY_1 + 3/2Y_1(Y_1 - 1) + 3/2X_1Y_1 \text{ then } \mathcal{O}(Y_2) = n^2.$$

Recursively, we have the same complexity on Y_i as on X_i: $\mathcal{O}(n^{2^{i-1}})$. After t steps (for $t \geqslant 2$), the total number of transformations of x or y has the following complexity: $\Sigma_{i=0}^{t} n^{2i} = [(n^2)^t - 1]/[n^2 - 1]$. The maximum number of regressions performed by EF is therefore: $d \times [[(n^2)^t - 1]/(n^2 - 1)]^2$.

3.5. Example

Let us consider the mathematical function describing an ellipse: $4x^2 + y^2 = 30$. There are 13 data points, with random relative errors on Y lower than 1%. EF first creates at the first and unique level 5 new x-terms and 8 new y-terms, which lead theoretically to 162 ($3 \times 6 \times 9$) terms but in reality, due to the elimination of equivalent terms, only to 85 models. Five models are found acceptable among which two are "non-zero acceptable", that is to say with non-zero values of the coefficient at the term with the highest power of x in the polynomial. The three best models are listed below, such that EF presents them:

Y = (SQRT(+A(*BXX)))|(AB) = (29.89 – 3.974) + – (0.212 0.031)χ^2 = 0.20795 Q = 0.9999 (which is indeed the best model).

Y = (SQRT(+A(*BX)(*CXX)))|(ABC) = (29.89 2.85E-4 – 3.9743567)
  + –(0.212 0.009 0.031)χ^2 = 0.207 Q = 0.9999

Y = (SQRT(+A(/BX)(/CXX)))(ABC) = (–2.9008 0.054 39.34)
  + –(0.064 0.063 0.381)χ^2 = 5405.036Q = 0.0D0

The results are printed in the following form: the model of the law, the values of the coefficients of the law, the absolute uncertainties of these coefficients, along with the rounded χ^2 and Q values. These results are quite easy to interpret, and the coefficients uncertainty can help very efficiently the user to measure the confidence rate he can have in the solution. In the example, the uncertainties allow us to eliminate the B coefficient of the second equation, and this way lead to the same model as the first one, confirming it as a good solution.

3.6. Additional Comments

The possibility of looking at only the three best models is very useful, since it provides a first selection to the user. We can suppose that facing more than a dozen models, even an expert would have some difficulties to make a choice. However, if among the three best models, like in the example above, there are only one or two non-zero acceptable models, the other models are just given for information and the user must be careful and take care of the Q value. This underlines the role of the statistical criterion and the difficulty of interpreting the solution.

Another possible limit for a user is that EF does not stop at the first acceptable model, but tries all the models inside the depth limit. It does not indeed have any heuristic to guide the search, and when it finds an acceptable model, it is not able to decide if it is the best one. Without additional knowledge, there is no other possibility than to test all possible models. In counterpart, that makes the execution time completely dependent on the depth parameter, instead of the task complexity. If the system does not find any solution for some depth values, the user needs to increase them in order to test if there are deeper solutions. In this case, the system is not able to take into account its previous results but tests again the same models. That underlines again that such exhaustive search does not really take care of the number of tests, but relies on efficient computational force.

4. LAGRANGE

LAGRANGE (Dzeroski and Todorovski, 1993) tests a list of models created automatically, like EF, following the increasing complexity, but LAGRANGE manages more than two variables. Moreover, it computes numerical derivatives and it can this way discover differential equations.

4.1. Input

a. Nature of the Data. Only numerical attributes are dealt with, and no uncertainty is considered. Furthermore, when looking for differential equations, the user gives the time interval separating two experiments.

b. Available Knowledge. For each angular variable, its sine and cosine are automatically added to the current set of variables, but no other unit is considered. Neither does it make any distinction between dependent and independent variables. The main drawback of this choice occurs when there is really a dependent variable and when the user wants to predict the value of the dependent variable on further experiments and when the selected law cannot however be written such that the dependent variable is a function of the others.

c. Parameters. Three parameters control LAGRANGE. A first parameter o indicates to which order derivatives must be created. For instance, if $o = 2$, it creates all first and second derivatives of the initial variables.[1] When $o = 0$, no differential equations but only algebraic ones can be discovered, and in this case LAGRANGE can be compared to the two other discovery systems. We have therefore set o to 0 in our experiments.

In addition, regression models are controlled by two parameters: r (by default $r = 2$) gives the maximum number of independent terms, and d stands for the maximum of the initial variables in the products (by default $d = 2$). When $d = 2$ and $r = 2$, it creates products of two variables, and it can apply regression on models like: $x_1 = ax_2 + b$, $x_3 = ax_1 + bx_2 + c$, or $x_3x_2 = ax_1x_2 + bx_2 + c$ where x_1, x_2, and x_3 are initial variables.

The equations found are evaluated by two significance thresholds t_R and t_S: by default $t_R = 0.01$ and $t_S = 0.01$. t_R is the statistical multiple correlation coefficient and the t_S value is also a function of the chi-square, which is used to differentiate two equations with similar t_R and when there is a perfect fit which causes the division by zero in the calculation of V_R. An equation is found acceptable if its rate V_R (usually called R) is higher than $1 - t_R$ and then if its V_S (usually called *sigma*) is lower than t_S. The higher is V_R and lower is V_S, the better is the solution.

4.2. Algorithm

When the parameter o is not zero, derivatives up to the oth order are created for all initial variables. Experiments are measured at the same interval of time and time derivatives are introduced by numerical derivation, computing the derivative $\dot{x}_i$ of a variable x at a point i from the neighbors x_{i-1}, x_{i-2}, x_{i+1}, x_{i+2}. Due to errors, Dzeroski and Todorovski (1993) advise measuring the derivatives directly whenever the measurement error is too high. Once the derivatives have been built, LAGRANGE then introduces new variables by applying multiplication to initial variables and their derivatives. Finally, it generates models and tests them by linear regressions. Like *Equation Finder*, LAGRANGE does not stop at the first adequate solution but applies regression on all possible models inside the given limits, and reports any satisfying law.

4.3. Scope

LAGRANGE applies regression between at most r independent terms which can be products of at most d initial variables. A law has therefore the following form:

$$x_{01} \cdots x_{0k} = \alpha x_{a1} \cdots x_{am} + \cdots + \sigma x_{s1} \cdots x_{sn} + \zeta,$$

[1] A specific variable is the time, but it is not used in the search.

where $k, m, n \leqslant d; a, s \leqslant r; x_{ij}$ stands for any initial variable or its derivative, and $\alpha, \ldots, \sigma, \zeta$ are parameters of the polynomial.

In the particular case where $o = 0$, $d = 2$, $r = 3$, the most complex model is: $x_i x_j = \alpha x_k x_1 + \beta x_m x_n + \gamma x_o x_p + \delta$.

The ability of the system is highly dependent on the parameters' values. For instance, if $d \leqslant 4$, LAGRANGE can find $x^3 = y$ but not $x^4 = y$.

4.4. Complexity

The total number of variables that LAGRANGE combines is $(o + 3)n$ and $k^{d*(r+1)}$ is the number of possible regressions one can perform from k initial variables with r dependent variables and at most d variables in a term. The complexity of the number of regressions that LAGRANGE performs is therefore:

$$\mathcal{O}((o + 3)n^{d(r+1)})$$

where n is the number of initial variables.[2] The authors emphasize that small values of these parameters are usually sufficient for dynamical systems. However, let us consider the previous case: $o = 0$, $d = 2$, $r = 3$, where neither derivatives nor trigonometric functions are considered. These values are the minimum necessary to discover a simple law like: $y = ax_1x_2 + bx_1 + cx_3$, and if we suppose that there are $n = 4$ initial variables, LAGRANGE applies more than $12^8 = 430$ millions of regressions!

Moreover, sums are more complex to discover than products: the model $z = x + y$ requires at least $d = 1$ and $r = 2$, therefore at least n^3 regressions, while the model $z = xy$ requires at least $d = 2$, $r = 1$, therefore n^2 regressions. Ratios are also more complex than products: the model $z = \alpha xu + \beta yv + \gamma$ needs $r = 2$, $d = 2$, while $z = \alpha x/u + \beta y/v + \gamma$ is found under the form $zuv + \beta\gamma uv$, if $r = 3$, $d = 3$. Finally, LAGRANGE can discover polynomials but a polynomial of degree k requires that $d = k$ and $r = k - 1$. Even if only two variables are considered, LAGRANGE performs $2k^2$ regressions in order to discover it. Comparing to *Equation Finder*, LAGRANGE is obviously more complex, and we will verify experimentally this result for different polynomial degrees in section 7.

4.5. Example

In the example of an ellipsoid law: $4x^2 + y^2 = 30$, here are the results provided by LAGRANGE:

Order of the system (o): 0
Maximal depth of new terms (d): 2

---

[2]The value 3 corresponds to the case when the sine and cosine of each variable are created.

```
Maximal num. of regr. vars (r): 2
Tolerance (R): 0.01 Tolerance (S): 0.01
eqn1: y*y = 29.9687 - 3.9885*x*x(*R = 0.999, sigma = 0.00236*)
eqn2: y*y = 29.9687 - 3.9885*x*x(*R = 0.999, sigma = 0.00236*)
neqns = 2(*Time elapsed [s]: 0.074 - Memory used [K]: 0*)
```

These two equations correspond to the same and good solution, but they have been probably found by two different ways. We already note on this example that the system does not eliminate redundant equations, and that it does not indicate the uncertainty of the coefficients.

4.6. Additional Comments

The main drawbacks of the system are: its incapacity to deal with uncertainties, its incapacity (like EF) to manage with background knowledge, its incapacity to stop at the first adequate solution, and the possibly huge number of solutions proposed, without being sorted according to the selection criteria, and with redundant solutions.

5. A DESCENDANT OF BACON: ARC.2

The BACON system is important for the history of quantitative law discovery because it sets out the main principles of inductive discovery: an iterative search in the solution space guided by very simple heuristics. An important basic notion is *monotonic dependency*, that is to say, the way a variable varies relative to a second one when all other variables remain fixed: the dependency is positive (resp. negative) if the values of y are increasing (resp. decreasing) when the values of x are increasing.

Our system, ARC.2 is a direct offspring of BACON.3 and of ABACUS (Falkenhainer and Michalski, 1990), and is also based on the study of monotonic dependencies. The first version, ARC, improves on the ABACUS system especially by dealing with uncertainties (Moulet, 1991), and by providing a new efficient evaluation function. The second version that we describe in this chapter, ARC.2, integrates polynomial regression in ARC and significantly extends the number of reachable laws.

5.1. Input

a. Nature of the Data. ARC.2 finds a relation between numerical attributes but, as in BACON, *symbolic* attributes can be provided which are possibly used to cluster the data and to characterize the scope of each law. For instance, given the nominal attribute *battery*, which is the name of the battery used, and given the attributes U and I, it can find Ohm's law in the following form:

$$\text{if battery} = \text{A then } UI = 5;$$
$$\text{if battery} = \text{B then } UI = 10;$$
$$\text{if battery} = \text{C then } UI = 15$$

which corresponds to a classification of the dataset.

We will see also in 5.6 that data must follow some characteristics but the algorithm needs firstly to be described.

b. Available Knowledge. As in ABACUS, units can be given. They help pruning sums and differences and since they allow early elimination of terms, they reduce the search very efficiently. Uncertainties can also be provided, associated either in a relative form to all the values of a variable or in an absolute form to specific values. These uncertainties are propagated along all subsequent computations and are mainly used to precisely select the solution.

Like ABACUS, ARC does not make any distinction between dependent and independent variables. ARC.2 can make such a distinction but, in distinction to EF, ARC.2 does not require it. If regression is used, and if the dependent variables is not indicated, then the solution found risks not to be invertible, like in LAGRANGE, that is, to be such that the dependent variable cannot appear as a function of the remaining ones.

c. Parameters. A dozen parameters help the user to control the system, but only two main parameters are very important: **constancy** (0.9) represents the minimum percentage of examples which need to be covered by the laws. That constrains in another sense the maximum percentage of outliers.

uncertainty (0.02) gives the maximum relative error on numerical data. It is used as a default value but each variable uncertainty can be specified as we have noted above.

Other main parameters are:

l-operators (* − +/regression), which is the list of operators which help to combine variables. When **l-operators** = (+ − */), the system is ARC. *unuseful-vars* (nil) can help to discard momentarily some variables, they are not still considered for the search. **max-nodes**(200), **search-limit** (3) control the search depth. They indicate the maximum number of combinations to create, and the maximum number of iterations in the combination phase, as in EF. **cluster-on** (nil) controls the clustering process: if it is set to t, the system can find a partition of the data such that each cluster verifies one different law.

5.2. Algorithm

The algorithm is composed of three main phases: first, iterative combinations are built from the initial variables with only the four arithmetical operators, selecting at each step the most promising combinations. Second, when the first phase does not lead to a solution, a polynomial regression phase is performed. Last, if regression does not succeed either, the iterative search is continued with the remaining combinations. The *first phase* is similar to ABACUS' algorithm:

- For each pair of variables x and y, ARC.2 studies their *monotonic dependency*. ARC.2 computes this dependency on the groups of examples for which all the variables except x and y remain constant (according to their respective uncertainties). This set is called the *projection* on x and y (denoted PROJ(x, y)) and plays an important role in all the following steps.
- The monotonic dependency determines for each pair of variables the arithmetical combinations to apply: $x - y$ and $\frac{x}{y}$ if the dependency is positive, $x + y$ and xy if it is negative. Sums or differences of variables with different dimensions are discarded.
- When all the combinations of the first generation have been created, they are sorted according to a specific evaluation function, and they are divided into two sets: the *active nodes*, which are first studied, and the *suspended nodes*, which are developed only in the third phase, when no solution has been found in the two previous phases. We describe below the selection criterion.
- The active nodes are in turn combined with initial variables and among themselves. That provides another generation which is in turn divided in two groups, keeping only the group of active nodes, and so on. This first process stops whenever no active node is found, or when **max-nodes** (200) combinations have been created, or when the number of iterations goes under **search-limit** (3), or when all possible arithmetical combinations have been created. On the other hand, the process succeeds when it finds a combination which is constant within the given percentage **constancy*(0.90)* of examples. In any case, the search stops at the first solution found, but continues as long as the law proposed is not satisfactory.

Evaluation function. The evaluation of a combination created by ARC.2 is based on the following idea. Let us consider for instance the law $xy = f(w, z)$. If w and z are fixed, then xy is constant while $x + y$ has no reason to be constant. Therefore, the combinations of x and y which appear in the law can be recognized by testing their constancy rate on the projection PROJ(x, y). The active nodes are successively the initial variables, any combination of two variables such that the final law can be written with this combination as a function of the other variables, the combinations of $3, 4, \ldots, n - 1$ variables which belong in the same way to the final law and lastly the good combination which is constant on the examples.

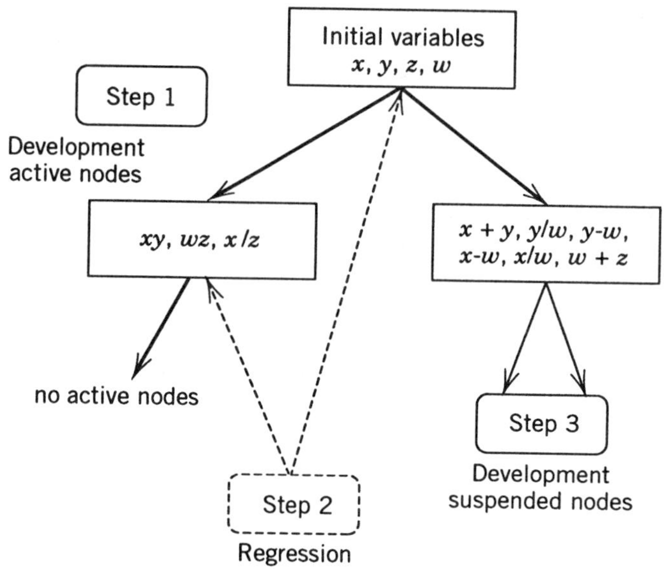

Figure 13.1. *Integration of regression in* ARC.2's *process.*

For instance, during the search for the law $xy/wz = constant$, the active nodes are: xy, xw, x/z, y/w, y/z, wz, then xy/w, xy/z and y/wz. In this way, irrelevant variables are automatically eliminated since a combination with such a variable has little chance to be constant on the corresponding projection, and then to be selected as an active node. Moreover, this criterion selects very few combinations, that significantly improves the system's efficiency, and finally, it also allows to reduce the number of models tested during the regression phase.

The *second phase* introduces regression. When no other active node exists, one more chance is given to all previous active nodes, including initial variables, by testing if a relation other than a simple arithmetical combination like a polynomial function relates them. ARC.2 determines heuristically on which variables to apply regression as well as the degree of the polynomial (for further details, see Moulet (1994).

ARC.2's three major steps are illustrated in Figure 13.1.

Selection of pairs of variables for regression. Among all the possible pairs of variables, a selection can be made, from the following remark: regression between two combinations like xy and yz, such that a same variable belongs to both of them is not worth being tried since regression from simpler variables like x and z would lead to a similar model and to a more precise law. If Y is selected as the first variable for regression, ARC.2 eliminates therefore as second variable the combinations which integrate Y or compose Y, or involve the same initial variables as Y.

Iterative generalization of the polynomial found. When regression is applied between two variables X and Y and if the initial variables are not all included in X and Y, then ARC.2 generalizes the model to the remaining variables (the ones not yet involved in X or Y), following an iterative strategy similar to that of BACON.3. First, regression is applied separately on each group of examples belonging to the projection of X and Y, that is, such that the remaining variables have fixed values. Second, it tries to relate each coefficient of the polynomial found to the remaining initial variables. For instance, ARC.2 discovers the law $x = 5t^2 + tV + x_0$ first by applying regression between $x - x_0$ and V on the model $x - x_0 = aV + b$, on groups such that t is constant, that leads to a list of polynomials, one for each value of t, and secondly by discovering iteratively the relations $a = t$ and $b = 5t^2$.

5.3. Scope

The space explored by ARC.2 is theoretically an infinite space of equations bounded by parameters, different from those explored by EF or LAGRANGE. When polynomial regression is not used, ARC.2 is reduced to ARC, which is able to reach laws of the following form:

$$f_1(x_{11}, \ldots, x_{1i}) \text{ op } f_2(x_{21}, \ldots, x_{2j}) = \textit{constant},$$

with $op = \{+, -, \times, /\}$, x_i being any initial variable, and f_i being in turn any initial variable or the composition of two other functions f_j and f_k, or the sine or cosine of a combination f_k. ARC can find for instance the momentum conservation's law: $m_1.v_1 + m_2.v_2 = m'_1.v'_1 + m'_2.v'_2$, by finding constant the combination $m_1.v_1 + m_2.v_2 - m'_1.v'_1 - m'_2.v'_2$ or the ideal gas law: $pv = kNt$, by finding constant the combination pv/Nt.

In this first phase, only linear functions of x with integer coefficients can be discovered. Thanks to polynomial regression, ARC.2 extends the scope of ARC to polynomial laws of the following form:

$$f'_1(x_{11}, \ldots, x_{1i}) = \mathscr{P}(f_2(x_{21}, \ldots, x_{2j})) = \alpha_p f_2^p + \alpha_{p-1} f_2^{p-1} + \cdots + \alpha_1 f_2 + \alpha_0$$

where $\mathscr{P}$ is a polynomial of degree p and α_i is a real or is in turn a function of the initial variables taking the form f_1 or f'_1.

Since all combinations are based on monotonicity, powers like x^3y cannot be created directly but only through successive products of xy and x. The square x^2 cannot be created alone and laws like the ellipsoid law $y^2 + 4x^2 = 30$ cannot be discovered.

5.4. Example

Let us consider the law $x = 5t^2 + Vt + x_0$, for which examples $e_i, i = 0, \ldots, 8$ are given in Table 13.1. There are four initial variables: $\{x, x_0, V, t\}$.

TABLE 13.1. Data for the Law $x = 5t^2 + Vt + x_0$

	x	x_0	V	t
e1	8	1	2	1
e2	25	1	2	2
e3	53	2	2	3
e4	55	4	2	3
e5	36	10	3	2
e6	64	10	3	3
e7	102	10	3	4
e8	106	10	4	4

(i) First Phase: active nodes search. There are only two non empty projections on initial variables:

$$\text{PROJ}(x, x_0) = \{\text{subsets of examples } V \text{ and } t \text{ are fixed}\} = \}\{e3, e4\}\}$$

$$\text{and PROJ}(x, t) = \{\{e1, e2\}\{e5, e6, e7\}\}.$$

Monotonic dependencies are found positive and the first generation of new combinations is: $G_1 = \{x - x_0, x/x_0, x - t, x/t\}$, where only $x - x_0$ is found an interesting node (i.e., it is constant on the two examples of PROJ(x, x_0)). ARC.2 continues then the search by studying the relation between only this active node and the initial variables. The variables $(x - x_0)/V$, $x - x_0 - V$, $(x - x_0)/t$ and $x - x_0 - t$ are created in the same way, but none is evaluated as an interesting node. The first phase stops then at this point and regression is called in order to detect polynomial relations.

(ii) Search by regression. The last set of active nodes is $\{x - x_0\}$. Since there is only one variable in this set, ARC.2 tries to relate it to previous active nodes, which are, in this case, initial variables. ARC.2 selects V and t as the first possible regression variables, since they do not integrate the same initial variables as $x - x_0$. The study of the degree of the possible polynomial relations between $x - x_0$ and then leads the system to try first to relate $x - x_0$ to V, on the model $y = ax + \beta$. Regression is applied separately on each subset of $\text{PROJ}(x - x_0, V) = \{\{e_2, e_5\}, \{e_3, e_4, e_6\}\{e_7, e_8\}\}$, where t is constant:

$$\text{For } t = 2, \quad x - x_0 = 2V + 20;$$

$$\text{For } t = 3, \quad x - x_0 = 3V + 45;$$

$$\text{For } t = 4, \quad x - x_0 = 4V + 80$$

ARC.2 tries then to generalize this solution to the remaining variable t, in order to include as many variables as possible, and to cover most examples. In this way, the system tries to relate each coefficient of the model $y = \alpha x + \beta$ to t, by calling recursively ARC.2 on each couple (α, t) and (β, t). First, it easily finds that $\alpha = t$. Second, it needs to call again regression to find a polynomial law between β and t: $\beta = 5t^2$.

Together, these two relations lead to the final law $x - x_0 = 5t^2 + Vt$ which covers all examples and is therefore proposed by ARC.2 as a possible solution. Let us note that ARC.2 could have found the solution also by starting to relate $x - x_0$ to V. In this example, only eight combinations are necessary and only two regressions are performed.

5.5. Complexity

Active node building. Let n be the number of initial variables. The number of combinations created is near to the maximum when (i) no unit is given, (ii) no projection is empty, (iii) there exists a monotonic dependency between any couple of variables. However, since only heuristics are used to make the selection, one cannot measure precisely this maximum. We give below only an approximation of ARC.2's complexity.

In the extreme case, at the first generation, the number of active nodes is $\binom{n}{2}$, which corresponds to all possible combinations of two variables that appear in the law. For instance, if the law is $x_1x_2x_3 = constant$ there are three "good" combinations: x_1x_2, x_1x_3, x_2x_3. In the same way, at the second step, when previous good combinations of two variables have been combined either with initial variables or together, at most $\binom{n}{3} + \binom{n}{4}$ combinations will be selected as active nodes. Consequently, the maximum number of selected variables as active nodes is the sum of all the left-hand-side nodes of the tree in Figure 13.2 (we just draw the three first levels, that corresponds to $n \leqslant 8$), where N'_i stands for the number of active nodes selected by the system and N_i is the total number of variables created at the ith iteration:

$N_0 = n$, and $N_i = 2\binom{N_{i-1}}{2}$, since, at most, two operators are applied to all pairs of previous variables. The tree's maximum depth is ceiling($\log_2 n$), that is to say the first integer equal or just higher than $\log_2 n$ but the search stops as

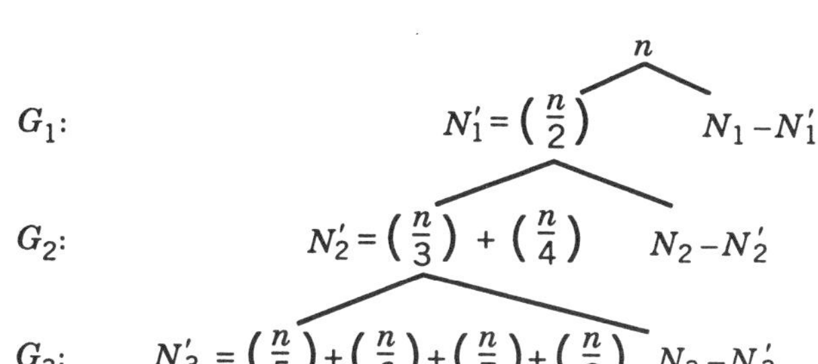

Figure 13.2. ARC.2's *decomposition of nodes into active and suspended nodes.*

soon as a good combination is found on the n variables. In the same way, the maximum depth can be reached but not the number N_i: if $n = 6$ then at most $N'_3 = \binom{n}{5} + 1$.

The number of active nodes is then $\Sigma_{k=1}^{n} \binom{n}{k} = 2^n - 1$. The total number of variables which are meanwhile created has the complexity of $\mathcal{O}(e^{\alpha n - \beta})$. Fortunately, the number of variables created in reality is much less than the theoretical one computed above because there does not always exist a monotonic dependency between any couple of variables. For instance, consider the law $S = (x + y)(z + w)$. Even if all dependencies can be computed, and even if no unit is given, 34 new terms are sufficient to lead to a good combination instead of the theoretical number of 870. The created nodes are decomposed as follows: $N_0 = 5$, $N_1 = \binom{5}{2} = 20$, only two combinations are selected as active nodes ($x + y$ and $w + z$): $N'_1 = 2$ instead of 15 and $N_2 = 14$ instead of 51, finally $N'_2 = 3$ ($(x + y)(w + z)$, $S/(x + y)$ and $S/(w + z)$), and the solution is found.

Regression phase. Regression is performed only with some variables among the $2^n - 1$. The active nodes are considered in inverse order of creation with the last created the first considered. Each variable Y is regressed against its "complementary" active nodes: the variables which only include the k initial variables which do not already include Y. This way, combinations of $n - 1$ variables are regressed against one initial variable, combinations of $n - 2$ variables are regressed against combinations of two variables, and so on. Therefore, we only consider levels n down to $n \bmod 2$ or $n \bmod 2 + 1$ if n is odd.

Once we have chosen a combination Y of $(n - k)$ initial variables, the independent variable X is successively: one combination of k variables, k combinations of $k - 1$ variables, $\binom{n}{2}$ combinations of $k - 2$ variables, and so on, finally the k initial variables which do not compose X. That makes at most $2^k - 1$ possible regressions. Since there are $\binom{n}{i}$ active nodes of i variables, the maximum number of regressions done is: $\Sigma_{i_{min}}^{n-1} \binom{n}{i}(2^{n-i} - 1)$ with $i_{min} = n \bmod 2$ if n is even or $i_{min} = n \bmod 2 + 1$ if n is odd. For instance, if there are 5 initial variables, at most 31 active nodes can be created, and regression would be at most performed $65 = \binom{5}{3}(2^2 - 1) + \binom{5}{4}(2 - 1)$ times.

The cost of a regression done by ARC.2 is however different from the usual regression done by EF or LAGRANGE: if two active nodes chosen for regression do not integrate all initial variables, ARC.2 tries to generalize the polynomial found by the remaining variables, by calling itself recursively on each possible pair of coefficients of the polynomial and of remaining variable. The most complex case holds when regression is applied between 2 initial variables, and the $n - 2$ remaining variables must be introduced one by one, each one relating to a coefficient of the previous polynomial by another polynomial. The complexity of the generalization of a polynomial to k remaining variables is therefore at most: $\Sigma_{i=1}^{k} p_i \,\mathrm{Comp}(2)$, where p_i is the degree of the polynomial relating the ith variable to the previous law and Comp(n) stands for the complexity of ARC.2 for n variables, i.e., the number of nodes plus the number of regressions done. When $n = 2$, at most two nodes are created until regression is called, and only regression between the two initial variables is possible.

In summary, the complexity of ARC.2 is the sum of two complexities: the number of terms created during the development of the active nodes (exponential function of the number n of initial variables) and the number of regressions done (itself a function of the number of active nodes). The complexity could be summarized roughly in the exponential term, but this complexity stands for the worst case which is not representative of real situations, while in the case of the two other systems, the complexity corresponds to the complexity really reached.

5.6. Additional Comments

In order to compute monotonic dependencies, the data must verify some important constraints, most of the data should be collected with a minimum of organization. In the opposite case where the data is varying a lot, or where there are numerous variables, projections may all be empty and the system may not find a solution. However, the failure will not take too much time since the system will stop at once.

This constraint is only applicable to multi-variate data, and even if there seems to be very little structure, if in counterpart examples are sufficiently numerous, one can expect that the projections will not be all empty and that the system will succeed.

Moreover, that does not mean that the user must respect a complete experimentation plan since the system does not need all projections, but only some of them. For instance, in order to find $xy = 3wz + 2$, *ARC*.2 just needs to compare the pairs (x, y) and (w, z). This way, only four examples are necessary to find the solution. It is however difficult to compute how many projections or examples are necessary. This constraint is however the necessary condition for the evaluation function to be active and for the heuristic search to limit the number of tested models. It turns out to be similar to other machine learning systems: the more noisy are the examples, the more numerous they must be in order to find a solution.

6. THEORETICAL COMPARISON: A SYNTHESIS

6.1. Inputs

Let us summarize in Table 13.2 the systems' inputs: the type of data they can manage, whether they are able to take into account initial uncertainties, and which knowledge can be provided. In the last row we indicate the set of main parameters controlling the system and their default values are given into parentheses.

TABLE 13.2. Summary of the Inputs' Characteristics

	ARC.2	EF	LAGRANGE
Type	numeric, symbolic	numeric	numeric
Uncertainty	yes	only for one variable	no
Knowledge	dimensions, dependent variable	dependent variable	angle dimensions
Parameters	**max – nodes** (200) **constancy** (0.9) **uncertainty** (0.02)	*d*: max degree (3) *n*: nb of operators (4) *t*: nb of iterations (1)	*o*: order of derivatives (1) *d*: max depth of terms (2) *r*: max regr vars (2)

Concerning the type of the data and the available knowledge. ARC.2 seems the most complete system: it takes into account symbolic attributes, the uncertainty and the unit of each variable. Physical dimensions allow the search to be pruned and uncertainties contribute to control the adequacy of both the combinations and the solution found. On the other hand, the system is very sensitive to uncertainties and may not discover a law if they are not well specified.

As opposed to EF and LAGRANGE, which can deal with any data, BACON-like systems and in particular ARC.2, present a major limitation that we presented in 5.6: all variables must be controlled and a minimum number of experiments must be run. That means that during some experiments, all variables except two must remain constant, while the two under study are varying, and this must be verified for different pairs of variables.

6.2. Algorithm's Characteristics

In Table 13.3 we summarize some major characteristics of the algorithms: the operators used to combine variables, a general form of the reachable laws. The last row presents the algorithms' complexity which is given as a function of: (i) the parameters of the system and (ii) the number n of initial variables for LAGRANGE and ARC.2, or the number k of possible operators in EF.

In ARC.2, we cannot measure directly the complexity as a function of the parameter's values. However, they influence it, in particular **constancy** and **uncertainty**. Moreover, these parameters can determine the success of the law discovery process. For instance, for too high a value of **constancy**, there may be no solution, and for a too low value, ARC.2 may stop at an uninteresting solution.

To summarize, *Equation Finder* is the most complete system concerning bivariate laws, including polynomials. LAGRANGE is essentially adapted to simple multivariate linear laws, when one wants to try simple models, but is not adapted to polynomials or complex forms with powers. On the other hand, ARC.2 is useful in those cases when the user wants to test complex laws or when he does not have any idea of the model.

TABLE 13.3. Summary of the Algorithms' Characteristics

	ARC.2	EF	LAGRANGE
Operators	$+, -, *, /$, sin x, cos x, polynomial regression	$\frac{1}{x}$, $*$, log x, e^x, $\sqrt{x}$, polynomial regression	$*$, multivariate linear regression
Search Space	$f_1(x_{11}, \ldots, x_{1i}) = \mathscr{P}(f_2(x_{21}, \ldots, x_{2j})$ P coefficients: real or functions	$y' = P(x')$, $y' = f(x, y)$ $x' = f(x)$ and $degree(P) \leqslant d$	$x_{01} \cdots x_{0k} = \alpha x_{11} \cdots x_{1m} + \cdots + \zeta$, $k, m, n \leqslant d, s \leqslant r$
Complexity	$\mathcal{O}(e^{\alpha n - \beta})$	$\mathcal{O}\left(d \times \left[\frac{(k^2)^t - 1}{k^2 - 1}\right]^2\right)$	$\mathcal{O}((o+3)n^{d(r+1)})$

Much care must be given to the parameter settings. In each of these three systems like in any parametric search, setting the parameters too small prevents the system from discovering a law. Moreover, when the parameters are set too large, it would be preferable that the search is not longer than necessary. While LAGRANGE explores all the space delimited by the parameters, and EF stops only at the level including a first solution, ARC.2 stops as soon as a solution is found (but it can continue if the user asks for it). Therefore, if the parameters are sufficiently high, the space really explored is proportional to the complexity of the law, and not to the parameters. When the law is unreachable, the role of the parameters is to prevent the system from exploring too large a space.

On the other hand, LAGRANGE explores all the space delimited by the parameters. In order to reduce the search, the user must set parameters to the most appropriate values, and therefore needs to have a good idea of the complexity of the solution. Moreover, the number of solutions increases with the value of the parameters and makes the discrimination task more difficult.

7. EMPIRICAL COMPARISON

We have separated experiments in two steps: a first set of experiments concerns only bivariate laws, on which all three systems can be compared (7.1), and a second set concerns multivariate laws that only ARC.2 and LAGRANGE are able to discover (7.2). In the first step, data can be created automatically and randomly, while a set of predefined data and laws have been used in the second step. All experiments have been run on a SUN SPARC 10. EF and ARC.2 have been run on *Harlequin Common Lisp*, while LAGRANGE is a C compiled program.

TABLE 13.4. Execution Time (in seconds) for Discovering Polynomials

Degree	ARC.2	EF	LAGRANGE
1	0.35	0.05	0.08, 2 eqns
2	0.58	0.054	0.054, 2 eqns
3	0.518	0.058	d = 3 and r = 3 0.15, 23 eqns
4	0.84	on 43 examples 3.085	d = 4 and r = 4 5.525, 1015 eqns

7.1. Comparison of All Three Systems

Discovering polynomials. In these experiments, we test successively the influence of the degree and the influence of the number of examples on the systems' complexity. For each configuration, we use 50 random datasets. By default, each dataset contains 20 examples described on two variables x and y, and these data have been randomly chosen with a relative error randomly chosen between 0% and 2%. Default values have been chosen for parameters, except the parameter o of LAGRANGE, which is set to 0 (see 4), and except when it is indicated. Table 13.4 provides the results of testing the three systems on polynomials of degree 1 up to 4, which seems to be a reasonable limit for experimental science.

EF takes the same time for any degree up to 3, since it performs in each case only three regressions. ARC.2 and LAGRANGE are more sensitive to the polynomial degree, and in particular LAGRANGE, since the higher is the degree the higher must be the parameters values, which lead to more search and regressions.[3]

Tests on fourth-degree polynomials. In about half of the cases tested by ARC.2, regression on a fourth degree polynomial model does not lead to correct coefficients (due to noise) and less than 90 percent of the examples are covered by the law. ARC.2 discovers however the solution of the form $Y/X^4 = constant$ in the third step when developing the suspended nodes. EF does not find either the solution by regression between X and Y (which is limited by default to the degree 3), but finds a solution of the form $Y = \alpha X^2 + \beta X^4$, by regression between Y and X^2. Since some terms are missing compared with the best solution, the fourth degree polynomial law found does not appear at the top of the three best models.

Influence of the number of examples. In the next experience, we use only second degree polynomials, but we vary the number of examples. In each case, 50 laws are automatically generated with an uncertainty rate which is

[3] In two cases, ARC.2 needs, due to noise, that **constancy** is set to 0.8, in order to find a solution.

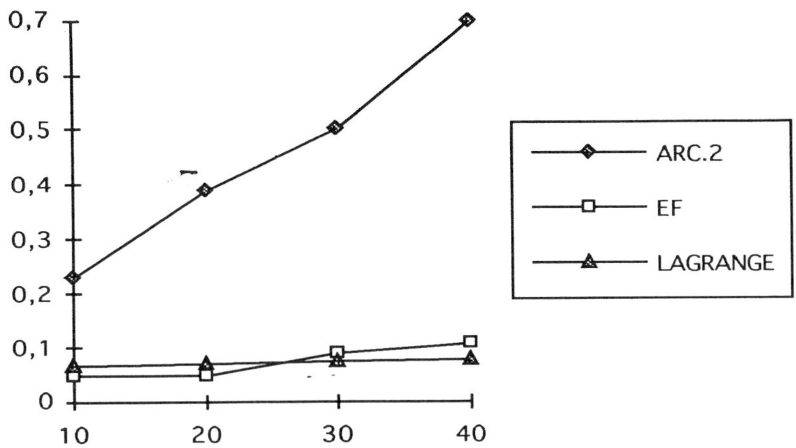

Figure 13.3. *Influence of the number of examples on performances.*

randomly selected between 0% and 2%. Figure 13.3 contains the results of these tests.

We can note that all three systems are sensitive to the number of examples, which is quite normal for a machine learning system. LAGRANGE stays the fastest algorithm, and the heuristic system ARC.2 is the slowest one. However, since all these systems present only a linear increase with respect to the number of examples, this is reasonable.

A non-polynomial law. All three systems can be also compared on simple bivariate laws, but we give only one real example, Kepler's third law: $d^3 = p^2$.

1. ARC.2 finds this law in the third phase of the algorithm, by finding that d^3/p^2 is constant through the creation of $d/p, d^2/p$. This discovery just requires the creation of seven new terms, and running the algorithm takes 0.54 second.

2. Concerning EF, among 135 models, 11 are found acceptable, 117 non-zero, and 7 non-zero-acceptable. The three best models contain an approximation of the expected law and this search requires 2.3 seconds.

$$P^2 = 2.5E - 4 - 0.0012D + 0.002D^2 + 0.991D^3$$

3. With default parameters, LAGRANGE does not find any equation. If we set d and r to the value 3, it finds 48 equations in only 0.47 second, the third one corresponding to Kepler's law:

$$\text{eqn3: } D^3 = 1.12716D + 0.99939P^2 (*R = 0.99, \, sigma = 0.020*)$$

However, 47 other equations are found among which some are even better than this third equation (according to the statistical criterion R), like the 44th equation:

$$\text{eqn44: } DP^2 = 347.71 + 1183.48P^2 - 1205.81D^3 + 9.83D^2P(*R = 1,\ sigma = 0.0017*)$$

We can note here that LAGRANGE does not make a distinction between very complex and simple laws, since only statistical criteria are taken account of. Conversely, EF and ARC.2 guarantee in a certain way that the first solution found will be the simplest (if we do not take possible simplifications into account), since the minimum number of transformations are applied.

7.2. Comparison of ARC.2 and LAGRANGE on Multivariate Laws

We have chosen to compare these two systems on five examples that we used currently for developing ARC.2. This is a first comparison, and should be extended in the future. However, they can help the reader to understand the limits of these two systems. Table 13.5 provides the results on these simple tests.

In some cases, for example for the law $a = hm_1 + lm_2/m_1 + m_2$, ARC.2 cannot find a solution because there is no monotonic dependency between sufficient pairs of variables.

In another case of the centrifugal force law: $f = mv^2/r \cos(\theta)$, LAGRANGE is not able to discover the expected law, even with high values of the parameters ($d = 3$ and $r = 3$). The reasons of this failure are difficult to find

TABLE 13.5. Execution Time of ARC.2 and LAGRANGE on Simple Laws

Law	ARC.2	LAGRANGE
$xy = 3wz + 2$	0.36s	$d = 2$ and $r = 2$ 0.114s, 5 eqns
$S = (x + y)(z + w)$	0.724s	$d = 4$ and $r = 4$ stop after 5 hours 1104 eqns
$m_1 . v_1 + m_2 . v_2 = m'_1 . v'_1 + m'_2 . v'_2$	3.66s	$d = 3$ and $r = 3$ 46mn, 235 eqns
$S = xy + z^2$	0.74s	$d = 3$ and $r = 3$ 197s, 144 eqns
$a = \dfrac{hm_1 + lm_2}{m_1 + m_2}$	no solution	$d = 2$ and $r = 3$ 92.029s, 1 eqn
$f = \dfrac{mv^2}{r} \cos(\theta)$	11.856s	no solution

since there is no possibility to have a trace during running, but we can suppose that errors are such that the statistical thresholds are not reached. ARC.2 discovers this law by regression between $fr/m \cos(\theta)$ and v.

8. CONCLUSION: STATISTICS VERSUS HEURISTICS

Statistical methods are obviously the quickest, when dealing with low degree polynomials or simple laws. But they do not allow to select precisely the best solution, since in general many equations verify the statistical criterion. *Equation Finder* is able to build a lot of new terms from only two variables, but is unable to evaluate them in order to have more information about the sought after law. These terms are considered as independent potential models. Like LAGRANGE, the search is reduced to testing a large set of models without any variable selection nor incremental discovery. On the other hand, the systematic search done by EF is enough and may be quicker than heuristic search when the law is simple, that is, first explored with regard to the search space. LAGRANGE does not present this advantage since it explores all the space bounded by the parameters. Its efficiency directly depends on the parameter settings.

ARC.2 combines many simple heuristics that help the system to find the law very efficiently, provided that it is reachable: first, the study of monotonic dependencies, units and initial uncertainties guide the building of new terms; then an evaluation function analyzes the behavior of these new terms in order to determine which among them are the most promising. This step is an important one and is similar to the variable selection phase, which is crucial in data analysis. Moreover, even in the case of failure, the selected combinations may mean a lot for an expert, who can then look for the model in other ways. Finally, the incremental construction of the law, using results of previous computations to guide the next search can be compared to the way human analysts usually proceed.

However, while EF and LAGRANGE are applicable to any domain, an important limit of ARC.2 lies in its requirements on the nature of the data: in order to compare a minimum of pairs of variables, some groups of examples must vary only on the two studied variables. This roughly requires that nearly all the variables are under control, and this reduces the applicability of our system to experimental domains. However, experimental domains constitute an important part of real applications, and the help provided by these restrictions confers an important advantage on heuristic search relative to the exhaustive approach.

We can debate the role to be played by a numeric law discovery system: must it be the best adapted to real problems, must it be able to discover as many laws as possible? Both of them, of course! However, we argue that only one system could not reach this goal alone: it needs several smaller modules, each one being adapted to a subset of problems or to a subset of functions.

The study here constitutes a necessary first step for a complete comparison but additional work is required in order to completely reach this goal, and in particular an experimental comparison on some real-world data.

ACKNOWLEDGMENTS

The author would like to thank Y. Kodratoff for his support, and A. Cornuéjols and C. Rouveirol for their comments. She also gratefully thanks Jan Zytkow for his support and his helpful comments. This work has been partially supported by the French PRC-IA.

REFERENCES

Bertrand, P. (1986). *Pratique de l'Analyse Statistique des Données.* Presses Universitaires du Québec.

Dzeroski, S., and Todorovski, L. (1993). Discovering dynamics. In *Proceedings of the Tenth International Conference on Machine Learning*, pp. 97–103, Morgan Kaufmann, San Mateo, CA.

Falkenhainer, B., and Michalski, R. (1986). Integrating quantitative and qualitative discovery in the ABACUS system. *Mach. Learn. J.* **4**(1), 367–401.

Falkenhainer, B., and Michalski, R. (1990). Integrating quantitative and qualitative discovery in the ABACUS system. In *Machine Learning: An Artificial Intelligence Approach* (Y. Kodratoff and R. Michalski, eds.), Vol. 3. Morgan Kaufmann.

Langley, P. (1979). Rediscovering physics with BACON.3. In *Proceedings of the Sixth International Joint Conference on Artificial Intelligence*, pp. 505–507. Morgan Kaufmann, San Mateo, CA.

Langley, P., Zytkow, J., Simon, H., and Bradshaw, G. (1986). The search for regularity: Four aspects of scientific discovery. In *Machine Learning: An Artificial Intelligence Approach* (R. Michalski, J. Carbonell and T. Mitchell, eds.), pp. 425–469. Morgan Kaufmann, San Mateo, CA.

Moulet, M. (1991). Using accuracy in numerical law discovery. In *Proceedings of the Fifth European Working Session on Learning* (Y. Kodratoff, ed.). Lect. Notes Artif. Intell., pp. 118–136. Springer-Verlag, Berlin.

Moulet, M. (1993). Découverte empirique de lois numériques ou ABACUS Revu et Corrigé. Ph.D. Thesis, Thèse de Doctorat, Université de Paris-Sud, France.

Moulet, M. (1994). Iterative model construction with regression. In *Proceedings of the Eleventh European Conference on Artificial Intelligence* (T. Cohn, ed.), pp. 448–452. Wiley, New York.

Schaffer, C. (1990). *Domain-Independent Scientific Function Finding*, Tech. Rep. No. LCSR-TR-149. Rutgers University, New Brunswick, NJ.

Wu, Y. (1988). Reduction: A practical mechanism of searching for regularity in data. In *Proceedings of the Fifth International Conference on Machine Learning*, pp. 374–380. Morgan Kaufmann, San Mateo, CA.

Wu, Y., and Wang, S. (1991). Discovering functional relationships from observational data. In *Knowledge Discovery in Databases* (G. Piatetsky-Shapiro and W. Frawley, eds.), pp. 55–70. MIT Press, Cambridge, MA.

Zembowicz, R., and Zytkow, J. (1991). Automated discovery of empirical equations from data. In *Methodologies for Intelligent Systems* (Z. Ras and M. Zemankova, eds.). Springer-Verlag, Berlin.

Zembowicz, R., and Zytkow, J. (1992). Discovery of equations: Experimental evaluation of convergence. In *Proceedings of the Tenth National Conference on Artificial Intelligence*, pp. 70–75. MIT Press, Cambridge, MA.

CHAPTER 14

Analogy as Minimization of Description Length

A. CORNUÉJOLS

1. INTRODUCTION: ANALOGY, INDUCTION, AND COMPUTATIONAL MODELS

1.1. Analogy and Induction

Peculiarities of analogy that make it different from other, traditionally considered as more typical forms of induction, are that (i) it works from the specific to the specific apparently without relying on vast numbers of observtions, and (ii) it does not necessarily involve general rules that would apply both to the 'source' of the analogy and to its 'target.' Models of analogy must therefore propose mechanisms allowing for specific to specific inferencing and this possibly between different domains. Analogy thus appears quite remote from the realms of statistics.

1.2. Current Models of Analogical Thought

A survey of existing computational studies shows that for solving the specific to specific inference problem, the hypothesis of analogy as a mapping process between the representations of the analogues has been unaminously adopted, following Winston (1980). Recognizing then that a combinatorial number of possible mappings between graph-representations exists, the problem becomes one of taming this complexity to allow focussing on the most relevant and promising ones. To this effect, a number of intuitively appealing hypotheses have been made, but we have no place to describe here (see Burstein, 1986;

Machine Learning and Statistics, The Interface, Edited by G. Nakhaeizadeh and C. C. Taylor.
ISBN 0-471-14890-3

Clement and Gentner, 1991; Falkenheimer, 1987; Gentner, 1983; Gick and Holyoack, 1980, 1983; Holyoack and Thagard, 1989; Kedar-Cabelli, 1985). These models, however, still leave place for dissastifaction.

First, there is no formal ground for choosing the analogy as a mapping paradigm. Second, all models mentioned so far are highly dependent on the a priori designed representation scheme (see Keane, 1988; Chalmers et al., 1992 for a critique and illustrations). The latter is a serious drawback, even more so that at the same time it seems quite undisputable that analogy is not only mapping between representations but also and foremost making these representations appropriate for a given context through perception. Finally, it has not been proved that existing models can account for observed properties of analogy. It seems that a well-founded theory of analogy must explain these properties.

1.3. Properties of Analogy

We list here six properties of analogy that we feel are telltale symptoms and accordingly important test beds for any would-be theory of analogy.

1. Analogy is *pervasive* in cognition.
2. *Abstract* analogies are generally preferred over more literal ones.
3. There is generally a good *agreement in the ratings of analogies* among different people.
4. Analogy is *non-involutive* ($f \circ f \neq Id$). In other words, analogy is non symmetric. One can readily find examples where transferring properties from the source to a target and then back from the initial target to the initial source, brings discrepencies between the initial properties of the source and the ones which are transferred back.
5. Analogy is *non-idempotent* ($f \circ f \neq f$). If A is used to infer analogically properties about B, and then B is used to augment the knowledge about C, the result may differ from the one obtained by directly using A analogically on C. It must be noticed here that properties 4 and 5 call into question the feasibility of Case Based Reasoning where the case library is supposed to grow incrementaly with the system's history. Indeed it follows from 4 and 5 that inconsistent case bases might be created.
6. Analogy making *can be improved with practice.* At least some domain-dependent forms of analogical reasoning can be learned.

Trying to account for these properties is part of the reason why the point of view on analogy reported in this paper has gradually emerged. The overall spirit of this view is presented in the next section. Then, in section 3, its formal foundations are reviewed and explained. Section 4 illustrates the application of this approach to examples drawn from a classical analogy domain. Finally, the conclusion section sums up the current findings and underlines the deep relationships between inference and statistics.

2. ANALOGY AS AN ECONOMY PRINCIPLE

2.1. The Classical Schema of Analogy

We limit ourselves here to one component of analogical reasoning: the transfer of some properties of the source to the target, ignoring the problem of mentally conjuring up a good analog as well as the difficulties related to the validation and tuning of the result in the target domain. We will assume for the time being that one can legitimately isolate such components.

Analogy making is, in part, *perceiving* some aspects of the structure of two situations the essences of those situations, in some sense as identical. By perception, we mean here the process by which some conceptual primitives and structures are summoned up and used in order to make sense of the raw data and to represent what is judged relevant in them. When perception is tuned to analogy making, the conceptual structures that arise amount to the filtering of the salient part of the data that make them analogous

In a way, one might consider the essences of the analogues as the necessary information that, when provided, allows one to easily transform part of one situation into part of the other, these parts being the relevant ones in the context of analogy. But how to draw out these essential aspects?

Suppose we try to transform, by way of symbolic manipulations, one analogue, the source, belonging to universe U_1, into the other, the target,

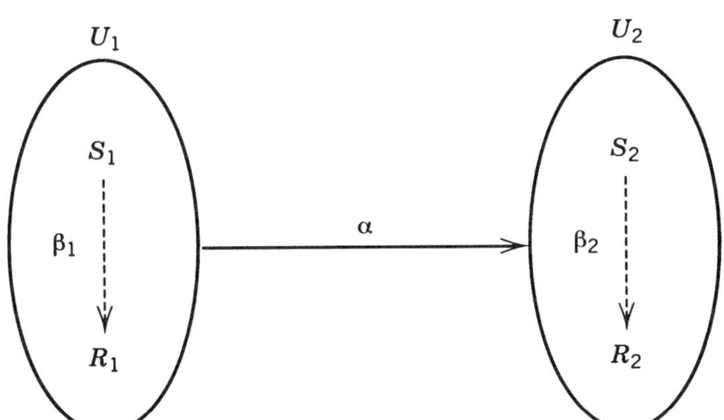

Figure 14.1. *The classical schema of analogy. The* source *in universe U_1 is translated into the target in universe U_2 by way of the transformation α. The simpler the transformation α, the more similar are the source and the target.*

belonging to universe U_2. We know that if we find an economic transformation, this means that there are reasons to believe that we hit on an appropriate information supply, that is some essential commonalties between the analogs.

In the classical schema (see Figure 14.1) where $(S_1 \Rightarrow R_1)$ is the *source* and S_2 is the *target* situation or problem with R_2 being the unknown, analogy making is trying to find ways of representing S_1, R_1, and S_2 so that (i) an economic transformation α using these representations exists between S_1 and S_2, and (ii) β_1 the relevant dependency $S_1 \Rightarrow R_1$ be represented.

It is then a simple matter to translate β_1 into β_2 by way of the mapping $\beta_2 = \alpha(\beta_1)$.

2.2. Analogy and Kolmogorov Complexity

In its naive version, induction consists in drawing general rules from specific observations. These general rules or hypotheses are subject to three constraints: (i) to be more general than the original observations, otherwise they have little value; (ii) to be testable so that later observations may falsify them; and (iii) to inspire confidence, that is to be the most probable given the observations.

A general principle that dates back at least to William of Occam (1290?–1349?) seems to verify these three criteria to the best. This is the Principle of Simplicity, which asserts that the "simplest" hypothesis or explanation is the most reliable. During the last thirty years, "simplicity" has been equated with "shortest effective description", so that if there are alternative explanations for a given body of observations, one should select the one with the shortest description. In the first formalization (Chaitin, 1987; Kolmogorov, 1965; Solomonoff, 1964) called algorithmic or Kolmogorov complexity, hypotheses were considered as the effective computable procedures able to generate these observations as output. The best hypothesis was thus the shortest procedure as measured with the Turing machine encoding scheme, that is the one that allowed maximum compressibility of the data.

We are going to use the same principle here. The conceptual primitives and structures used in the perception and representation of the analogues will stand for the hypotheses, and the raw data will stand for the observations. This way, we will be able to compare the quality of various analogies between some source and target by measuring the corresponding Kolmogorov complexities. The analogies with the lowest algorithmic complexities will be deemed the best.

As has been noted in the introduction, analogy is a special form of induction in which the data consist in a single compound object: the source and its target. It is therefore a great advantage for our purpose, and one that distinguishes it from statistical approaches, that algorithmic complexity applies equally well to sets of observations and to single objects. However, compressibility through Turing machines is not a method that is to be used lightheartedly in Artificial Intellgence. Indeed, one usually looks for explanations expressed within some model classes, and not explanations expressed as abstruse bit strings. This is where the Minimum Description Length Principle (MDLP) (Rissanen, 1989; Wallace & Boulton, 1968) intervenes.

2.3. The Minimum Description Length Principle (MDLP)

Rissanen (1989) starts from the observation that scientific theories often involve two steps. First, the formulation of a set of possible alternative hypotheses (for which he does not offer any mechanism), and, second, the selection of one hypothesis as the most likely one. Rissanen proposes that this selection mechanism obeys the Minimum Description Length Principle which states that:

The best theory to explain a set of data is the one which minimizes the sum of

- The length, in bits, of the description of the theory; and,
- The length, in bits, of data when encoded with the help of the theory.

There is a deep relationship between the MDLP and the Bayesian approach. Indeed, one can derive the former from the latter by observing that from Bayes' Rule:

$$P(H \mid D) = \frac{P(D \mid H) \times P(H)}{P(D)}$$

where H is an hypothesis, and D is the set of observed data, and where we are looking for the hypothesis H that maximizes $P(H \mid D)$ it follows:

$$-\log P(H \mid D) = -\log P(D \mid H) - \log P(H) + \log P(D)$$

Minimizing this expression is then equivalent to minimizing:

$$-\log P(D \mid H) - \log P(H)$$

since the length of D is fixed for any H. This yields the MDLP.

2.4. Practical Problems: Encoding Scheme and Intractability

The problems when trying to apply the MDLP are twofold. First, the length of an explanation will depend on the languages or codes used for describing both the theory and the data. Second, Kolmogorov has proved that searching for the shortest description of an object is NP-complete.

The first problem is known as the *encoding problem*. An answer to it is that, following the equivalence just mentioned between the MDLP and Bayes' Rule, the code used should reflect, when possible, our prior expectations about the environment: that is descriptions of common or important concepts should be shorter than descriptions of unusual or unimportant ones. We will therefore require that the coding schemes be efficient, that is, that they provide optimal encodings of the theories with respect to their a priori probability of occurrence, and of the data with respect to each theory. This means that what constitutes a good theory will always be dependent on our expectations about the world. If this seems disappointing, it is exactly these expectations that make the induction problem tractable.

To the second problem, the absence of any effective way of calculating the best theory, there is no other answer except to be content with searching for satisfying theories only.

3. ANALOGY AND MDLP

3.1. Analogy as an Economical Perception of the Source and Target

Looking at an analogy problem, we can consider the data as the descriptions of, and the theories as, the conceptual constructions that allow us to represent S_1, S_2, R_1 and R_2 as well as the transformations and β_1. Now what makes α_S and β_1 interdependent is that they are built using as many common parts or abstractions as possible. This allows the economical description of the analogy. Because there does not always exist a common generalization of the analog (they might belong to different domains), it might be necessary to introduce different sets of abstractions or models, M_1 and M_2. These models correspond to the ontologies or primitive descriptions that allow the "perception" of the relevant features of the initial data S_1, S_2 and R_1 (see Figure 14.2).

Given the data S_1, R_1 and S_2, the information needed to derive them from the model M_1 is given by the formula:

$$I(S_1 \mid M_1) + I(\beta_1 \mid M_1) + I(M_2 \mid M_1) + I(S_2 \mid M_2)$$

where $I(x \mid y)$ is the information needed to describe x when y is given (the conditional information). We have chosen here to underline the dependency relation β_1 that allows us to derive R_1 from S_1. If we insist instead on R_1 itself,

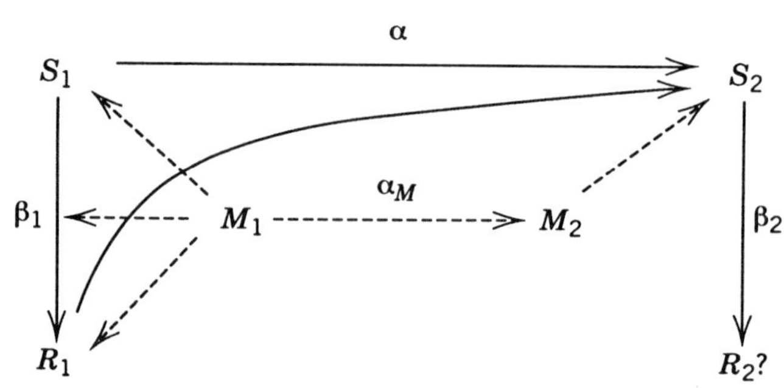

Figure 14.2. *The interplay between conceptual structures M_1 and M_2 and the representation of S_1, S_2 and R_1 as well as α_S and α_M.*

we could use as well the formula:

$$I(S_1 \mid M_1) + I(R_1 \mid M_1) + I(M_2 \mid M_1) + I(S_2 \mid M_2)$$

Therefore, in application of the MDLP, the best model for a completed analogy (where R_2 is hypothesized) is the one that minimizes the following formula:

$$L(Analogy) = L(M_1) + L(S_1 \mid M_1) + L(\beta_1 \mid M_1) + L(\alpha_M) + L(S_2 \mid M_2) + L(\beta_2 \mid M_2)$$

It must be noted that $\alpha_M = I(M_1 \mid M_2)$ the information required to generate M_2, when knowing M_1.

3.2. Overview of the Task Domain and the Experiments

As a task domain, we have chosen the microworld developed by Hofstadter et al. for the COPYCAT project (Mitchell, 1993). In this domain, the analogy problems consists in finding how a letter string is transformed given, as an example, another string and its transform. For instance, given that **abc** ⇒ **abd**, what becomes of **iijjkk** ⇒ **?**.

This microworld offers several useful features. It is simple in its definition and it is reasonably straightforward to find adequate knowledge representation primitives for it. It nonetheless provides much of the richness of the analogy problem. And, finally, it is easy to rate the relative merits of the possible solutions for the test riddles.

As a first step toward a complete account of analogy as a Minimum Description Length problem, a preliminary check for the feasibility of the project is to build an encoding scheme for the abstractions, rules and concepts than can enter analogy making, and then see if the ensuing description lengths of possible solutions for given test problems reflect the expectations about their quality. For instance, it is expected that, to the afore-mentioned problem, **iijjll** is a better solution than **iijjld**. Is the description length associated with the first solution accordingly shorter than the description length for the second one?

3.3. Encoding Scheme

In an analogy problem, we can consider that the data are the descriptions of the source and the target, and the theories are the conceptual constructions that allow us to express the transformation of the source into the target and the relevant properties of the source that need to be transferred (see Figure 14.2). We now have to describe the primitives that allow us to represent both the data the theories. At the same time we have to assign them a coding length that obeys as much as possible the efficiency criteria underlined in section 2.3.

The knowledge representation primitives are mostly the ones used in Mitchell (1993). Table 14.1 lists these primitives and the grammar like structures that were selected for this microworld. In order for the quality

TABLE 14.1. List of Representation Primitives and of Grammar Like Conceptual Structures

• Features describing the conceptual structures:	
– orientation ($\rightarrow / \leftarrow$)	1 *bit*
– cardinality or number of elements: n	$log_2(n) + 1$ *bits*
– type of elements	length: see below
– length: 1	$log_2(l) + 1$ *bits*
– starting or ending with element = x	$L(x)$ *bits*
• **Letter**	(1/2)
Particular letter (e.g. 'd')	(1/2.26)
• **String** (orientation,elements)	(1/8)
$L = 3 + L(orientation) + \Sigma\, L(elements)$ e.g. L('a3bd' with orientation = $\rightarrow$) $= 3 + 1 + log_2((1/2.26)^3 + L(3)$ $= 3 + 1 + 18 + 3 = 25$ *bits*	
• **Set** (type of elements, cardinality, elements)	(1/8)
$L = 3 + L(type) + L(\text{cardinality}) + \Sigma\, L(elements)$	
• **Group** (type of elements, number of elements, elements)	(1/8)
$L = 3 + L(type) + L(nbofel.) + \Sigma\, L(elements)$	
• **Sequence** (orientation, type of elements, succession law or number of elements, length, starting or ending with)	(1/8)
$L = 3 + L(orient.) + L(type) + L(law)\, or\, L(nbel.) + L(length) + L(start/\text{end})$	
• Description and length of a **succession-law** succ(type-of-el.,n,x) ≡ the nth successor of the element x of type type-of-el. $L = L(type) + L(n(\text{see below})) + L(x)$ $L(n) = \begin{cases} L(1/6) & \text{if n} = 1 \text{ or } -1 \text{ (first successor or predecessor)} \\ L(1/3) & \text{if n} = 0 \text{ (same element)} \\ L((1/3).(1/2)^p) & \text{otherwise (with p} = \text{n if n} = 0\text{, p} = -\text{n otherwise)} \end{cases}$	
• **First / last**	1 *bit*
• **nth**	*nbits*

criterion to be computable, each construct is associated with a number, that corresponds either to a prior probability from which it is easy to draw the related length using the relation $L = -\log_2(P)$ (e.g. the concept of set is associated with the prior 1/8, hence, is of length 3 bits), or directly to a length in bits (e.g., the concept of *n*th requires *n* bits). These numbers can be modified either manually or through learning to yield various biases corresponding to a variety of contexts or prior knowledge.

Using these primitives, we define templates for the descriptions of the letter strings and for the descriptions of the transformations between letter strings. These templates constitute the theories that are to be compared with respect to the simplicity of the description they will allow. The templates or conceptual primitives that yield the shortest description of the analogy will be selected as the best ones. By using them, we single out the most relevant aspects of the analogy.

3.3.1. Description for the Letter Strings. As an illustration of the use of this scheme, the string 'abc' could be described as:

'abc' ≡ String (1/8)
orientation: → (1/2)
1st = 'A', 2nd = 'B', 3rd = 'C' $(1/4.26)^3$
TOTAL Length: **21** *bits*

or alternatively as:

'abc' ≡ Set (1/8)
'A', 'B', 'C' $(1/4.26)^3$
TOTAL Length: **20** *bits*

or else as:

'abc' ≡ Sequence
orientation: → (1/8)
type of elements = letters (1/2)
succession-law: (1/2)
succ(elt(letter = x) = elt(succ(letter,1,x)
L(*letter*) + *L*(1*stsucc*) + *L*(*x*) = *L*(1/2.1/6.1)
= *L*(1/12) = 4 *bits*
length = 3 3 *bits*
starting with element(letter = 'A') (1/26)

TOTAL Length: **17** *bits*

It is apparent that the last description, which more fully represents the structure of the string 'abc', is the most economical one, even though it describes it more completely than the second description which corresponds to the perception of a set of three letters.

3.3.2. Description for the Transformations. A transformation between letter strings or concepts or abstractions must specify what has to be changed in the description so as to obtain the transformed object from the original one. This corresponds to the conditional information needed, in addition to the information contained in the original object, to get the new one. The best induction is obtained when this conditional information is maximally compressed. Here the application of the MDLP results in the choice of certain abstractions and constructions thereof.

For instance, following the work of Mitchell (1993), we will restrain the description of the *dependence relation* β to the mould "*Replace_by_*" where the _ can stand for diverse concepts of various levels of abstraction, for example, *letter*, *unit-base* (...), "A", succ(letter,i,'A').

The *resemblance relation* α, on the other hand, will state the correspondence for each modified attribute of a description. For instance, between the descriptions of **abc** and **ijk**, the α relation may be stated as:

$$\alpha = \textit{starting-with}\ (\textit{unit-base} = \text{"A"}) \\ \mapsto \textit{starting-with}\ (\textit{unit-base} = \text{"I"})$$

It could as well be stated as:

$$\alpha = \text{"A"} \mapsto \text{"I"};\ \text{"B"} \mapsto \text{"J"};\ \text{"C"} \mapsto \text{"K"}$$

Figure 14.3 gives an example of an analogy using these description schemes.

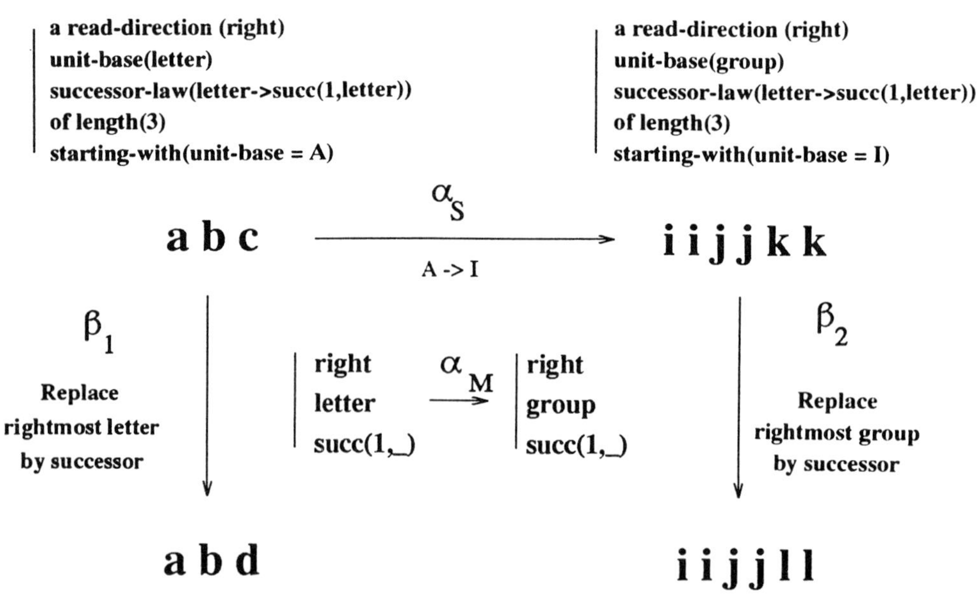

Figure 14.3. *A simple analogy described using a certain set of primitives giving prominence to the concepts of* group *and of* succession-law.

3.4. Illustration

We illustrate this with the following example: **abc** ⇒ **abd**; **iijjkk** ⇒ **?**

While limited space prevents us from giving a detailed account of the derivation of all the complexity figures in the experiments, we will nonetheless describe the overall method and explain the numbers for the first experiment. First, some model M_1 must be hypothesized stressing the importance of chosen aspects of the source and target. For instance, the model M_1 below is the one used in solution 1 for the problem described in section 4.

M_1 = Sequence	(1/8)
orientation: → (right)	(1/2)
type of elements = letters	(1/2)
succession-law:	
succ(elt(letter = x) = elt(succ(letter,1,x)	4 *bits*
Last (rightmost)	1 *bits*
	TOTAL Length: **10** *bits*

The second step consists in constructing representations for the different data and hypotheses intervening in the analogy, and computing the relevant differences in information between them. For instance, the representation of **abc** below and associated with (Problem 1; Solution 1) differs from M_1 in two aspects: the *length* of the sequence and the *starting element*. These amount to 8 *bits*, hence the value of $L(S_1 \mid M_1)$.

'abc' ≡ Sequence	
orientation: →	(1/8)
type of elements = letters	(1/2)
succession-law:	(1/2)
succ(elt(letter = x) = elt(succ(letter,1,x)	
$L(letter) + L(1stsucc) + L(x)$	$= L(1/2.1/6.1)$
	$= L(1/12) = 4$ *bits*
length = 3	3 *bits*
starting with element(letter = 'A')	(1/26)
	TOTAL Length: **17** *bits*

β_1 (*"Replace the rightmost letter by its successor"*) is supposed to directly follow from M_1, and therefore, its cost is fixed to 0 *bit*. The same reasoning yields the other values.

We then get using an extended version of the encoding scheme described above:

$$
\begin{aligned}
L(M_1) &= -\log(1/8) - 2\log(1/2) + 4 + 1 = 10.00 \textbf{ bits} \\
L(\beta_1 \mid M_1) &= \qquad = 0 \\
L(\alpha_M) &= -\log(1/8) - \log(1/2) + 1 \qquad = 5 \\
L(\beta_2 \mid M_2) &= \qquad = 0 \\
L(S_1 \mid M_1) &= 3 - \log(1/26) \qquad = 8 \\
L(S_2 \mid M_2) &= 3 - \log(1/26) \qquad = 8
\end{aligned}
$$

So that the overall complexity of this analogy is : **31** *bits.*

4. EXPERIMENTAL RESULTS

In the work partially reported here, we examined three questions. First, does the *analogy* = *economical perception* perspective seem to have some validity? Second, what exactly is transferred in an analogy? And last, what is the optimal level of abstraction, in the context of an analogy, for the intermediary concepts M_1 and M_2?

To this aim, we have set to manually translate miscellaneous analogy problems in the letter microworld with various competing solutions, and then to compute their corresponding complexity. We then compared the results obtained with COPYCAT's behavior so as to test if the fact that we share with this system a lot of the same representation primitives induces a bias towards similar results.[1] The expectation, if, our model is correct is that less complex analogies should correspond to the preferred ones. At the same time, we inspected the abstractions used in the best analogies with regard to the second and third questions above. For lack of space, only sketchy details of some experiments are given thereunder, more can be found in Cornuéjols (1996).

Problem 1: **abc ⇒ abd; iijjkk ⇒ ?**

Solution 1: "Replace rightmost group of letters by its successor"	**iijjkk ⇒ iijjll**
Solution 2: "Replace rightmost letter by its successor"	**iijjkk ⇒ iijjkl**
Solution 3: "Replace rightmost letter by D"	**iijjkk ⇒ iijjkd**
Solution 4: "Replace third letter by its successor"	**iijjkk ⇒ iikjkk**
Solution 5: "Replace Cs by Ds"	**iijjkk ⇒ iijjkk**

[1]Recall that COPYCAT, because it relies on a stochastic algorithm, is not deterministic, and may give different responses on different runs. This is why the results indicate the percentage that the different responses were obtained.

Only 'solution 1' implies the perception of a *sequence*, while the other solutions do not perceive either **abc** or **iijjkk** as *sequences*, but as *strings* (solutions 2, 3 and 4), or only as a background from which elements like 'C' and 'D' stand out (solution 5).

Problem 2: **abc ⇒ abd; aababc ⇒ ?**

Solution 1: "Replace rightmost group by its successor"	**aababc ⇒ aababcd**
Solution 2: "Replace 'abc' by 'abd'"	**aababc ⇒ aababd**

This problem underlines the difference that can result from the perception of the deep structure of the letter string **aababc**, seen as a sequence of three successive groups of letters: 'a', 'ab' and 'abc', compared with a myopic one that only detect the presence of the string 'abc' in the source situation, unrelated to the rest of the string.

Problem 3: **abc ⇒ abd; xcg ⇒ ?**

Solution 1: "Replace rightmost letter by its successor"	**xcg ⇒ xch**
Solution 2: "Replace Cs by Ds"	**xcg ⇒ xdg**
Solution 3: "Replace rightmost letter by D"	**xcg ⇒ xcd**

The following charts were obtained for the complexity measure corresponding to the three problems and the solutions above.

	P1;S1	P1;S2	P1;S3	P1;S4	P1;S5
$L(M_1)$	10 bits	9	11	11	12
$L(beta_1 \mid M_1)$	0	0	0	0	1
$L(alpha_M)$	5	0	0	0	0
$L(beta_2 \mid M_2)$	0	0	0	0	0
$L(S_1 \mid M_1)$	8	18	18	18	22
$L(S_2 \mid M_2)$	8	36	36	36	42
TOTAL	**31** bits	**63**	**65**	**65**	**77**
Rank	1	2	3	3	5
COPYCAT	81%	16.5%	0.3%	0%	0%
Rank	1	2	3	4	4

	P2;S1	P2;S2	P3;S1	P3;S2	P3;S3
Complexity	26	69	40	57	46
Rank	1	2	1	3	2
COPYCAT	N/A	N/A	97.4%	1.4%	1.2%
Rank	1	/1	1	2	3

The results obtained using our formula for the MDLP and the representation primitives with their attached complexity numbers are intuitively satisfactory. They rank the different solutions to the analogy problems in a way that corresponds to their intuitive qualities, and to the percentages measured on COPYCAT. On the other hand, the encoding scheme used here leaves place for arbitrariness in some places, and the absolute values of the complexity numbers should be taken with a grain of salt. Furthermore, producing an analogy involves various search processes whose effects on the final output depend heavily on their computational complexities. Often, the fastest one wins the price even if its output is not the best one. The MDLP does not take this dynamic complexity into account. This is why we do not intent here to suggest that the results obtained in the toy problems may serve as a validation. "Proving" analogies requires a lot of care and generally implies to make finely controlled experiments with human subjects. Results may vary from one culture to another and are always open to arguments.

The experiments reported here are only meant to show that the view exposed in this paper: analogy as economical perception, is reasonable and produces interesting results. Computer implementation of this idea and large testing on a variety of analogy tasks remain to be done.

5. CONCLUSION

This chapter has presented a new perspective where analogy is seen as the result of an economical perception of the analog. A formal account for this was given implying a form of the Minimum Description Length Principle. Experimental tests on toy problems reveal: (i) that the encoding problem, particularly assigning description lengths to conceptual primitives, is difficult, and (ii) that nonetheless first results confirm that less complex analogies are also the preferred ones. In addition, the proposed model offers some bases for explaining the properties of analogy listed in section 1.3. Analogical thinking is pervasive because it is intrinsically an economical mode of thinking. Abstract analogies are preferred because they correspond to the most economical way of perceiving two things as similar. Good agreement in the ratings of analogies by different people would result from the sharing of the same optimal way of perceiving things. Analogy is noninvolutive and non-idempotent because of the asymmetry between the source and the target. And finally, analogy making can be improved in our model by the learning of prior probabilities of the perceptual primitives, therefore altering their description length, and hence the overall relative complexities of the various possible analogies. On the other hand, this last property is also responsible for the sensitivity to the coding prescription that renders it a delicate task.

This research reveals the deep interdependencies that may exist between statistics and forms of inference like analogy that seem to imply only symbolic manipulations of the descriptions of single situations. If we consider indeed, as we did here, any inferencing as the search for theories that provide economic encodings for the evidences, and if we recognize that an efficient coding scheme must reflect our prior expectations about the world, then it follows that inferencing will at the end rely on statistics. This is why symbolic reasoning and statistical analysis are two sides of the same coin that cannot be separated in Artificial Intelligence.

REFERENCES

Burstein, M. (1986). *Machine Learning: An Artificial Intelligence Approach*, Vol. 2. Morgan Kaufmann, San Mateo, CA.

Chaitin, G. (1987). *Information, Randomness and Incompleteness—Papers on Algorithmic Information Theory*. World Scientific Press, Singapore.

Chalmers, D., French, R., and Hofstadter, D. (1992). High-level perception, representation and analogy: A critique of artificial intelligence methodology. *J. Exp. Theor. Artif. Intell.* **4**, 185–211.

Clement, C., and Gentner, D. (1991). Systematicity as a selectional constraint in analogical mapping. *Cogn. Sci.* **15**, 89–132.

Cornuéjols, A. (1996). *Analogy as a Minimization Principle in Operation: The Details*, Tech. Rep., Laboratoire de Recherche en Informatique, Université de Paris-Sud, France.

Falkenheimer, B. (1987). An examination of the third stage in the analogy process: Verification-based analogical learning. In *Proceedings of the Tenth International Joint Conference on Artificial Intelligence*, pp. 260–263.

Gentner, D. (1983). Structure mapping: A theoretical framework for analogy. *Cogn. Sci.* **7**, 155–170.

Gick, M., and Holyoack, K. (1980). Analogical problem solving. *Cogn. Psychol.* **12**, 306–355.

Gick, M., and Holyoack, K. (1983). Schema induction in analogical transfer. *Cogn. Psychol.* **15**, 1–38.

Holyoack, K., and Thagard, P. (1989). Analogical mapping by constraint satisfaction. *Cogn. Sci.* **13**, 295–355.

Keane, M. (1988). Analogical mechanisms. *Artif. Intell. Rev.* **2**, 229–250.

Kedar-Cabelli, S. (1985). Purpose-directed analogy. In *Cognitive Science Society Conference*. In *Proceedings of the Cognitive Science Society, Irvine, CA*, pp. 150–159.

Kolmogorov, A. (1965). Three approaches to the quantitative definition of information. *Probl. Inf. Transm. (Engl. Transl.)* **1**(1), 1–7.

Mitchell, M. (1993). *Analogy-Making as Perception*. MIT Press, Cambridge, MA.

Rissanen, J. (1989). *Stochastical Complexity and Statistical Inquiry*. World Scientific Press, Singapore.

Solomonoff, R. (1964). A formal theory of inductive inference. *Inf. Control* **7**(1), 224–254.

Wallace, C., and Boulton, D. (1968). An information measure for classification. *Comput. J.* **11**, 185–195.

Winston, P. (1980). Learning and reasoning by analogy. *Commun. ACM* **23**(12), 689–702.

INDEX